For Joe Boeddeker,

At the start
of Libraria Inc,

Bay OBrin
Via Ron Pasari

THE
COMBINATORIAL
INDEX

THE COMBINATORIAL INDEX

Barry A. Bunin

Arris Pharmaceutical
San Francisco, California

ACADEMIC PRESS

San Diego   London   Boston   New York   Sydney   Tokyo   Toronto

Academic Press
a division of Harcourt Brace & Company
525 B Street, Suite 1900, San Diego, California 92101-4495, USA
http://www.apnet.com

Academic Press Limited
24-28 Oval Road, London NW1 7DX, UK
http://www.hbuk.co.uk/ap/

Library of Congress Card Catalog Number: 97-81368

International Standard Book Number: 0-12-141340-3

PRINTED IN THE UNITED STATES OF AMERICA
98 99 00 01 02 03 QW 9 8 7 6 5 4 3 2 1

■ CONTENTS

4 Combinatorial Solid-Phase Synthesis 77

5 Analytical Methods for Solid-Phase Synthesis 213

6 Preparation of Solution Libraries and Combined Approaches at the Solution/Solid-Phase Interface 237

■ FOREWORD

The idea that the drug discovery process can be accelerated by parallel and combinatorial synthesis is now entering its second decade. Pioneering work in the field was done in Australia by H. Mario Geysen and co-workers in the mid 1980s. They used the power of parallel synthesis of small peptides, in conjunction with ELISA screening of the resulting products, to identify small antigenic regions of large proteins. Seminal work in the area of split synthesis was accomplished by Árpád Furka in the late 1980s with the description of methods now known as split synthesis, used to generate pools of large numbers of compounds for biological screening.

The power of combinatorial methods for the generation of molecular diversity was clearly demonstrated by these early chemical methods. Additional research into biological diversity, such as phage libraries, further showed the power of the field. However, along with this early success came growing recognition of the limitations of peptides as orally available therapeutic agents. Interest of the pharmaceutical industry in this now-burgeoning field was sparked by a 1992 publication of Barry Bunin and Jonathan Ellman describing a solid-phase synthesis for 1,4-benzodiazepines. The field has grown such that nearly every large pharmaceutical company actively conducts research in the area, and many small companies devoted exclusively to combinatorial technology have been founded in recent years.

One of the main barriers facing a synthetic chemist contemplating a combinatorial approach to a medicinal chemistry problem is information. Prior to 1994, very few synthetic methods for parallel and combinatorial synthesis were available in the literature. Today, with the explosion of solid-phase methods, the opposite problem now confronts a researcher: Where to begin?

The Combinatorial Index is an answer to that problem. Presented in the pages that follow is an easy-to-access compendium of reliable synthetic transformations culled from classical and recent literature. There is a strong emphasis on solid-phase methods, as most of the combinatorial libraries reported to date have been constructed using some form of a solid support. The popularity of this approach results from the ability to drive a reaction to completion by the use of excess reagents and the ease of purifications between chemical steps, as noncovalently bound material may simply be rinsed away. Resin synthesis presents its own difficulties, and a chapter on analytical methods helps to illuminate the "black box" that support-bound intermediates can become. A chapter on different linkers will greatly assist in the judicious choice of the linking strategy for a combinatorial synthesis. Finally, the use of solution-based synthesis for the generation of molecular diversity is also covered.

This book offers a practical approach to combinatorial techniques, and any chemist contemplating a parallel or combinatorial synthesis will find *The Combinatorial Index* an invaluable resource. Looking to the future, the use of parallel and combinatorial synthesis methods will become even more tightly integrated into the toolbox of the medicinal chemist. *The Combinatorial Index* is an important and significant contribution to that end.

Matthew J. Plunkett
Berkeley, California

ACKNOWLEDGMENTS

It is a pleasure to acknowledge my friends and colleagues who provided suggestions during the preparation of this compilation. First, I thank Professor Jonathan Ellman for demonstrating the feasibility and utility of the combinatorial synthesis of small molecules. I especially thank my old labmate, Guang-cheng Liu, for providing copies of articles on all the classical (and more obscure) linkers for solid-phase synthesis. I am indebted to Ellen Kick, Brad Backes, Ruo Xu, Frank Woolard, Matt Plunkett, and Doug Livingston for proofreading various sections of the Index. Ron Zuckermann's insightful suggestions regarding the organization of the material and the structural appendixes were particularly valuable and are much appreciated. I also thank my family, especially Anna for her help with the home computer. *The Combinatorial Index* is dedicated to my good dog Lucky.

INTRODUCTION

In the preface to *The Practice of Peptide Synthesis*[1] Bodanszky and Bodanszky mentioned that, "In the library of the Institute we noted that the volumes of *Houben-Weyl's Handbuch der Organischen Chemie* dealing with peptide synthesis were so much in use that they were ready to fall apart because the researchers of the Institute consulted them with amazing regularity. They were looking for references, but even more for experimental details that could be adapted to the particular problem they happened to face. In planning a new synthetic endeavor they tried to lean on the experience of others in *analogous* situations. This suggested to us that a smaller and hence more tractable book may be needed, a volume that could be kept on or near the bench to make examples of fundamental methods readily available in the laboratory. Such a collection could save numerous short trips to the library, a point particularly important where a library well equipped with the sources of the literature of peptide synthesis is not near at hand." In preparing *The Combinatorial Index*, I have applied Bodanszky and Bodanszky's philosophy and format to combinatorial chemistry.

Combinatorial chemistry is a useful tool for rapidly optimizing molecular properties, particularly ones that are difficult to design a priori. The combinatorial approach can be used to generate a large number of variables in a format that allows the selection of an optimal subset of variables. Nature uses a combinatorial approach to generate diverse functional macromolecules, such as the large number of antibodies that recognize non-self molecules. Combinatorial chemistry builds on this evolutionary approach by generating diversity in a controlled setting and then applying it to different problems such as drug discovery or catalysis development. Although this approach is relatively new, combinatorial chemistry is already a fundamental approach to the identification of novel molecular properties.

Combinatorial chemistry is an interdisciplinary field currently undergoing rapid growth. The goal of this book is to provide practical information about the combinatorial synthesis of small molecules for a range of individuals interested in the field, including bench chemists, lab managers, medicinal chemists, agrochemists, academic chemists, computational chemists, and even biologists. Regardless of one's experience in combinatorial chemistry, many questions inevitably arise, often simple questions such as: What has already been done? How was it done? Where can the necessary materials or instruments be found? With an explosion in information, the organization of the information becomes crucial. *The Combinatorial Index* provides a compilation of the synthetic information for constructing combinatorial libraries of small molecules.

A number of different formats are used to construct combinatorial libraries (e.g., pooled, spatially separate, encoded), and the method of choice varies from laboratory to laboratory. Combinatorial chemistry has been applied most commonly to lead identification and optimization for the drug discovery process. Combinatorial approaches have also been applied to the optimization of catalysts, novel materials, and receptors. However, this book is intentionally limited to a discussion of synthetic methods related to library generation and analytical methods to assist synthetic efforts. A number of different structures have been prepared by combinatorial synthesis. Initially, libraries of peptides and oligonucleotides were the focus of combinatorial chemistry. However, peptides and oligonucleotides generally have poor oral activities and rapid *in vivo* clearing times; therefore, their utility as bioavailable therapeutic agents is limited. As a result of such limitations, the synthesis and screening of libraries of nonbiological oligomers (see Appendix 1) and nonpolymeric organic compounds ("small molecules") have rapidly become the focus of intensive research efforts. *The Combinatorial Index* describes the methods reported in the literature for the rapid preparation of functionally diverse small molecules. Points of interest, representative examples, and literature procedures summarize each method. In addition to focusing on small molecules, *The Combinatorial Index* is a compilation derived exclusively from journal articles (i.e., patent literature is not included).[2] Comments on the original studies are made; however, the majority of the information presented is factual and is thus reported as described in the primary literature. References are provided at the beginning of each new heading. Although this compilation provides an overview of the combinatorial synthesis of small molecules, it is primarily a resource book.

Many combinatorial libraries are currently constructed using solid-phase synthesis. Therefore, following a brief introduction and background, three chapters are devoted to the different linkers, reactions, and analytical techniques for solid-phase synthesis. Although many reactions on support are high yielding, significant optimization is often required before they are efficient and general enough to be used to construct combinatorial libraries. It is hoped that providing examples of different reactions along with the specific conditions that were necessary for optimization will assist related combinatorial studies. Representative examples are given to help assess the generality and limitations of the different methods. Reliable methods for reactions such as substitutions, cyclizations, condensations, and Suzuki couplings are included specifically because they can be used in different contexts.

Doing organic chemistry on solid support has been likened to working with a blindfold on because of the limited analytical and purification techniques available relative to those available in solution. This is one of the arguments in favor of building libraries in solution. The most direct way to evaluate the fate of a particular set of reaction conditions on support is to cleave the material off of support and rigorously characterize the products. Unfortunately, this is not always possible because intermediates are often unstable to cleavage conditions. Furthermore, particularly in a multistep sequence, there are often faster methods for determining whether a particular reaction worked (i.e., Fmoc quantitation). Detailed procedures for a range of different quantitative and qualitative analytical methods for solid-phase synthesis are described in Chapter 5.

A growing number of reports on solution libraries have appeared in the literature. Criteria and methods for building solution libraries are discussed in Chapter 6. Whether a library is prepared on support or in solution is often dictated by the type of chemistry being developed (or vice versa, the type of chemistry being developed is often dictated by whether a library is prepared on support or in solution). High-throughput purification techniques such as solid- and liquid-phase extractions are often critical for preparing solution libraries that are useful for screening. The challenges associated with the construction of solution libraries (e.g., solubility and purification) can be quite different from the challenges associated with solid-phase combinatorial synthesis (e.g., monitoring reaction progress and scale up). Although there are many differences between the solution- and solid-phase strategies for generating libraries, in both cases the synthetic challenge is to develop reaction conditions that are general and high yielding.

Often I have chosen to be inclusive, rather than selective, in citing examples from the literature. Even so, due to the large body of literature, many examples are cited in related studies. For example, in the sections discussing subjects such as the formation of amides and esters, only selected examples are provided for obvious reasons. I have made every effort to be both fair and informative about the strengths and weaknesses of the various methods in this compilation. I apologize for any topics that were underrepresented or misrepresented. I would appreciate being notified at bunin@combinatorial.com of new or incomplete information for incorporation into future editions or online supplements.

REFERENCES

1. Bodanszky, M.; Bodanszky, A. *The Practice of Peptide Synthesis:* Springer-Verlag: New York, 1984.
2. For a detailed review of recent solid-phase organic reactions including the patent literature, see Hermkens, P. H. H.; Ottenheijm, H. C. J.; Rees, D. *Tetrahedron* **1996**, *52*, 4527–4554.

2
BACKGROUND

A salient feature of combinatorial synthesis is that a large amount of diversity can be generated from a relatively small number of building blocks. A representative example of a simple combinatorial library prepared on solid support from three sets of building blocks, A, B, and C, is illustrated below. From only 10 derivatives of each building block, a library of 1,000 trimers can be generated; with 100 derivatives of each building block, 1,000,000 compounds can be accessed. With rapid access to such large numbers of compounds, new issues arise such as which compounds are the most useful to make and how to keep track of the large amount of information that is generated.

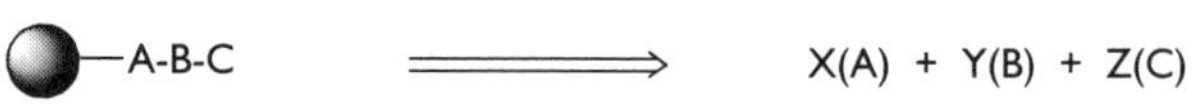

●—A-B-C $\Longrightarrow$ X(A) + Y(B) + Z(C)

X + Y + Z = Total number of variables used as inputs in the library

(X)(Y)(Z) = Total number of compounds generated from the library synthesis

Currently, there are a number of distinct approaches for generating combinatorial libraries *in vitro*. The compounds can be synthesized in a spatially separate format or as pooled mixtures. A number of methods for identifying active compounds in a mixture have been developed. Obviously, identifying an active compound is straightforward when the compounds are synthesized in a spatially separate format. A brief overview of the different methods for preparing synthetic combinatorial libraries follows. More detailed discussions can be found in a number of review articles.[1–4]

Methods for Generating Combinatorial Libraries

A. Spatially Separate Synthesis. The most straightforward approach to library analysis is to keep the different compounds (or other variables) spatially separate in a parallel array. The primary advantage of keeping the compounds spatially separate is that it removes some of the ambiguities associated with pooling compounds. When the compounds are spatially separate, direct structure–activity relationships are obtained from biological evaluation. Analytical evaluation of the chemical integrity of the compounds is also straightforward when the compounds are spatially separate. The primary disadvantage of spatially separate libraries is that the number of compounds that can be synthesized is more limited.

The first combinatorial library was prepared in a spatially separate format by Geysen and co-workers in 1984.[5] They developed functionalized pins for solid-phase peptide synthesis and epitope analysis. The pins were configured to be compatible with 96-well microtiter plates. The pin technology has been improved using different polymers, as well as higher loading levels and functional linkers to accommodate other chemical applications.[6] Fodor and co-workers at Affymax have developed photolithographic methods for building large libraries on a silicon wafer.[7] Large spatially separate libraries (100,000 compounds) can be prepared with this method. However, because it requires photolabile protecting groups and support-bound biological assays, the technology is primarily being applied to DNA diagnostic tests.[8] A number of new technologies for the preparation of spatially separate libraries on resin and in solution are currently being developed.

B. Split Synthesis. There are a number of different pooling strategies. The earliest of these, developed independently by Furka,[9] Lam,[10] and Houghten,[11] employ a split and mix procedure to generate mixtures of peptides. In a split synthesis, a quantity of resin is split into equal-sized portions in separate reaction vessels and reacted with different monomers. After the reactions are complete, the resin is pooled together and thoroughly mixed. A common protecting group can be removed, or a common transformation can be performed, in a single reaction vessel. For the coupling of a second monomer, the resin is split again, and the process is repeated until the end of the combinatorial synthesis. To couple different building blocks, such as activated amino acids, the resin must be split into separate reaction vessels to allow reactions with different rates to be driven to completion.[12]

There are a number of techniques for identifying biologically active components from a library prepared by a split synthesis. The active components in a mixture can be isolated by deconvolution studies such as an iterative resynthesis and evaluation of smaller pools. A portion of the resin can be saved at each step to facilitate the iterative resynthesis. In addition, orthogonal,[13] positional,[14] and indexed[15] libraries all use pooling strategies that minimize the amount of deconvolution required.

The combinatorial methods initially developed for peptide synthesis have also been applied to the combinatorial synthesis of unnatural biopolymers and small molecules. In one early example, high-affinity ligands to 7-transmembrane G-protein-coupled receptors (7TM/GPCR) were identified from the split synthesis of a diverse peptoid library.[16]

At the end of any split synthesis, each individual bead theoretically contains a single product, since all of the sites on any particular bead have been exposed to the same synthetic reagents. "One-compound, one-bead" approaches have been developed to identify the active components in a biological assay without resorting to a time-consuming iterative resynthesis. With certain assays of support-bound compounds, an active compound from a single resin bead is identified after it binds with a radiolabeled or fluorescent-labeled receptor.[17] After active components on support are detected and isolated, the chemical structure can be determined using a method such as Edman degradation for the identification of support-bound peptides. Methods for the partial release of compounds off the support have been developed for biological evaluation in solution. After biological evaluation, the compound that remains on the resin beads can be used for structural identification.[18]

A conceptually different approach to deconvoluting active components from a library prepared by split synthesis involves a molecular tagging scheme. In this approach, readable tags that encode the reaction sequence are attached to resin. DNA was an obvious choice for encoding,[19] since that is what Nature uses. Unfortunately, DNA is not chemically stable under many of the reaction conditions frequently used in organic synthesis. To circumvent this problem, encoding has been performed with peptides prepared from amino acids that have relatively unreactive side chains[17] or GC–EC tags that are inert to most of the reaction conditions typically employed.[20] The advantages of the GC–EC tags, developed by Still and co-workers, are that they can be both detected at less than 0.1 pmol and attached directly to polystyrene via carbene chemistry. Thus, the method does not require an orthogonal protecting strategy. Radiofrequency tagging strategies have also been developed as an alternative method for encoding libraries on resin.[21,22] Alternative approaches to generating combinatorial libraries and optimizing biological activity, such as genetic algorithms, are currently being investigated.[23,24]

At least as important as the format in which libraries are prepared are the classes of compounds that are accessible. This compilation describes synthetic methods and analytical techniques to assist in the development of chemistry for combinatorial libraries.

REFERENCES

1. (a) Thompson, L. A.; Ellman, J. A. *Chem. Rev.* **1996**, *96*, 555–600. (b) Choong, I. C.; Ellman, J. A. *Annu. Rep. Med. Chem.* **1996**, *31*, 309–318.
2. (a) Gallop, M. A.; Barrett, R. W.; Dower, W.; Fodor, S. P. A.; Gordon, E. M. *J. Med. Chem.* **1994**, *37*, 1233–1251. (b) Gallop, M. A.; Barrett, R. W.; Dower, W.; Fodor, S. P. A.; Gordon, E. M. *J. Med. Chem.* **1994**, *37*, 1385–1401.
3. (a) Jung, G.; Becksickinger, A. *Angew. Chem., Int. Ed. Engl.* **1992**, *31*, 367–383. (b) Früchtel, J. S.; Jung, G. *Angew. Chem., Int. Ed. Engl.* **1996**, *35*, 17–42.
4. Rinnova, M.; Lebl, M. *Collect. Czech. Chem. Commun.* **1996**, *61*, 171–231.

5. Geysen, H. M.; Meloen, R. H.; Barteling, S. J. *Proc. Natl. Acad. Sci. U.S.A.* 1984, *81*, 3998–4002.

6. Maeji, N. J.; Valerio, R. M.; Bray, A. M.; Campbell, R. A.; Geysen, H. M. *React. Polym.* 1994, *22*, 203–212.

7. Fodor, S. P. A.; Read, J. L.; Pirrung, M. C.; Stryer, L.; Lu, A. T.; Solas, D. *Science* 1991, *251*, 767–773.

8. Abbott, A. *Nature* 1996, *379*, 392.

9. Furka, A.; Sebestyen, F.; Asgedom, M.; Dibo, G. *Int. J. Pept. Protein Res.* 1991, *37*, 487–493.

10. Lam, K. S.; Salmon, S. E.; Hersh, E. M.; Hruby, V. J.; Kazmierski, W. M.; Knapp, R. J. *Nature* 1991, *354*, 82–84.

11. Houghten, R. A.; Pinilla, C.; Blondelle, S. E.; Appel, J. R.; Dooley, C. T.; Cuervo, J. H. *Nature* 1991, *354*, 84–86.

12. Houghten has found that the relative rates of amide bond formation on solid support with different amino acids were primarily dependent on the activated amino acid in solution. By adjusting the concentration of the amino acids to compensate for their relative rates, different amino acids can be pooled in the same reaction vessel.

13. Deprez, B.; Williard, X.; Bouirel, L.; Coste, H.; Hyafil, F.; Tartar, A. *J. Am. Chem. Soc.* 1995, *117*, 5405–5406.

14. Pinilla, C.; Appel, J. R.; Blanc, P.; Houghten, R. A. *Biotechniques* 1992, *13*, 901–905.

15. Pirrung, M. C.; Chen, J. *J. Am. Chem. Soc.* 1995, *117*, 1240–1245.

16. Zuckermann, R. N.; Martin, E. J.; Spellmeyer, D. C.; Stauber, G. B.; Shoemaker, K. R.; Kerr, J. M.; Figliozzi, G. M.; Goff, D. A.; Siani, M. A.; Simon, R. J.; Banville, S. C.; Brown, E. G.; Wang, L.; Richter, L. S.; Moos, W. H. *J. Med. Chem.* 1994, *34*, 2678–2685.

17. Vetter, D.; Tate, E. M.; Gallop, M. A. *Bioconjugate Chem.* 1995, *6*, 319–322.

18. Lebl, M.; Patek, M.; Kocis, P.; Krch(ák, V.; Hruby, V. J.; Salmon, S. E.; Lam, K. S. *Int. J. Pept. Protein Res.* 1993, *41*, 201–203.

19. Needels, M. C.; Jones, D. G.; Tate, E. H.; Heinkel, G. L.; Kochersperger, L. M.; Dower, W. J.; Barrett, R. W.; Gallop, M. A. *Proc. Natl. Acad. Sci. U.S.A.* 1993, *90*, 10700–10704.

20. Ohlmeyer, M. H.; Swanson, R. N.; Dillard, L. W.; Reader, J. C.; Asouline, G.; Kobayashi, R.; Wigler, M.; Still, W. C. *Proc. Natl. Acad. Sci. U.S.A.* 1993, *90*, 10922–10926.

21. Moran, E. J.; Sarshar, S.; Cargill, J. F.; Shahbaz, M. M.; Lio, A.; Mjalli, A. M. M.; Armstrong, R. W. *J. Am. Chem. Soc.* 1995, *117*, 10787–10788.

22. Nicolaou, K. C.; Xiao, X. Y.; Parandoosh, Z.; Senyei, A.; Nova, M. P. *Angew. Chem., Int. Ed. Engl.* 1995, *34*, 2289–2291.

23. Weber, L.; Wallbaum, S.; Broger, C.; Gubernator, K. *Angew. Chem., Int. Ed. Engl.* 1995, *1*, 2280–2282.

24. Sheridan, R. P.; Kearsley, S. K. *J. Chem. Inf. Comput. Sci.* 1995, *35*, 310–320.

3
LINKERS FOR SOLID-PHASE SYNTHESIS

Solid-phase synthesis is typically performed on linkers attached to various solid supports (or resins). A number of different solid supports, ranging from membranes to photolithographic chips to polyethylene pins, have been used for combinatorial synthesis. The most common resins currently in use are the polystyrene/divinylbenzene and poly(ethylene glycol)–polystyrene/divinylbenzene resins. Along with the identification of the appropriate solvent, concentration, and temperature, the selection of solid support can be a critical step in solid-phase reaction optimization.

Choosing the correct linker is also a crucial step for combinatorial synthesis on solid support. The linker must be compatible with all the synthetic steps, yet labile under cleavage conditions that do not cause decomposition of the compounds generated in the library. The appropriate linker is often dictated by the functionality present in the specific class of molecules of interest. For example, either peptide acids or amides can be obtained by selecting the appropriate Rink ester or amide linker for solid-phase peptide synthesis. Many of the linkers commonly used for combinatorial synthesis are borrowed from previous research on the solid-phase synthesis of peptides. For example, the solid-phase synthesis

X = NH$_2$, Rink amide linker
X = OH, Rink acid linker

of 1,4-dihydropyridines, a class of small molecules that have found commercial utility as calcium channel blockers, was performed on both PAL and Rink resins that were initially developed for solid-phase peptide synthesis.

The solid-phase synthesis of 1,4-dihydropyridines on peptide resins.

Small molecules have been linked to support through carboxylic acid derivatives, phenols, alcohols, amines, and even carbon–carbon bonds for combinatorial applications. Therefore, although linkers developed for solid-phase peptide synthesis will continue to play an important role in combinatorial synthesis, new linkers and new modes of cleavage are particularly important for the solid-phase synthesis of small molecules. Small molecules can be cleaved from resin by a number of different mechanisms. The most common linkers, such as the acid-labile Rink or Wang linkers, provide molecules with a specific functional group. The auxiliary functional group can have a positive, negative, or negligible effect on biological activity. Thus if a certain functional group on a particular scaffold is known to be important for the desired activity in the library, it is an attractive site for attaching the linker. Alternatively, compounds may be cleaved via a cyclic mechanism in which no memory of the linker is present. Hydantoins have been cleaved from resin via a cyclic mechanism under either acidic or basic conditions.

The cleavage of hydantoins from resin under either acidic or basic conditions.

Other linkers, such as the highly activated acylsulfonamide linker, allow the incorporation of additional diversity during the cleavage step. Treatment of the activated acylsulfonamide with limiting amounts of an amine nucleophile results in complete consumption of the amine to provide the amide product in pure form. This is particularly useful for combinatorial synthesis because additional diversity can be incorporated during the cleavage step. By adding a limiting amount of an equimolar mixture of amines to the resin, an equimolar mixture of extremely pure amide products is produced.

1. $(R^1CO)_2O$, catalytic DMAP
2. ICH_2CN, DIEA

A highly activated acylsulfonamide linker than can be cleaved with a limiting quantity of a mixture of amines (R^2R^3NH).

Usually small molecules prepared by solid-phase synthesis contain an auxiliary functional group upon cleavage from support. Sometimes, classes of compounds that do not contain an auxiliary functional group are desired. In addition to cyclic cleavage, "traceless linkers" that replace the linker with an innocuous carbon–hydrogen bond after cleavage have been developed.

A traceless linker for the solid-phase synthesis of 1,4–benzodiazepines. Upon completion of the synthesis the carbon–silicon bond is replaced with a carbon–hydrogen bond.

This chapter provides a description of the strengths and limitations of various linkers, with a particular emphasis on linkers that are appropriate for combinatorial synthesis. Additional details, particularly regarding the commercially available linkers, can be found in the synthetic notes of catalogs for purchasing resins for solid-phase peptide or combinatorial synthesis.

3.1 RESIN DERIVATIZATION

TABLE 3.1 Functionalization of Polystyrene Resins

Linker name	Page	Preparation conditions and comments
Bromination and lithiation	12	Br_2, Tl(OAc)$_3$ then s-BuLi, TMEDA and the appropriate quench for functionalization
Chloromethylation	13	MOMCl, Lewis acid
Aminomethylation	14	TfOH, N-(chloromethyl)phthalimide; then hydrazine
PEG–PS		Various PEG-PS resins have been prepared by different methods (i.e., alkylation, polymerization, carbon–carbon bond formation).[a]

[a] Bayer, E. *Angew. Chem., Int. Ed. Engl.* **1991**, *30*, 113–216.

3.1.1 Bromination and Lithiation: Two Important Steps in the Functionalization of Polystyrene Resins[1]

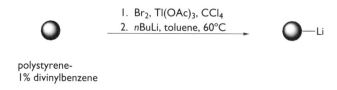

polystyrene-
1% divinylbenzene

Points of Interest

1. The bromination of polystyrene resins proceeded best with catalytic Tl(OAc)$_3$.

2. Brominated macroreticular resins reacted completely with n-BuLi in THF. In contrast, the direct lithiation reaction of 1% cross-linked polystyrene resins was dependent on both the loading level and the solvent employed. The lithiation was incomplete in THF or cyclohexane but occurred quantitatively in benzene or toluene.

3. Direct lithiation of cross-linked polystyrene resins proceeded best with n-BuLi and TMEDA in cyclohexane. However, the two-step procedure (bromination, followed by lithium/halogen exchange) is probably the method of choice if control of the position of lithiation and of the degree of functionalization is desired, since predictable results can be obtained in the bromination over a very broad range of degrees of functionalization.

4. The lithiated resins were used to prepare a wide range of polymers containing carboxylic acid, thiol, sulfide, boronic acid, amide, silyl chloride, phosphine, alkyl bromide, aldehyde, alcohol, or trityl functional groups for applications in polymer-assisted synthesis.

5. Many derivatized polystyrene resins are commercially available.

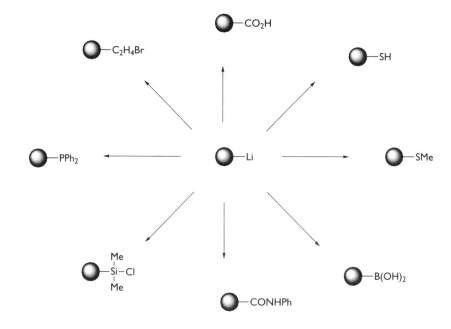

Literature Procedures

Prewashing Commercially Available Resins.   The resins used in this study[2] were a solvent-swellable 1% divinylbenzene–styrene copolymer, Bio-Beads SX 1, purchased from Bio-Rad Laboratories, and a macroreticular resin, Amberlite XE-305, purchased from British Drug House. The resins used were washed routinely to remove surface impurities. The following solutions were used at 60–80°C with, in each case, a contact time of 30–60 min with the resin: 1 N NaOH, 1 N HCl, 2 N NaOH/dioxane (1:2), 2 N HCl/dioxane (1:2), H_2O, DMF. The resins were then washed at room temperature with the following: 2 N HCl in MeOH, H_2O, MeOH, MeOH/DCM (1:3), MeOH/DCM (1:10). The resins were dried under reduced pressure at 50–70°C. In general, the washing was accompanied by a loss of weight of up to 10%.

Bromination of Cross-Linked Polystyrene Resins.   To a suspension of 20 g of washed resin in 300 mL of carbon tetrachloride was added 1.18 g of thallic acetate. The reaction mixture was stirred in the dark for 30 min, and 13.6 g of bromine in 20 mL of carbon tetrachloride was added slowly. After 1 h of stirring at room temperature in the dark, the mixture was heated to reflux for 1.5 h. The reaction mixture, which had lost all the coloration due to free bromine, was collected by filtration and washed with carbon tetrachloride, acetone, acetone/water (2:1), acetone, benzene, and methanol. After drying under vacuum, 26.3 g of resin containing 3.10 mequiv of bromine per gram was obtained (24.8% Br). Thus the resin obtained in this preparation had functional groups on 43% of the aromatic rings.

Lithiation and Quenching of 1% Cross-Linked Brominated Resin.   The lithiation was carried out by heating a suspension of 2.04 g of brominated resin (2.89 mequiv/g) in 30 mL of dry benzene (or toluene) with 10 mL of 1.6 M *n*-BuLi and stirring the suspension at 60°C for 3 h. After quenching with powdered carbon dioxide in THF, washing, and drying, 1.86 g of a polymer containing 2.9 mequiv of $-CO_2H$ per gram was obtained (by titration). The infrared spectrum of the polymer included very broad hydroxyl and carbonyl absorptions. Detailed procedures for the preparation of thiol, sulfide, boronic acid, amide, silyl chloride, phosphine, bromide, aldehyde (and its oxime), alcohol, and trityl resins were provided.

3.1.2   Zinc Chloride-Catalyzed Chloromethylation of Resins for Solid-Phase Peptide Synthesis[3]

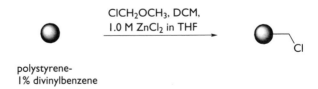

polystyrene-
1% divinylbenzene

Points of Interest

1. Zinc chloride was found to catalyze chloromethylation of resins for solid-phase peptide synthesis, producing low levels (0.07–1.27 mmol/g) of

chloromethylation more accurately and under more convenient conditions than $SnCl_4$.

2. The preparation of a stable solution of $ZnCl_4$ catalyst in THF was described and data were presented detailing the effects of varying temperature, time, and the concentrations of catalyst and chloromethyl methyl ether (*Caution:* toxic!) on the chloromethylation reaction.

3. Resin beads that had undergone chloromethylation swelled in organic solvents to the same extent as nonchloromethylated controls, suggesting little or no cross-linking occurred between polymer chains.

4. (Chloromethyl)polystyrene–divinylbenzene resins (Merrifield resins) are available from a number of commercial sources.

Literature Summary

For the Preparation of a 1.0 M $ZnCl_2$ Solution in THF.   A flask containing $ZnCl_2$ (1.5 g), 3 drops of catalytic concentrated HCl, and 5 drops of H_2O was heated until all solids dissolved. The water was evaporated by gradually increasing the temperature, and the remaining solids were melted by stronger heating. When the $ZnCl_2$ became a clear mobile liquid with no further evolution of bubbles, the flask was cooled in a desiccator, and the resulting solid was dissolved in freshly distilled THF (10 mL). The exact concentration of the resulting solution was determined by pipetting a sample into water containing several drops of HNO_3 and titrating with 0.1 M $AgNO_3$.

Chloromethylation of Polystyrene Beads (10-g Sample).   The resin (10 g of Bio-Beads, SX 1, 200–400 mesh, Bio-Rad Laboratories), $ClCH_2OCH_3$ (60 mL; **CAUTION:** methoxymethyl chloride is toxic!), and DCM (58 mL) were stirred for 5 min. Then 2.0 mL of the 1.0 M $ZnCl_2$ solution was added, and the resulting suspension was stirred for 3 h at 40°C. The resin was then filtered and washed with THF (4×), 3:1 THF/H_2O (3×), 3:1 THF/3 N HCl (3×), 3:1 THF/H_2O (3×), THF (5×), and MeOH (3×). The resin was dried under high vacuum at room temperature overnight and the sample was analyzed potentiometrically, giving Cl = 0.18 mmol of Cl/g of resin.

3.1.3   High-Capacity (Aminomethyl)polystyrene Resin[4]

polystyrene-
1% divinylbenzene

Points of Interest

1. (Aminomethyl)copolystyrene–1% divinylbenzene resins of high purity and a variety of substitution levels were synthesized by an improved method. After examination of different catalysts for the preparation of phthalimidomethyl resins, the use of ferric chloride and N-(chloromethyl)phthalimide was found to be suitable for obtaining a pure and uniform resin. The substitution level was up to 96% of the available benzene rings.

2. (Aminomethyl)copolystyrene–1% divinylbenzene resins can conveniently be derivatized with a large number of different linkers.

3. (Aminomethyl)polystyrene resins are commercially available from a number of sources.

4. Ferric chloride can also be used as a Friedel–Crafts catalyst (0.3–0.4 equiv) with 2-bromopropionyl chloride, *p*-nitrobenzoyl chloride, and *o*-chlorobenzoyl chloride to give the corresonding 2-bromopropionyl resin, *p*-nitrobenzophenone resin (for the oxime resin, see p 50), and *o*-chlorobenzophenone resin (for the Cl-trityl resin, see p 24). In the case of the preparation of the 2-bromopropionyl resin, orange specks of polymer were not observed (in contrast to an earlier preparation of 2-bromopropionyl resin).

Literature Summary

Copolystyrene–1%-divinylbenzene resin (1 g), *N*-(chloromethyl)phthalimide (0.43 g, 2.2 mmol), and 20 mL DCM were mixed briefly with an overhead stirrer, and then $FeCl_3$ (0.1 g, 0.62 mmol) was added. After 2 h of stirring, the resin was filtered, rinsed with dioxane, 1 N HCl/dioxane (1:1), dioxane, and MeOH, and dried under high vacuum at room temperature. The phthalimido group was removed by treatment with 5% hydrazine hydrate in ethanol under reflux to yield resin with a substitution of 2 mmol of N/g. Resins were tested for the absence of the carbonyl group at $1720 \ cm^{-1}$ by IR spectroscopy to verify that hydrazinolysis was complete. Washing the insoluble byproduct, 2,3-dihydro-1,3-phthalazinedione, required hot ethanol rinses, especially from the high-loading resins.

REFERENCES FOR SECTION 3.1

1. Farrall, M. J.; Frechet, J. M. J. *J. Org. Chem.* **1976,** *41* (24), 3877–3882.
2. The initial study was reported in 1976; commercial sources of underivatized resins can change over time.
3. Feinberg, R. S.; Merrifield, R. B. *Tetrahedron* **1974,** *30,* 3209–3212.
4. Zikos, C. C.; Ferderigos, N. G. *Tetrahedron Lett.* **1995,** 3741–3744.

3.2 LINKERS FOR CARBOXYLIC ACIDS

TABLE 3.2   Acid-Labile Linkers for Carboxylic Acids (Note: The Linkers Are Organized by Increasing Acid Lability)

Linker name	Page	Cleavage conditions and comments
Glycoamidic ester linker	17	Complete acid stability, cleaved with NaOH
Sheppard linker **3**	18	Complete acid stability, cleaved with amines
Sheppard linker **2**	18	HF (***Caution!***)
PAM linker	19	16% HBr in AcOH/TFA (75 min, rt), **or** anhydrous HF (***Caution!***)/anisole (1 h, 0°C)
tert-Butyl linker	20	TFA/thioanisole (3 h), **or** 1 N HBF_4/thioanisole/TFA (4°C, 1 h)

(Continues)

TABLE 3.2 (*Continued*)

Linker name	Page	Cleavage conditions and comments
Sheppard linker **1**	18	TFA (30 min)
Wang linker	21	50% TFA/DCM (30 min)
Sheppard linker **4**	18	1% TFA/DCM
SASRIN linker	22	0.5–1% TFA/DCM
HAL linker	23	0.1% TFA/DCM (5 min)
2-Chlorotrityl linker	24	1:1:8 AcOH/TFE/DCM
Rink acid linker	25	10% AcOH/DCM (1.5 h)

TABLE 3.3 Silyl-Based Linkers for Carboxylic Acids

Linker name	Page	Cleavage conditions and comments
Silyl acid (SAC) linker	26	95:2.5:2.5 TFA/thioanisole/phenol
Barany's Pbs linker	27	TBAF, thiophenol, DIEA, DMF (2 min) **or** TFA
Ramage's silyl linker	28	TBAF/DMF (5 min)

TABLE 3.4 Photolabile Linkers for Carboxylic Acids

Linker name	Page	Cleavage conditions and comments
Rich's *o*-nitrobenyl linker	28	350-nm *hv* (18–24 h)
α-Methylphenacyl ester linker	29	350-nm *hv* (72 h)

TABLE 3.5 Safety catch linkers for carboxylic acids

Linker name	Page	Cleavage conditions and comments
Reductive acidolysis safety-catch linker	31	SiCl$_4$/thioanisole/anisole/TFA (3 h)
Safety-catch linker cleaved by intramolecular activation	31	Activation and cleavage by $^-$OH (pH 10)
Safety-catch linkage for direct release into aqueous buffers	33	Activation with 0.1% HCl, cleavage in 0.01 M phosphate buffer

TABLE 3.6 Other carboxylic acids linkers

Linker name	Page	Cleavage conditions and comments
Allylic linker	34	Pd(PPh$_3$)$_4$, morpholine, THF
Fluorene linker	35	20% morpholine, DMF (2 h)
Base-labile *β*-elimination linker	35	30:9:1 dioxane/MeOH/4 N NaOH (30 min)
Oxidation-labile linker	36	0.5 M CuSO$_4$, pyridine/AcOH, DMF (16 h)

3.2.1   Glycolamidic Ester Linker [1]

X = OH, Br

Points of Interest

1. The glycolamidic ester linkage is compatible with both Boc and Fmoc strategies, yet labile to cleavage with NaOH, alkoxides, or primary amines under the appropriate conditions. The glycolamidic anchorage is stable to piperidine.

2. The glycolamidic ester linkage is extremely stable to acids; thus all of the standard amino acid side-chain protecting groups can be removed by treatment with strong acids with no concomitant cleavage from the resin.

3. The glycolamidic ester linkage is both more suitable for hindered acids and more labile than the similar 4-(hydroxymethyl)benzamido linker [2] (resin 3; see p 18).

Literature Procedure

Bromoacetic acid (0.70 g, 5.04 mmol) was dissolved in 20 mL of DCM, and after the mixture was cooled to 0°C, DCC (0.519 g, 2.52 mmol) in 5 mL of DCM was added. The mixture was stirred at 0°C for 15 min and the precipitated dicyclohexylurea was eliminated by filtration. The solution was added to 2.0 g of aminomethyl resin (0.84 mmol of NH_2) that had been prewashed with DCM (4×), 5% DIEA in DCM (2×), and DCM (4×). The suspension was shaken at room temperature for 15 min, DIEA (0.143 mL, 0.84 mmol) was added, and the shaking was continued for 60 min. The bromoacetamido resin was washed with DCM (4×), DMF (4×), DCM (4×), and diethyl ether (4×) and then dried *in vacuo* for 24 h (yield 2.1 g). The bromoacetamido resin (2.1 g) was treated with the cesium salt of Fmoc-L-Tyr(tBu)OH (2.48 g, 4.2 mmol) dissolved in DMAC (40 mL) after shaking for 48 h, and the resin was drained and washed with DMAC (10×), DMF (4×), DCM (4×), and DMAC (4×). The resin was treated with 20% piperidine in DMAC for 10 min to remove the Fmoc group. After peptide synthesis, cleavage from support was performed as follows. The peptide resin (either in the side-chain-protected or unprotected form, 456 mg) was suspended in 20 mL of 2-propanol/water (70:30, v/v), and 0.266 mL of 1 M NaOH was added. The mixture was stirred for 3 h and then the resin was drained and washed with water (2×), ethanol (2×), and methanol (2×). The filtrates were pooled and the peptides were isolated after acidification.

REFERENCES FOR SECTION 3.2.1

1. Baleux, F.; Calas, B.; Mery, J. *Int. J. Pept. Protein Res.* **1986**, *28*, 22–28.
2. Atherton, E.; Logan, C. J.; Sheppard, R. C. *J. Chem. Soc., Perkin Trans. 1* **1981**, *1*, 538–546.

3.2.2   Sheppard's Linkers for Fmoc-Amino Acid Synthesis [1]

Points of Interest

1. Sheppard's linkers are derived from Wang resin (*vide infra*); however, the linkers are readily attached to aminomethyl resin via an amide bond.

2. Resin **2** is an orthogonal linker for standard Fmoc-amino acid synthesis.

3. Bpoc-amino acids were loaded as the first residue due to the potential lability of Fmoc-amino acids to catalytic DMAP which is used in the esterification of the resin. Other researchers have reported successfully loading Fmoc-amino acid derivatives in the presence of catalytic DMAP.[2]

4. Carboxylic acids linked to resin **3** can be cleaved from support with a variety of nucleophiles including primary amines, secondary amines, hydrazines, and methanol or reduced to the corresponding alcohols.[3]

5. Sheppard's linkers are commercially available from a number of sources (for example, NovaSyn TG HMBA resin is linker **3** on PEG–PS).

6. The experiments and reasoning responsible for the development of Fmoc-amino acid synthesis as an alternative to Boc-amino acid synthesis are discussed in detail in *Solid-phase Peptide Synthesis: A Practical Approach*. The reinvestigation of solid-phase peptide synthesis provides a truly fascinating example of innovative research.[4]

Literature Procedures

For Loading onto Linkers.   To approximately 1 g of the washed poly-(dimethylacrylamide) resin was added (dimethylamino)pyridine (2 mmol) in

7.5 mL of DMF followed after 5 min by a solution of Bpoc-glycine (from the dicyclohexylammonium salt) and DCC (2 mmol) in 7.5 mL of DMF. After 17 h, the resin was thoroughly washed with DMF. The Bpoc group was removed with 0.09 M HCl–AcOH and neutralized prior to standard Fmoc-amino acid synthesis.

For Cleavage.   Resin **1**[5] is slightly less acid labile than Wang resin; the esters of the benzylic hydroxyl group are cleaved under the same acidic conditions used for *tert*-butyl derivatives (e.g., trifluoroacetic acid, 30 min).

Resin **2** requires more drastic hydrogen fluoride treatment associated with benzyl ester derivatives for cleavage.

Resin **3** displays almost complete acid stability (unaffected even by liquid hydrogen fluoride), but enhanced nucleophilic lability. Thus resin **3** has been successfully used for the preparation of peptide amides. Peptide amides were initially formed with cold methanolic ammonia; more recently, Story and Aldrich have reported that ammonia in 2-propanol is the preferred solvent for cleavage because it reduces the quantity of impurities due to transesterification.[6]

Resin **4**[2] may be cleaved with 1% TFA/DCM and is thus suitable for the preparation of peptide fragments with a *tert*-butyl-protected side chain.

REFERENCES FOR SECTION 3.2.2

1. Atherton, E.; Logan, C. J.; Sheppard, R. C. *J. Chem. Soc., Perkin Trans. 1* **1981**, *1*, 538–546.
2. Rink, H. *Tetrahedron Lett.* **1987**, *28* (33), 3787–3790.
3. Atherton, E.; Sheppard, R. C. In *Solid-phase Peptide Synthesis: a practical approach*; IRL Press: Oxford, UK, 1989; pp 152–154.
4. Atherton, E.; Sheppard, R. C. *Solid-phase Peptide Synthesis: a practical approach*; IRL Press: Oxford, UK, 1989.
5. Sheppard, R. C.; Williams, B. J. *Int. J. Pept. Protein Res.* **1982**, *20*, 451–454.
6. Aldrich, J. V.; Story, S. C. *Int. J. Pept. Protein Res.* **1992**, *39*, 87–92.

3.2.3   PAM Linker: Phenylacetamidomethyl (PAM) Resin[1]

PAM Linker

PAM Linker loaded with Boc amino acid

Points of Interest

The PAM linker is typically used for peptide synthesis with Boc-amino acids.

1. The primary advantage of the PAM linker is that it is 100 times more stable than the peptidyl benzyl ester used initially for solid-phase peptide synthesis. The relative stability was examined under standard conditions for Boc deprotection: 1:1 TFA/DCM.

2. Peptides synthesized using the PAM linker are cleaved from support with HBr in acetic acid/TFA or with anhydrous HF.

3. For the preparation of (aminomethyl)polystyrene–divinylbenzene resin, see p 14.[2]

4. Peptides can be cleaved from PAM resins by transesterification (DBU/LiBr) without epimerization at the C-terminal stereocenter.[3]

Literature Procedures

For Loading onto the Linker.   A solution of Boc-valine (0.413 g, 1.90 mmol) and carbonyldiimidazole (0.276 g, 1.70 mmol) in DCM (10 mL) was kept at –5°C for 30 min and then added to a reaction vessel containing PAM-derivatized resin (1.00 g, 1.89 mmol/g loading). The residual solution was washed in DCM (2 mL), and the suspension was shaken for 8 h at room temperature. The resin was filtered, washed with DCM, and acetylated by being shaken with 1:1 pyridine–acetic anhydride (20 mL) for 40 min. Subsequent filtration, washing with DMF, DCM, 2-propanol, and DCM, and drying under vacuum afforded resin which contained 0.647 mmol of Val per gram of substituted resin (0.760 mmol of Val per gram of PAM-polystyrene resin).

For Cleavage.   After synthesis of a model tetrapeptide using standard Boc-amino acids, a portion (212 mg, 28.4 mmol) of the peptide-resin was shaken in a mixture of 16% HBr in 1:1 v/v acetic acid–trifluoroacetic acid for 75 min at room temperature. The model tetrapeptide was obtained in 35% yield and 98.0% purity. When the tetrapeptide PAM-resin was treated with anhydrous HF (10 mL) containing anisole (1 mL) for 1 h at 0°C, 11.6 mmol (86%) of the model tetrapeptide was released in 87% yield.

REFERENCES FOR SECTION 3.2.3

1. Mitchell, A. R.; Erickson, B. W.; Ryabtsev, M. N.; Hidges, R. S.; Merrifield, R. B. *J. Am. Chem. Soc.* **1976**, *98*, 7357–7362.
2. Zikox, C. C.; Ferderigos, N. G. *Tetrahedron Lett.* **1995**, *36*, 3741–3744.
3. Seebach, D.; Thaler, A.; Blaser, D.; Ko, S. Y. *Helv. Chim. Acta* **1991**, *74*, 1102–1119.

3.2.4 *tert*-Butyl Resin: 4-(1′,1′-Dimethyl-1′-hydroxypropyl)phenoxyacetyl (DHPP) Handle[1]

Points of Interest

1. DHPP resin significantly suppressed diketopiperazine formation of C-terminal dipeptides relative to traditional Wang resin.

2. Esterification of the resin was performed with Fmoc-amino acid chlorides.

3. The *tert*-butyl alcohol type resin was initially introduced by Wang and Merrifield to prepare protected peptide fragments.

4. The *tert*-butyl alcohol type resin is a useful alternative to Wang resin when strong bases are employed for solid-phase organic synthesis.[2]

Literature Procedures

For Loading onto the Linker. The DHPP resin was quantitatively esterified with Fmoc-Pro-Cl (5 equiv) in 40% pyridine (the Fmoc protecting group is stable to pyridine even after 48 h) in DCM after 10 h at room temperature (by amino acid analysis vs internal standard). Less than 1.5% racemization was detected after esterification. In contrast, esterification of the support-bound tertiary alcohol with DIC/DMAP was unsuccessful.

For Cleavage. Peptides were quantitatively cleaved from resin with either TFA/thioanisole (3 h) or 1 N HBF_4/thioanisole/TFA (4°C, 1 h).

REFERENCES FOR SECTION 3.2.4

1. Akaji, K.; Kiso, Y.; Carpino, L. A. *J. Chem. Soc., Chem. Commun.* **1990**, 584–586.
2. Chiron Mimotopes notes on handles available on the Multipin System (Chiron Mimotopes, Melbourne, Australia).

3.2.5   Wang Linker: *p*-Alkoxybenzyl Alcohol Resin [1]

Points of Interest

1. The first amino acid was loaded onto resin via esterification as the symmetric anhydride. Wang resin was initially used for the synthesis of protected peptides employing highly labile Bpoc-amino acids.

2. The Wang resin was used primarily for the preparation of Bpoc-protected peptide fragments until it was introduced as the linker of choice for Fmoc solid-phase synthesis by Sheppard and coworkers.[2]

3. Dipeptides prepared on Wang resin are prone to diketopiperazine formation, especially with C-terminal proline residues.

4. C-terminal cysteine and histidine residues are prone to racemization on Wang resin.

5. Peptides can be cleaved from Wang resins by transesterification (DBU/LiBr) without epimerization at the C-terminal stereocenter.[3]

6. Hydroxylamines were attached to Wang resin using Mitsunobu loading conditions in the solid-phase synthesis of hydroxamic acids.[4]

Literature Procedures

For Loading onto Linker.   Alkoxybenzyl alcohol resin (5 g, 4.4 mmol) was washed several times with DCM and then allowed to react with 2.5 g of Bpoc-L-Phe (6.3 mmol) and 1.3 g of DCC in the presence of 0.51 mL of pyridine for 150 min. After washings, 5.8 g of Bpoc-Phe-resin was obtained. The resin was then treated with 1.65 mL of pyridine and 1.95 mL of benzoyl chloride in 58 mL of DCM at 0°C for 15 min (alternatively remaining hydroxyl groups may be capped with acetic anhydride with triethylamine and a catalytic amount (0.05 equiv) of 4-(dimethylamino)pyridine). The resin was characterized by IR spectroscopy amino acid analysis, and CHN microanalysis indicated a loading of 0.41 mmol of phenylalanine per gram of resin. Less than 0.1% racemization was detected.

For Cleavage.   Peptides were removed from resin by treatment with 1:1 TFA/DCM for 30 min.

REFERENCES FOR SECTION 3.2.5

1. Wang, S. *J. Am. Chem. Soc.* **1973**, *95*, 1328–1333.
2. Atherton, E.; Logan, C. J.; Sheppard, R. C. *J. Chem. Soc., Perkin Trans. 1* **1981**, *1*, 538–546.
3. Seebach, D.; Thaler, A.; Blaser, D.; Ko, S. Y. *Helv. Chim. Acta* **1991**, *74*, 1102–1119.
4. Floyd, C. D.; Lewis, C. N.; Patel, S. R.; Wittaker, M. *Tetrahedron Lett.* **1996**, *37*, 8045–8048.

3.2.6   SASRIN (Super Acid Sensitive ResIN) [1]

Points of Interest

1. The SASRIN resin is similar to the dialkoxybenzyl linker developed by Sheppard for aminomethyl resin.
2. Between 0.5 and 0.7 mequiv/g of amino acid was loaded under optimized conditions. Racemization could be suppressed during loading except for the Cys and His derivatives.
3. Carboxylic acids are cleaved from SASRIN resin with 1% TFA in DCM.

Literature Summary

For Loading onto the Linker.   Alkoxide resins were esterified with Fmoc/*tert*-butyl-protected amino acids, DCC, and catalytic DMAP in a 1:4 mixture of DMF/DCM at 0°C overnight; unreacted groups are capped with benzoyl chloride/pyridine. Fmoc-amino acid resin is obtained in about 80% coupling yield.

For Cleavage. Cleavage was performed with 0.5–1% TFA in DCM.

REFERENCES FOR SECTION 3.2.6

1. Mergler, M.; Tanner, R.; Gosteli, J.; Grogg, P. *Tetrahedron Lett.* **1988**, *29*, 32, 4005–4008.

3.2.7 Hypersensitive Acid-Labile (HAL) Tris(alkoxy)benzyl Ester Anchor[1]

Points of Interest

1. Peptide acids are cleaved from dialkoxyphenyl linkers (see p 22) with 1% TFA/DCM, which can also remove *tert*-butyl side-chain protecting groups. Peptide acids are cleaved from more labile linkers, such as the Rink ester (see p 25) and the chlorotrityl resin (see p 24), with 10% acetic acid/DCM; however, with these linkers peptide acids can also be cleaved prematurely in the presence of a free C^α-carboxyl group of an incoming protected amino acid. The tris(alkoxy)benzyl ester (HAL) anchor has an intermediate acid lability.

2. The HAL linker is essentially the ester analog of the PAL linker used for the preparation of peptide amides.

Literature Summary

For Loading onto the Linker. The support-bound hydroxymethyl group was esterified, in 85–95% yield, by treatment with an Fmoc-amino acid derivative (5 equiv) and DIC (5 equiv) with catalytic DMAP (0.5 equiv) in DMF for 1 h. Less than 0.5% racemization was observed.

For Cleavage. The HAL linkage was stable to 0.1 M HOBt and Boc-amino acids in DMF, but the linkage was labile to 10% (v/v) acetic acid in DCM. Typically cleavage employed very dilute solutions of TFA in DCM. The model tripeptide linked to HAL was cleaved completely after 5 min with 0.1% (v/v) TFA or after 45 min with 0.05% TFA. Back-addition/alkylation was observed for peptides containing tryptophan when scavengers were omitted.

REFERENCES FOR SECTION 3.2.7

1. Albericio, F.; Barany, G. *Tetrahedron Lett.* **1991**, *32* (8), 1015–1018.

3.2.8   2-Chlorotrityl Chloride Resin [1]

Points of Interest

1. Unlike the standard trityl chloride resin, the 2-chlorotrityl chloride resin is sufficiently electron deficient to accommodate the formation of stable esters.

2. 2-Chlorotrityl chloride resin is rapidly esterified with Fmoc-amino acids and DIEA; for example, in 10 min an almost quantitative esterification with Fmoc-Ile was achieved (96%).

3. Esterification avoids electrophilic activation of the carboxyl group and consequently <0.05% racemization was detected. Loading via esterification is most efficient in DCM or DCE.

4. The linker is stable to DCC/HOBt and Bop/DIEA.

5. Carboxylic acids can be cleaved from the resin with AcOH/TFE/DCM or 0.5% TFA. These mild conditions allow cleavage from support in the presence of side-chain-protected amino acid derivatives.

6. The 2-chlorotrityl cation obtained after cleavage is mild and does not attack nucleophilic side chains of Trp, Met, and Tyr.

7. Zikos and Ferderigos have developed a 2-fluoro-4'-carboxytrityl linker that can be used for Fmoc-amino acid synthesis on any aminomethyl resin.[2]

Literature Procedures

For Loading onto the Resin.   2-Chlorotrityl chloride resin (1 g, 1.6 mmol of Cl⁻/g) and a limiting quantity of the Fmoc-amino acid in DCE (10 mL) were stirred for 5 min at room temperature. DIEA (0.44 mL, 2.5 mmol) in DCE (5 mL) was then added dropwise over 5 min and the mixture was stirred for another 20 min. The reaction (and excess support-bound 2-chlorotrityl chloride) was then quenched with methanol (1 mL) for 10 min. The resin was filtered, washed (3 × DCE, 2 × DMF, 2 × 2-propanol, methanol, and 2 × ether), and dried *in vacuo* for 24 h.

For Cleavage.   The amino acid or peptide-resin ester was suspended in a 1:1:8 mixture of AcOH/TFE/DCM. (Note: Trifluoroethanol is a very effective accelerator for cleavage from support; substituting methanol was far less effective.) The mixture was stirred at room temperature for about 1 h. The resin was separated by filtration and washed four times with the cleavage cocktail. For resin derivatized with Fmoc-amino acids or peptides, the cleavage reaction could be monitored by the ninhydrin test after the resin was subjected to Fmoc cleavage conditions (piperidine/DMF). This was only necessary for larger peptides that inhibited the rate of cleavage.

REFERENCES FOR SECTION 3.2.8

1. Barlos, K.; Chatzi, O.; Gatos, D.; Stavropoulos, G. *Int. J. Pept. Protein Res.* **1991**, *37*, 513–520.
2. Zikos, C. C.; Ferderigos, N. G. *Tetrahedron Lett.* **1994**, *35*, 1767–1768.

3.2.9 Rink Ester and Amide Resins [1]

$$X = OH, NH_2$$

Points of Interest

1. Loading the C-terminal Fmoc-Gly-OH (5 equiv) onto the support-bound secondary alcohol (Rink ester resin, 1 equiv) employing DCC (5.2 equiv), *N*-methylmorpholine (1.2 equiv), and catalytic DMAP (0.25 equiv) resulted in <1% dipeptide formation (indicating that premature Fmoc cleavage was not particularly problematic).

2. A potential limitation for peptide synthesis is the sensitivity of the Rink ester linkage to repeated use of HOBt; therefore Hunig's base was recommended to buffer the HOBt. Hunig's base was also recommended for peptide synthesis with the symmetric anhydrides of Fmoc-amino acids.

3. Peptides were cleaved from the Rink ester linkage under extremely mild conditions with acetic acid/DCM (1:9). Peptides were cleaved from the Rink amide linkage with TFA/DCM (50:50).

4. The Rink linkers are commercially available from a number of sources.

5. The Rink amide linker has also been used for the preparation of sulfonamides.[2]

Literature Summary

For Loading onto the Ester Linker. The support-bound secondary alcohol (2.00 g, 0.80 mmol) was esterified with Fmoc-Gly-OH (4.00 mmol) in 20 mL of DCE with DCC (4.20 mmol), DMAP (0.20 mmol), and NMM (1.00 mmol) at room temperature for 4 h. Unreacted hydroxy groups were acylated with benzoic anhydride (5 equiv) in pyridine/DMA (1:4). The Fmoc content of the derivatized resin was 0.36 mmol/g. Less than 1% Fmoc-Gly-Gly-OH (HPLC) was found in the cleavage product of the resin (0.1% TFA in DCM, 2 min, room temperature).

For Cleavage from the Ester Linkage. Cleavage from the resin was carried out with AcOH/DCM (1:9) for 1.5 h at room temperature. These conditions do not cleave Lys(Boc) or Tyr(*tert*-butyl) in detectable amounts (TLC). Cleavage from the resin was also performed by a 3-min treatment with 0.2% TFA in DCM at room temperature.

For Cleavage from the Amide Linkage.   Unoptimized conditions for cleavage of the peptide amide from the resin with concomitant removal of two *tert*-butyl groups employed TFA/DCM (1/1) for 15 min at room temperature and resulted in 80% yield of the expected peptide.

REFERENCES FOR SECTION 3.2.9

1. Rink, H. *Tetrahedron Lett.* **1987**, *28*, 3787–3790.
2. Beaver, K. A.; Siegmund, A. C.; Spear, K. L. *Tetrahedron Lett.* **1996**, *37*, 1145–1148.

3.2.10   SAC (Silyl ACid) Linker[1]

Points of Interest

1.   The SAC linker was prepared in three steps, and the preformed Fmoc-amino acid-SAC linker derivatives were coupled to support.
2.   Peptides were cleaved from the SAC linker with fluoride ions or 1% TFA/DCM.
3.   The SAC linker prevents diketopiperazine formation and tryptophan back-alkylation.

Literature Procedures

For Loading onto the Linker.   The preformed Fmoc-amino acid-SAC linker was attached to amino-functionalized support as the trichlorophenyl ester with HOBt/DIEA in DMF for 5 h. Peptide coupling reactions were carried out using 4 equiv of Fmoc-amino acid/BOP/DIEA for 60 min.

For Cleavage.   C-terminal tryptophan peptides were cleaved from support with 20 mL of TFA/thioanisole/phenol (95:2.5:2.5) solution for 60 min and isolated in 92% yield (70% purity).

REFERENCES FOR SECTION 3.2.10

1. Chao, H.; Bernatowicz, M. S.; Reiss, P. D.; Klimas, C. E.; Matsueda, G. R. *J. Am. Chem. Soc.* **1994**, *116*, 1746–1752.

3.2.11   Barany's Pbs Linker [1]

Points of Interest

1. Peptides were readily cleaved from the Pbs resin with TBAF in 2 min or with TFA in 1 h via a 1,6-elimination mechanism.
2. The Pbs linker is stable to Fmoc and Dts (dithiasuccinoyl) peptide synthesis strategies.
3. No racemization was observed in the synthesis of a tetrapeptide.
4. The linker was prepared via a 10-step synthesis and then prederivatized with C-terminal amino acids prior to loading onto the resin.

Literature Procedures

For Loading onto the Linker.   Preformed amino acid handles were prepared in solution with Fmoc-amino acids and dimethylformamide–dineopentyl acetal. The preformed handles were attached to support via their trichlorophenyl esters with DMAP and HOBt: the appropriate Fmoc-AA-Pbs handle (3 equiv) and HOBt (3 equiv) were dissolved in the minimal amount of DMF (ca. 0.1 M) which was added to the resin (1 equiv of free amino groups). DMAP (0.5 equiv) was added and the reaction mixture was shaken for 2 h. The resin was then washed with DMF (3×) and capped with acetic anhydride (3 equiv) and HOBt (3 equiv) in DMF.

For Cleavage.   The peptide resin (1 equiv of peptide based on amino acid analysis) was treated with a solution of tetrabutylammonium fluoride trihydrate (1 equiv), thiophenol (1.2 equiv), and (for protected peptides) DIEA (0.5 equiv) in DMF (1 mL per 100 mg of resin). After 2 min, the DMF solution was expressed under positive nitrogen pressure into a vessel containing both Dowex 1X 8-400 (hydroxide form) resin (2 equiv). Two further DMF washes of equal volume were also expressed into the mixed resin bed. After treatment for a total of 5 min, the resins were removed by filtration through a plug of glass wool. The DMF was removed at 2 mm and 35°C, and the resultant residues were triturated with diethyl ether.

REFERENCES FOR SECTION 3.2.11

1. Mullen, D. G.; Barany, G. *J. Org. Chem.* **1988**, *53*, 5240–5248.

3.2.12   Ramage's Silyl Linkers[1]

Points of Interest

1. Three structurally related silyl linkers were synthesized in three to six steps. Varying the substituents on the silyl group attenuates the reactivity of the linkers to fluoride. The trimethylsilyl linker shown was cleaved instantaneously and was thus the linker of choice.

2. Fluoride-induced fragmentation affords the tetrabutylammonium salt of the peptide, leaving the bis-quinone methide attached to the resin which can tautomerize to the stable resin-bound cinnamide.

3. Ramage's silyl linkers are not compatible with acid-labile protecting groups.

Literature Summary

For Loading onto the Linker.   Preformed amino acid handles were attached to resin employing a standard DCC/HOBt coupling. Peptides were prepared with Fmoc-amino acids activated by the phosphinic-carboxylic mixed anhydride method.[2]

For Cleavage.   A protected hexapeptide was cleaved from support by treatment with Bu$_4$NF (2 equiv) in DMF for 5 min and isolated in 62% yield after HPLC.

REFERENCES FOR SECTION 3.2.12

1. Ramage, R.; Barron, C. A.; Bielecki, S.; Thomas, D. W. *Tetrahedron Lett.* **1987**, *28*, 4105–4108.
2. Ramage, R.; Atrash, B.; Hopton, D.; Parrot, M. J. *J. Chem. Soc., Perkin Trans. 1* **1985**, *1*, 1617.

3.2.13   The First *o*-Nitrobenzyl Resin for Solid-Phase Synthesis of Boc-Peptide Acids[1]

Points of Interest

1. Peptide acids are removed from resin by photolysis at 350 nm under anaerobic conditions.

2. Boc and benzyl groups, as well as aromatic amino acids, are stable to the photolytic cleavage conditions.

3. Boc amino acids can be attached to the nitrobenzyl bromide resin under reflux with TEA or DIEA. Alkylation of tetramethylammonium salts or with other strong bases caused some decomposition.

4. A modified (2′-nitrobenzhydryl)polystyrene has been developed that provides increased acid stability during peptide synthesis with Boc-amino acids.[2]

Literature Procedures

For Loading onto *o*-Nitrobenzyl Bromide Linker. The 3-nitro-4-bromomethyl resin (4.0 g, 0.3 mmol of Br/g) was added slowly to the solution of Boc-glycine (0.70 g, 4.0 mmol) in 20 mL of ethyl acetate. Diisopropylethylamine (0.52 g, 4.0 mmol) was added and the suspension was gently heated at reflux for 48 h. The resin was collected by filtration, washed with EtOAc, MeOH, DCM, and MeOH (3 × 25 mL for 2 min), and dried *in vacuo* to give the desired product. The resin contained 0.3 mmol/g of Boc-glycine and no detectable bromine by the Dorman method.[3]

For Cleavage. A suspension of resin in anhydrous MeOH or EtOH was placed in a flask surrounded by a jacket containing a 40% $CuSO_4$ solution. Dissolved air was removed from the suspension by passing prepurified, oxygen-free nitrogen for 2 h through the solution, which was under a slight vacuum. The suspension was then irradiated at 350 nm for 18–24 h. Upon completion of photolysis, the suspension was filtered and the resin washed three times for 2 min with 20-mL portions of each of the following solvents: EtOH, DCM, DCM/EtOH, EtOH. The filtrate and washings were evaporated *in vacuo*. The crude product was purified by chromatography over a Sephadex LH-20 column in methanol.

REFERENCES FOR SECTION 3.2.13

1. Rich, D. H.; Gurwara, S. K. *J. Am. Chem. Soc.* **1975**, *97*, 1575–1579. (For other examples of the use of the *o*-nitrobenzyl group, see: Amit, B.; Zehavi, U.; Patchornik, A. *J. Org. Chem.* **1974**, *39*, 192.)
2. Ajayaghosh, A.; Pillai, V. N. R. *J. Org. Chem.* **1987**, *52*, 5714–5717.
3. Dorman, L. C. *Tetrahedron Lett.* **1969**, *28*, 2314.

3.2.14 *α*-Methylphenacyl Ester Linker[1]

Points of Interest

1. The linker is prepared in a single step from polystyrene–2% divinylbenzene and 2-bromopropionyl chloride. Alternatively, a similar 4-(2-bromopropionyl)phenoxyacetic acid linker can be used with (aminomethyl)polystyrene.[2] Both linkers are commercially available.

2. Peptides were cleaved from α-methylphenacyl linker by irradiation at 350 nm with a half-life of 5 h.

3. The reactivity of the α-methylphenacyl linker was compared with that of o-nitrobenzyl and benzyl linkers.

4. The anchoring bond of the α-methylphenacyl linker is completely stable to TFA/DCM (1:1) but labile to hydrazinolysis.

5. Related multidetachable p-alkoxybenzyl resins that undergo 1,6-elimination after photolytic cleavage have been reported.[3]

6. A photolabile α-sulfanyl-substituted phenyl ketone linker has been reported; cleavage of a C–S bond provided a mixture of disulfides or tolyl derivatives in varying, substrate-dependent ratios.[4]

Literature Summary

For Loading onto the Resin. The linker was prepared on support by Friedel–Crafts acylation of polystyrene–2% divinylbenzene beads (200–400 mesh) with 2-bromopropionyl chloride in the presence of $AlCl_3$ as catalyst. Alternatively, for the use of $FeCl_3$ as the catalyst, see p 14. The derivatized resin contained 0.94 mmol of Br/g. Boc-Gly-OH (4.38 g, 25 mmol) was dissolved in a mixture of 40 mL of EtOH and 10 mL of water and titrated to pH 7.0 with 20% Cs_2CO_3. The solution was evaporated to dryness and reevaporated twice with DMF (35°C). The cesium salt was stirred in DMF (80 mL) with 20 g (18.8 mmol) of derivatized resin for 17 h. The esterified resin was then washed and dried to give 20.5 g of derivatized resin. Amino acid analysis indicated 0.49 mmol of glycine per gram of resin. Virtually no residual bromide was left (0.13%).

For Cleavage. A protected tetrapeptide resin (10 g, 4.1 mmol) was suspended in 250 mL of DMF that had been treated with argon gas (2 mL/s) for 15 min inside a jacketed Pyrex tube (3.5 × 30 cm). The suspension was further flushed with argon for an additional 60 min with gentle magnetic stirring. The reaction mixture was then tightly stoppered and irradiated at 350 nm (16 × 24 W) in a Rayonet photochemical reaction chamber for 72 h with efficient water cooling (20°C). The released peptide was separated by suction filtration and the solvent removed at 40°C under reduced pressure to give 3.5 g of clear oil which solidified immediately on treatment with EtOAc. It was crystallized from THF and water: yield 2.42 g (77%). The residual resin (7.3 g) after photolytic cleavage contained 0.047 mmol of peptide according to amino acid analysis (thus photolysis was 92% complete).

REFERENCES FOR SECTION 3.2.14

1. Wang, S. *J. Org. Chem.* **1976,** *41,* 3258–3261.
2. Bellof, D.; Mutter, M. *Chimia* **1985,** *39,* 10.

3. Tam, J. P.; Dimarchi, R. D.; Merrifield, R. B. *Int. J. Pept. Protein Res.* **1980**, *16*, 412–425.
4. Soucholeiki, I. *Tetrahedron Lett.* **1994**, *35*, 7307.

3.2.15   Reductive Acidolysis Safety-Catch Linker[1]

Points of Interest

1. This safety-catch linker is conceptually similar to the SCAL linker described by Patek and Lebl (see p 50).

2. The 4-(2,5-dimethyl-4-(methylsulfinyl)phenyl)-4-hydroxybutanoic(DSB) acid linker was prepared in six steps in solution.

3. A similar safety-catch protecting group has been developed based on the 4-(methylsulfinyl)benzyl (Msob) group.

Literature Summary

For Loading onto the Resin.   The C-terminal amino acid was coupled to resin with the DIC/DMAP method with 1.7% racemization by the GITC method.[2]

For Cleavage.   The amount of cleavage by TFA/anisole for 24 h was 3.5%, whereas 90.5% of loaded amino acid was cleaved by reductive acidolysis [SiCl$_4$/thioanisole/anisole/TFA, 3 h]. γ-Endorphin was prepared using *N*-Boc-amino acids bearing reductive acidolysis-cleavable side-chain protecting groups in 62% yield after reverse phase FPLC (fast protein liquid chromatography).

REFERENCES FOR SECTION 3.2.15

1. Kiso, Y.; Fukui, T.; Tanaka, S.; Kimura, T.; Akaji, K. *Tetrahedron Lett.* **1994**, *35*, 3571–3574.
2. Nimura, N.; Ogura, H.; Kinoshita, T. *J. Chromatogr.* **1980**, *202*, 375.

3.2.16   Safety-Catch Linker Cleaved by Intramolecular Activation[1]

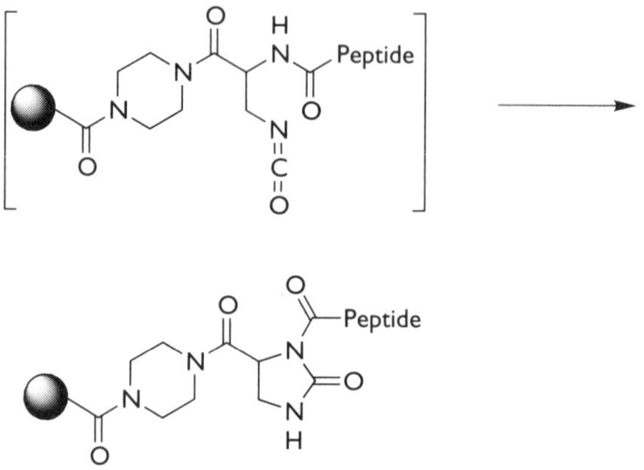

Points of Interest

1. The N^3-(phenyloxycarbonyl)-2,3-diaminopropionic acid residue behaves as a versatile linker [Dpr(Phoc)linker], displaying high stability under neutral and acidic conditions but undergoing activation under mild alkaline conditions for the release of peptide acids or amides by nucleophilic cleavage.

2. The Dpr(Phoc) residue is loaded to piperazine-derivatized resin; after peptide synthesis the isocyanate is generated at alkaline pH followed by fast and selective intramolecular reaction of the electrophile with the nearest primary amide group, leading to a five-membered ring. After cyclization, peptide acids or amides are formed from either hydrolysis or aminolysis.

3. The Dpr(Phoc) linker has also been employed for Fmoc-based solid-phase peptide synthesis of C-terminal peptides.[2]

4. The N^1-*tert*-butylcarbonyl-N^3-(phenyloxycarbonyl)-2,3-diaminopropionic acid residue was prepared via a Hoffmann rearrangement from Boc-Asn. The literature procedure[3] was modified by using I,I-diacetoxyiodobenzene instead of I,I-[bis(trifluoroacetoxy)]iodobenzene and by avoiding the use of pyridine. This modification to the original literature procedure has also been used for the Hoffmann rearrangement of Fmoc-Asn.[4]

Literature Procedures

For Loading onto the Linker.   The linker and other Boc-amino acids were assembled under the following reaction conditions on the hydrophilic polyacrylamide support Expansin, which was first derivatized with N-Boc-piperazine (0.75 mequiv/g). The Boc groups were removed with 6 M HCl for 35 min, followed by washes with water and DMF. Boc-amino acids were coupled in DMF as preformed HOBt esters using *in situ* neutralization. Most couplings were complete in 30 min, as monitored by a ninhydrin test.

For Cleavage.   The acylurea was formed from the protected phenyl carbonate by two successive treatments with 0.04 M NaOH (2 equiv, 1 h) in iPrOH/water (7:3). After neutralization, UV monitoring of the hydrolysate indicated the presence of phenol and the peptide in more than 90% yield. The peptides were purified by HPLC or silica gel chromatography.

REFERENCES FOR SECTION 3.2.16

1. Sola, R.; Saguer, P.; David, M.; Pascal, R. *J. Chem. Soc., Chem. Commun.* **1993**, 1786–1788.
2. Sola, R.; Mèry, J.; Pascal, R. *Tetrahedron Lett.* **1996**, *37*, 9195–9198.
3. Waki, M.; Kitajima, Y.; Izumiya, N. *Synthesis* **1981**, 266–267.
4. Bunin, B. A. Thesis, University of California at Berkeley, 1996.

3.2.17 Safety-Catch Linkage for the Direct Release into Aqueous Buffers[1]

Points of Interest

1. The Boc-protected imidazole linker is deprotected with TFA. Upon neutralization to pH 7, the imidazole residue acts as an intramolecular catalyst for cleavage via ester hydrolysis.

2. Linkages for the direct release of C-terminally modified peptides and peptide amides using diketopiperazine formation or ammonia vapor have been reported by Bray and coworkers.[2,3]

3. This safety-catch method is compatible with the N^α-Fmoc/*t*Bu solid-phase synthesis strategy; after the final acidic deprotection the linker is still blocked by protonation of the imidazole ring.

Literature Summary

For Loading onto the Linker. Loading was accomplished with MSNT/MeIm activation for esterification.[4]

For Cleavage into Aqueous Buffer. After final acidic deprotection and successive extractions of resin (3 × 10 min each) with MeOH/water (1:1) containing 0.1% HCl and 1 M acetic acid, the desired peptide acids were obtained by cleavage in 0.01 M phosphate buffer (pH 7.5) at 50°C for 25 min.

REFERENCES FOR SECTION 3.2.17

1. Hoffman, S.; Frank, R. *Tetrahedron Lett.* **1994**, *35*, 7763–7766.
2. Bray, A. M.; Maeji, N. J.; Geysen, H. M. *Tetrahedron Lett.* **1990**, *31*, 5811.
3. Bray, A. M.; Maeji, N. J.; Jhingran, A. J.; Valerio, R. *Tetrahedron Lett.* **1991**, *32*, 6163.
4. Blankemeyer-Menge, B.; Frank, R. *Tetrahedron Lett.* **1990**, *31*, 1701.

3.2.18 Allylic Linkers [1]

X = Br, Cl, OH

Points of Interest

1. Cleavage is carried out under practically neutral conditions with palladium: Boc and Fmoc protecting groups as well as glycoside bonds remain intact during cleavage.

2. An allylic linker for primary and secondary amino groups has also been reported. The allyl carbamate linkage was stable to acids and bases but labile to palladium-catalyzed hydrostannolysis.[2]

3. Allylic linkers have been cleaved under a number of different palladium-catalyzed reaction conditions. Palladium-catalyzed hydrostannolysis has been reported by Guibe and coworkers.[3] Albericio and coworkers reported that both the conditions described by Kunz and Guibe afforded peptides from resin in low yield.[4] In their hands, palladium-catalyzed cleavage of allylic linkers in 2:2:1 THF/DMSO/0.5 M HCl provided peptides in >90% yield. A number of problems, variable yields, and different solutions have been associated with the cleavage of allyl linkers and the deprotection of allyl carbamates on support. The combination of TMSN$_3$ and TBAF in conjunction with palladium initially developed by Shapiro and Buechler for the deprotection of allyl carbamates has proven reliable enough to be used in iterative encoding strategies.[5]

Literature Procedures

For Loading onto the Linker. The allyl halide resin was alkylated with the cesium salts of *N*-protected amino acids (molar ratio 1:1.5) to avoid possible racemization during esterification. Yields were 70–82% by elemental analysis.

For Cleavage. After standard Boc-amino acid synthesis, peptides and glycopeptides are cleaved from resin with 90 mg of Pd(PPh$_3$)$_4$ per mequivalent in 50 mL of THF and 5 mL of morpholine.

REFERENCES FOR SECTION 3.2.18

1. Kunz, H.; Dombo, B. *Angew. Chem., Int. Ed. Engl.* **1988**, *27*, 711–712.
2. Kaljuste, K.; Uden, A. *Tetrahedron Lett.* **1996**, *37*, 3031–3034.
3. Guibe, F.; Dangles, D.; Balavoine, G.; Loffet, A. *Tetrahedron Lett.* **1989**, *30*, 2641–2644.
4. Lloyd-Williams, P.; Jou, G.; Albericio, F.; Giralt, E. *Tetrahedron Lett.* **1991**, *32*, 4207–4210.
5. Shapiro, G.; Buechler, D. *Tetrahedron Lett.* **1994**, *35*, 5421–5424.

3.2.19 Fluorene-Derived Linker: *N*-[9-(Hydroxymethyl)-2-fluorenyl]succinamic Acid (HMFS) handle [1,2]

Points of Interest

1. The fluorene linker is compatible with the Boc strategy and cleaved with piperidine, although cleavage with morpholine afforded slightly purer products.

2. The HMFS handle was prepared in four steps in solution from 2-aminofluorene in 55% overall yield.

3. An appropriately derivatized HMFS handle can be used for oligonucleotide synthesis.[3]

Literature Summary

For Loading onto the HMFS Handle. The C-terminal Boc-amino acid (5 equiv) was anchored to the hydroxymethyl group of the resin-bound handle with a double coupling using DCC in the presence of DMAP (0.5 equiv) in DMF for 1 h at room temperature.

For Cleavage. The handle is stable to DMF for 24 h at room temperature and 5% DIPEA in DCM for 2 h at room temperature (<8% loss of product from resin was observed after 24 h at room temperature). The acylamido group was specifically designed to provide a linker with increased stability to DIPEA relative to previously developed fluorene linkers. The peptides were cleaved with 20% morpholine DMF for 2 h at room temperature in >93% purity by HPLC.

REFERENCES FOR SECTION 3.2.19

1. Rabanal, F.; Giralt, E.; Albericio, F. *Tetrahedron Lett.* **1992**, *33*, 1775–1778.
2. Rabanal, F.; Giralt, E.; Albericio, F. *Tetrahedron Lett.* **1995**, *51*, 1449–1458.
3. Eritja, R.; Robles, J.; Albericio, F.; Pederoso, E. *Tetrahedron* **1992**, *40*, 4171–4182.

3.2.20 Base-Labile Linker: For Cleavage via β-Elimination [1]

Points of Interest

1. The linker was compatible with both Fmoc and Boc cleavage conditions in model studies.

2. Although the sulfone was prepared in solution, this base-labile linker could potentially be used as a safety-catch linker by converting a support-bound thioether to a sulfone following solid-phase synthesis.

Literature Procedure

The solid-phase synthesis of leu-enkephalin was carried out with both Boc- and Fmoc-amino acids using the DCC/HOBt coupling method. The desired peptide (54–60% yield) was cleaved from the resin in 30 min with dioxane/ MeOH/4 N NaOH (30:9:1).

REFERENCES FOR SECTION 3.2.20

1. Katti, S. B.; Misra, P. K.; Haq, W.; Mathur, K. B. *J. Chem. Soc., Chem. Commun.,* **1992,** *11,* 843–844.

3.2.21   An Oxidation-Labile Linker[1]

Points of Interest

1. Cross-linked poly(2-hydroxyethyl methacrylate) derivatized with phenylhydrazine groups was used as a resin for peptide synthesis. It was obtained by a two-step modification of the commercially available arylamino-containing polymer (Spheron ArA 1000, LaChema, Czechoslovakia).

2. The phenylhydrazide linker is compatible with acid, base, and reduction-labile protecting groups and may thus be applicable for the synthesis of protected peptides.

3. The phenylhydrazide group was used for masking the carboxylic function. It was removed under mild oxidative conditions using copper(II) complexed to nitrogen-containing ligands as the catalyst.

4. Peptides containing Trp and Met were prepared as a model due to their sensitivity to oxidation. Cys residues were protected during cleavage.

Literature Procedure

The model peptide was prepared on the hydrazine resin with Boc-amino acids under standard conditions (DIC/HOBt). After completion of the synthesis the peptidyl-polymer (1 g) was suspended in a mixture containing 8 mL of DMF, 2 mL of 1 M pyridine/acetate buffer in DMF, 9 mL of 20% aqueous acetic acid, and 1 mL of 0.5 M aqueous $CuSO_4$. The mixture was stirred in the air for 16 h to accomplish the copper complex-catalyzed oxidation of phenylhy-drazide to the highly active, unstable phenyldiimide, which resulted in sponta-neous decomposition with evolution of nitrogen gas. One major peak was ob-served by HPLC (83%) and its structure was confirmed by NMR. No products resulting from the oxidation of Met or Trp were detected.

REFERENCES FOR SECTION 3.2.21

1. Semenov, A. N.; Gordeev, K. Y. *Int. J. Pept. Protein Res.* **1995**, *45*, 303–304.

3.3   LINKERS FOR AMIDES

TABLE 3.7   Acid-Labile Linkers for Amides (Note: The Linkers Are Organized by Increasing Acid Lability

Linker name	Page	Cleavage conditions and comments
Rink amide linker	38	1:1 TFA/DCM
PAL linker	39	90:5:3:2 TFA/thioanisole/EDT/anisole (2 h), but slightly more labile than Rink linker
CHA/CHE linkers	40	85:10:5 DCM/TFA/phenol, but slightly more labile than PAL
Sieber's Xan linker	41	2:98 TFA/DCE

TABLE 3.8   Silyl-Based Linkers for Amides

Linker name	Page	Cleavage conditions and comments
Silyl amide linker (SAL)	42	95:5:3:2 TFA/EDT/thioanisole/phenol (1 h)

TABLE 3.9   Photolabile Linkers for Amides

Linker name	Page	Cleavage conditions and comments
Photolabile primary amide linker	42	Rayonet hv, in 5:14:1 TFE/DCM/Aq NH$_4$OH (16 h)
Modified o-nitrobenzyl linker	43	365-nm hv, in PBS containing 5% DMSO (3 h)
Photolabile alkylamide linkers	44	350-nm hv, in 1:4 TFE/DCM (24 h)

TABLE 3.10   Safety-Catch Linkers for Amides

Linker name	Page	Cleavage conditions and comments
Oxidative phenol–sulfide linker	46	Activation with H$_2$O$_2$, cleavage with amines
Kenner's sulfonamide linker	47	Activation with diazomethane or iodoacetonitrile, cleavage with amines, hydrazine, or hydroxide
Safety-catch acid labile (SCAL) linker	50	1 M Me$_3$SiBr/thioanisole/TFA at 0°C (2 h)

TABLE 3.11   Other amide linkers

Linker name	Page	Cleavage conditions and comments
Oxime resin	50	hydrazines, amines, or anilines with AcOH catalyst

3.3.1   Rink Ester and Amide Resins [1]

X = OH, NH$_2$

Points of Interest

1. The C-terminal Fmoc-Gly-OH (5 equiv) was loaded onto the support-bound secondary alcohol (Rink ester resin, 1 equiv) employing DCC (5.2 equiv), N-methylmorpholine (1.2 equiv), and catalytic DMAP (0.25 equiv) and resulted in <1% dipeptide formation (indicating that premature Fmoc cleavage was not particularly problematic).

2. A potential limitation for peptide synthesis is the sensitivity of the Rink ester linkage to repeated use of HOBt; therefore Hunig's base was recommended to buffer the HOBt. Hunig's base was also recommended for peptide synthesis with the symmetric anhydrides of Fmoc-amino acids.

3. Peptides were cleaved from the Rink ester linkage under extremely mild conditions with acetic acid/DCM (1:9). Peptides were cleaved from the Rink amide linkage with TFA/DCM (50:50).

4. The Rink linkers are commercially available from a number of sources.

5. The Rink amide linker has also been used for the preparation of sulfonamides.[2]

Literature Summary

For Loading onto the Ester Linker.   The support-bound secondary alcohol (2.00 g, 0.80 mmol) was esterified with Fmoc-Gly-OH (4.00 mmol) in 20 mL of DCE with DCC (4.20 mmol), DMAP (0.20 mmol), and NMM (1.00 mmol) at room temperature for 4 h. Unreacted hydroxy groups were acylated with benzoic anhydride (5 equiv) in pyridine/DMA (1:4). The Fmoc content of the derivatized resin was 0.36 mmol/g. Less than 1% Fmoc-Gly-Gly-OH (HPLC) was found in the cleavage product of the resin (0.1% TFA in DCM, 2 min, room temperature).

For Cleavage from the Ester Linkage.   Cleavage from the resin was carried out with AcOH/DCM (1:9) for 1.5 h at room temperature. These conditions do not cleave Lys(Boc) or Tyr(*tert*-butyl) in detectable amounts (TLC). Cleavage from the resin was also performed by a 3-min treatment with 0.2% TFA in DCM at room temperature.

For Cleavage from the Amide Linkage.   Unoptimized conditions for cleavage of the peptide amide from the resin with concomitant removal of two *tert*-butyl groups employed TFA/DCM (1/1) for 15 min at room temperature and resulted in 80% yield of the expected peptide.

REFERENCES FOR SECTION 3.3.1

1. Rink, H. *Tetrahedron Lett.* **1987**, *28*, 3787–3790.
2. Beaver, K. A.; Siegmund, A. C.; Spear, K. L. *Tetrahedron Lett.* **1996**, *37*, 1145–1148.

3.3.2   PAL Linker: Fmoc-(Aminomethyl)-3,5-dimethoxyphenoxyvaleric Acid (PAL) Resin [1]

Points of Interest

1. Support-bound amides were cleaved twice as fast from the PAL linker as the Rink linker ($t_{1/2}$ 4 min vs 9 min in 95/5 TFA/phenol).

2. The (aminomethyl)-3,5-dimethoxyphenoxyvaleric acid (PAL) linker was prepared from 4-hydroxy-2,6-dimethoxybenzaldehyde in five steps in 74% overall yield. An efficient route to 4-hydroxy-2,6-dimethoxybenzaldehyde via the silyl-protected phenol has been reported by Landi and Ramig.[2]

3. The PAL linker is typically used with Fmoc-amino acid synthesis to provide peptide amides.

4. Boojamra and coworkers have employed a related linker for the solid-phase synthesis of 1,4-benzodiazepine-2,5-diones (see p 150).[3]

5. The PAL linker is commercially available.

Literature Procedures

For Loading onto the Resin.   Norleucyl ("internal reference") (amidomethyl)poly(styrene-*co*-1% divinylbenzene) resin was derivatized with the Fmoc-PAL handle under standard HOBt/DIC conditions (for most researchers this step is unnecessary since PAL-derivatized resins are commercially available). The Fmoc group was completely removed from the handle with piperidine–DMF (1:1 v/v), 1-min prewash + 10-min treatment. The Fmoc group on the protected PAL handle is somewhat harder to completely deprotect than a standard amino acid-bound Fmoc group (treatment with piperidine–DCM (1:1 v/v) resulted in only partial removal). Automated peptide synthesis was performed on Pepsyn K (Kieselguhr-encapsulated poly(dimethylacrylamide)) with Fmoc-protected amino acids using the following repetitive sequence on a MilliGen/

Biosearch Model 9600 Peptide Synthesizer: DCM/DMF (1:1 v/v) washes, 10 × 0.3 min; piperidine/DMF (3:7 v/v), 2 min then 8 min; DCM/DMF (1:1 v/v) washes, 10 × 0.3 min; Fmoc-AA/BOP/HOBt/NMM (1:1:1:1) in DMF for 2 h.

For Cleavage. Cleavage of the complete peptide-resin was achieved with exposure to freshly prepared reagent R, TFA/thioanisole/EDT/anisole (90:5:3:2) (1 mL used per 100 mg of peptide-resin), 2 h at room temperature. Alkylation of Trp residues is a commonly observed side reaction (note: this side reaction can be minimized by using the chlorotrityl resin).[4]

REFERENCES FOR SECTION 3.3.2

1. Albericio, F.; Kneib-Cordonier, N.; Biancalana, S.; Gera, L.; Masada, R. I.; Hudson, D.; Barany, G. *J. Org. Chem.* **1990**, *55*, 3730–3743.
2. Landi, J. L.; Ramig, K. *Synth. Commun.*, **1991**, *21*, 167–171.
3. Boojamra, C.; Burrow, K.; Ellman, J. A. *J. Org. Chem.* **1995**, *60*, 5742–5743.
4. Barlos, K.; Chatzi, O.; Gatos, D.; Stavropoulos, G. *Int. J. Pept. Protein Res.* **1991**, *37*, 513–520.

3.3.3 CHA and CHE Linkers [1]

Points of Interest

1. The CHE linker was designed for the preparation of side-chain-protected peptide amides.

2. Both the CHA and CHE linkers are more acid labile than PAL or Rink amide linkers. The following cleavage data were obtained for a support-bound pentapeptide in 25% TFA/DCM: $t_{1/2}(\text{Rink}) = 38$ min, $t_{1/2}(\text{PAL}) = 35$ min, $t_{1/2}(\text{CHA}) < 4$ min, $t_{1/2}(\text{CHE}) < 2$ min.

3. The CHA linker provides peptide amides in higher yield than the CHE linker, particularly for peptides containing acid-labile residues such as Trp and/or Tyr(SO_3H). The CHA linker is also more readily synthesized than the CHE linker.

Literature Procedures

For Loading onto the Linkers. Standard peptide synthesis with PyBOP, HOBt, and NMM (1:1:1.5 molar ratio) was performed on CHA- and CHE-derivatized aminomethyl-TentaGel S. D/L amino acid analysis revealed that there was no racemization.

For Cleavage. Cleavage was complete for Fmoc-Val after exposure to 10% TFA/5% phenol in DCM for 1 h.

REFERENCES FOR SECTION 3.3.3

1. Noda, M.; Yamaguchi, M.; Ando, E.; Takeda, K.; Nokihara, K. *J. Org. Chem.* **1994**, *59*, 7968–7975.

3.3.4 9-Xanthenyl (Xan) Linker: For Cleavage of Peptide Amides by Very Mild Acidolysis [1]

Points of Interest

1. Sieber's Xan linker is more labile than the corresponding Rink amide linker: peptide amides were cleaved from the Xan resin with 98:2 dichloroethane/trifluoroacetic acid.

2. Cleavage from support was faster (<8 min) than aminolysis of sterically hindered peptide esters to provide peptide amides as described by Sheppard (see p 18).

3. The linker is prepared via a five-step sequence; it is also commercially available from a number of sources.

4. The 9-xanthenyl group has also been used to protect Asn/Gln side chains.

5. The resin can be reductively alkylated to provide a support suited to the synthesis of secondary carboxamides.[3]

Literature Summary

For Loading onto the Linker. Fmoc cleavage and protected amino acid couplings were performed by standard methods (piperidine and HOBt/DCC).

For Cleavage. Cleavage of a simple model peptide was accomplished with TFA/DCE (2:98) or TFA/DCE/EDT (2:98:0.1).

REFERENCES FOR SECTION 3.3.4

1. Sieber, P. *Tetrahedron Lett.* **1987**, *28*, 2107–2110.
2. Han, Y.; Bontems, S. L.; Hegyes, P.; Munson, M. C.; Minor, C. A.; Kates, S.; Albericio, F.; Barany, G. *J. Org. Chem.* **1996**, *61*, 6326–6339.
3. Chan, W. C.; Mellor, S. *J. Chem. Soc., Chem. Commun.* **1995**, 1475–1477.

3.3.5   SAL (Silyl Amide Linker) [1]

Points of Interest

1. Upon exposure to TFA the α-silyl cation can undergo elimination to produce a vinyl phenyl ether on support.

2. The silyl amide linker minimizes irreversible back-alkylation onto the resin commonly observed for C-terminal tryptophan peptides prepared with linkers that form a stable carbocation upon cleavage. With scavengers in the cleavage cocktail, a peptide containing a C-terminal tryptophan residue was obtained in 90% yield after cleavage from SAL. Without scavengers, the peptide with a C-terminal tryptophan was isolated in 45% yield.

3. The silyl amide linker was synthesized in solution in six steps.

Literature Procedures

For Loading onto the Linker. The Fmoc group was cleaved under standard conditions (20% piperidine, 15 min). Peptide chains were assembled using 4 equiv of preformed pentafluorophenyl esters in the presence of equimolar HOBt/DIEA.

For Cleavage. After completion of the synthesis, the resin sample was treated with 15 mL of TFA/EDT/phenol/thioanisole solution (90:5:3:2) at room temperature for 1 h. The solution was collected by filtration and the resin washed thoroughly with 5 mL of the cleavage solution.

REFERENCES FOR SECTION 3.3.5

1. Chao, H.; Bernatowicz, M. S.; Matsueda, G. R. *J. Org. Chem.* **1993**, *58*, 2640–2644.

3.3.6   Photolabile Amide Linkers [1]

P = Boc, Fmoc, Dts

Points of Interest

1. A range of protected (aminomethyl)-*o*-nitrobenzyl linkers were prepared in solution. Dts was the optimal protecting group for the handle.
2. Peptide amides were obtained upon cleavage.

Literature Procedures

For Loading onto the Linker. The Dts-protected handle (56 mg, 0.18 mmol) was dissolved in 0.1 M HOBt/DMF (5 mL) and was added to an aminoacylamidomethyl resin (500 mg, 0.09 mmol). This was followed by DCC (41 mg, 0.2 mmol) or DIC (31 μL, 0.2 mmol), and after 2 h of shaking, ninhydrin tests were negative. The resin was then filtered, washed with DMF, DCM, and MeOH, and dried at 2 mm over P_2O_5. The Dts group was removed with β-mercaptoethanol (0.5 M)–DIEA (0.5 M) in DCM (3 × 2 min), followed by washing with DCM (5×). Stepwise solid-phase peptide synthesis proceeded under standard Fmoc conditions.

For Cleavage. Peptide resins were cleaved for 16 h with the Rayonet apparatus. Cleavage yields were highly solvent dependent. The highest yields (83% for a tetrapeptide amide) were obtained in TFE/DCM/aqueous NH_4OH (5:14:1 v/v/v). The corresponding peptide acid was always present, albeit in modest relative amounts (5−6%).

REFERENCES FOR SECTION 3.3.6

1. Hammer, R. P.; Albericio, F.; Gera, L.; Barany, G. *Int. J. Pept. Protein Res.* **1990**, *36*, 31−45.

3.3.7 A New *o*-Nitrobenzyl Photolabile Linker for Combinatorial Synthesis [1]

Points of Interest

1. The linker was prepared in solution in seven steps that do not require any chromatography.
2. Introduction of an α-methyl group onto the benzylic carbon (Pillai and coworkers also have described an α-methyl-*o*-nitrobenzyl linker prepared by nitration of resin[2]) greatly facilitates cleavage with >350-nm light and results in the formation of a nitroso ketone. Other *o*-nitrobenzyl linkers generate a nitroso aldehyde on support which can trap liberated compounds such as amines.
3. The linker was used for the synthesis and cleavage of a 4-thiazolidinone and a cholecystokinin peptide.

4. As with other photolabile linkers, this modified photolabile linker was stable to acidic or basic conditions for protecting group manipulations.

5. Photolytic cleavage was performed in an aqueous buffer with 5% DMSO to allow direct biological evaluation after cleavage.

Literature Procedures

For Loading onto the Linker.   Standard Fmoc amino acid chemistry was used for amide bond formation.

For Cleavage.   4-Thiazolidinone was obtained in >90% yield (95% purity) after photolytic cleavage in pH 7.4 phosphate-buffered saline (PBS) containing 5% DMSO (simulating a cleavage cocktail appropriate for transfer to a biological assay) by irradiating for 3 h with 365-nm UV light. A cholecystokinin peptide (H-Met-Gly-Trp-Met-Asp-Phe-NH$_2$) was cleaved with 1 h of irradiation in the presence of hydrazine as a scavenger in 87% purity (75% yield).

REFERENCES FOR SECTION 3.3.7

1. Holmes, C. P.; Jones, D. G. *J. Org. Chem.* **1995**, *60*, 2318–2319.
2. Ajayaghosh, A.; Pillai, R. *Tetrahedron* **1988**, *44*, 6661–6666.

3.3.8   *N*-Methyl- and *N*-Ethylamides of Peptides [1]

R = Me, Et

Points of Interest

1. Alkylamines were incorporated into the ((3-nitro-4-(bromomethyl)-benzamido)methyl)polystyrene resin prior to peptide synthesis.

2. Photolysis provides C-terminal-modified *N*-alkylamido peptides.

3. Amino esters have been incorporated into a (chloromethyl)-*o*-nitrobenzyl resin to prepare peptide esters.[2]

Literature Procedures

For Loading onto the Linker.   The ((3-nitro-4-(bromomethyl)benzamido)-methyl)polystyrene resin (10 g, 0.56 mmol of Br/g) was suspended in DCM (100 mL) in a 250-mL stoppered bottle and cooled to 0°C. Dry methylamine or ethylamine was bubbled through the suspension for 2 h. The reaction bottle was tightly stoppered and shaken for 12 h at room temperature. The resin was collected by filtration, washed with DCM (50 mL × 3 × 3 min), THF (50 mL × 2 × 2 min), water (50 mL × 6 × 3 min), and MeOH (40 mL × 3 × 3 min), and dried *in vacuo*. The resin had a substitution of 0.54–0.60 mmol of HN/g as indicated by Gisin's picric acid method (see p 217).[3]

The alkylamido resins were placed in a silanized glass reaction vessel clamped to a manually operated mechanical shaker. In separate experiments, symmetric anhydrides of Boc-amino acids were prepared by reacting 3 equiv of the respective Boc-amino acid in DCM with 1.5 equiv of DCC for 1 h at 0°C. The precipitated dicyclohexylurea (DCU) was removed by filtration, and the symmetric anhydride solution was added to the resin. After 3 h of shaking, the resin was collected by filtration and washed with DCM (15 mL × 3 × 3 min) and MeOH (15 mL × 3 × 3 min). Each coupling was carried out in duplicate to minimize error sequences. Each step of the coupling was thoroughly monitored by the semiquantitative ninhydrin method (see p 214). Coupling of a Boc-amino acid to the proline resin was monitored by the chloranil test.[4] After each Boc-amino acid was incorporated, the resin was heated with 4 N HCl–dioxane[5] (20 mL) for 30 min, filtered, and washed with dioxane (15 mL × 2 × 3 min) and DCM (15 mL × 3 × 3 min). The deprotected resin was neutralized with 10% DIEA in DCM (20 mL) by shaking for 10 min. The neutralized resin was washed with DCM (15 mL × 3 × 3 min). The coupling, deprotection, and washings were repeated until the desired sequence was achieved.

For Cleavage.   After the synthesis, peptides were removed from the supports by photolyzing. The peptide alkylamido resin (1.8 g) was suspended in a mixture of TFE/DCM (20%, 200 mL) and placed in an immersion-type photochemical reactor. The suspension was degassed with dry nitrogen for 1 h and irradiated with a Philips HPK 125-W medium-pressure mercury lamp at 350 nm for 24 h. The crude protected peptide *N*-methylamide was obtained in 66% yield. This was then treated with liquid HF (10 mL) for 30 min at 0°C in the presence of anisole (1 mL) to remove side-chain protecting groups. *Caution:* breathing HF can cause death! Even liquid HF reacts with glass; do not use HF without the appropriate training and equipment. The excess HF was blown off through a NaOH solution with dry nitrogen. The residue was dried *in vacuo* over KOH, and the free peptide was treated with ether for the removal of anisole. The solvent was removed to obtain the crude deblocked nonapeptide *N*-methylamide and purified on a Sephadex LH-20 column in 56% yield.

REFERENCES FOR SECTION 3.3.8

1. Ajayaghosh, A.; Pillai, V. N. R. *J. Org. Chem.* **1990,** *55,* 2826–2829.
2. Renil, M.; Pillai, V. N. R. *Tetrahedron Lett.* **1994,** *35,* 3809–3812.
3. Gisin, B. F. *Anal. Chim. Acta* **1972,** *58,* 248.
4. Christensen, T. *Acta Chem. Scand.* **1979,** *33,* 763.
5. The authors had previously demonstrated that the 2′-nitrobenzhydryl ester linkage was more stable to 4 N HCl–dioxane than TFA/DCM (1 : 1): Ajayaghosh, A.; Pillai, V. N. R. *J. Org. Chem.* **1987,** *52,* 5714–5717.

3.3.9 Oxidative Phenol–Sulfide Safety-Catch Linker[1]

Points of Interest

1. The phenol–sulfide linker was used for the preparation of protected peptide fragments.

2. When the sulfide is oxidized to the sulfone, the anchoring ester linkage is converted into an activated ester.

3. Cysteine, methionine, and tryptophan can also react with the peroxide that is employed to activate the phenol–sulfide linker.

Literature Procedures

For Loading onto the Linker.   Boc-Glu(OBz)OH (0.53 g, 1.5 mmol) was dissolved in 15 mL of DCM and added to 1.0 g of derivatized resin. DCC (0.31 g, 1.5 mmol) dissolved in 15 mL of DCM was then added and the mixture was stirred overnight. The resin was filtered, washed with ethanol, acetic acid, and ethanol, and dried. The amount of glutamic acid coupled to support was ca. 0.8 mmol/g.

For Cleavage.   The peptide polymer was treated with 2 mL of 30% H_2O_2 in 20 mL of acetic acid with stirring for 12 h at room temperature. After filtering and washing with ethanol, removal of the peptide from the polymer was accomplished by stirring for 24 h with 0.75 mmol of glycine (as the sodium salt) in a DMF/water solvent. The mixture was filtered and the filtrate was evaporated to dryness. The residue was dissolved in water, the pH was adjusted to 3.5, and the precipitate that formed was collected by decantation. The wet residue was dissolved in ethanol and evaporated to dryness. Protected peptides were further purified by recrystallization or chromatography.

REFERENCES FOR SECTION 3.3.9

1. Marshal, D. L.; Liener, I. E. *J. Org. Chem.* **1970**, *35*, 867–868.

3.3.10 Kenner's Safety-Catch Linker [1,2]

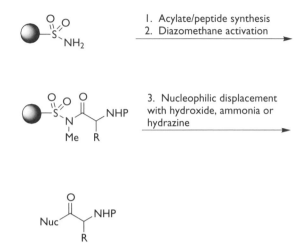

1. Acylate/peptide synthesis
2. Diazomethane activation

3. Nucleophilic displacement with hydroxide, ammonia or hydrazine

Points of Interest

1. The linker is completely stable to basic or strongly nucleophilic agents, as well as strongly acidic reaction conditions prior to activation.

2. About 7% of the sulfonated ion-exchange resin was not converted into the sulfonamide. An improved method for loading a sulfonamide via a standard amide bond has been developed.[2]

3. Activation by treatment with diazomethane provided the N-methylacyl-sulfonamide that can be cleaved with hydroxide or nucleophilic amines.

4. Although initially developed for preparation of peptide acids, amides, and hydrazides, more recently the linker was adapted for solid-phase organic synthesis of arylacetic acids. The linker was stable to basic reaction conditions required for acylsulfonamide enolate alkylation and Suzuki reactions.[2] After activation with diazomethane, cleavage conditions were extended to include a range of nucleophilic amines.

Literature Procedures

Original Method for Preparing and Derivatizing the Sulfonamide Resin. Fully sulfonated styrene–2% divinylbenzene copolymer was converted, by treatment with chlorosulfonic acid and then aqueous ammonia, into the polysulfonamide. About 7% of the sulfonic groups were inaccessible to these reagents and remained unchanged; their presence did not interfere with the subsequent steps. Treatment of the sulfonamide resin in DMF with the 2,4,5-trichlorophenyl esters of various protected amino acids (2 equiv based on total sulfonamide) and triethylamine (1 equiv based on sulfonamide + sulfonic acid) gave up to 25% incorporation of the amino acid (ca. 1 mmol/g) as the acylsulfonamides. Further acylation of the resin sulfonamide seemed inhibited by the electrostatic effects within the resin matrix, ionization of the more acidic acylsulfonamide groups inhibiting ionization (and hence acylation) of the residual sulfonamide.

A Method for Loading the Sulfonamide onto the Resin via a Standard Amide Bond. To a 250-mL three-neck round-bottom flask fitted with an overhead stirrer were added dry aminomethylated macroreticular resin (15 g, 5.9 mmol), 4-carboxybenzenesulfonamide (3.1 g, 16 mmol), DIC (2.4 mL, 16 mmol), HOBt (2.1 g, 16 mmol), and THF (100 mL). The slurry was stirred for 10 h after which the resin was filtered and washed with THF (3×) and DCM (3×). Reaction completion could be monitored by the standard ninhydrin test[3] (see p 214); a positive ninhydrin test for unreacted amine gives an intense blue color and a positive test for unreacted sulfonamide gives a pale red color.

For Activation and Cleavage.   The acylsulfonamide bonds were labilized by *N*-methylation with diazomethane in ether–acetone. The protected peptide could then be removed from resin by alkaline hydrolysis (1 equiv of 0.5 N NaOH), by aminolysis (an excess of 0.5 N ammonia in dioxane), or by hydrazinolysis (3 equiv of methanolic hydrazine).

REFERENCES FOR SECTION 3.3.10

1. Kenner, G. W.; McDermott, J. R.; Sheppard, R. C. *J. Chem. Soc., Chem. Commun.* **1971**, *636*– 637.
2. Backes, B. J.; Ellman, J. A. *J. Am. Chem. Soc.* **1994**, *116*, 11171–11172.
3. Kaiser, E.; Colescot, R. L.; Bossinger, C. D.; Cook, P. I. *Anal. Biochem.* **1970**, *34*, 595–598.

3.3.11   A Highly Reactive Acylsulfonamide Linker.[1,2]

Points of Interest

1. The cyanomethyl linker is highly labile to nucleophilic displacement, with $t_{1/2} < 5$ min for displacement with 0.007 M benzylamine in DMSO. For comparison, the $t_{1/2}$ for the corresponding *N*-methyl derivative under the same conditions is 790 min.

2. Cyanomethylation provides a highly activated linker which can be readily cleaved with amines. Treatment of the activated acylsulfonamide with limiting amounts of an amine nucleophile results in complete consumption of the amine to provide the amide product in pure form. This is particularly useful for combinatorial synthesis because additional diversity can be incorporated during the cleavage step. When a limiting amount of an equimolar mixture of amines is added to the resin, an equimolar mixture of extremely pure amide products is produced.

3. The highly activated linker reacts with relatively poor nucleophiles such as aniline and *tert*-butylamine.

Literature Procedures

For Loading onto the Linker. Sulfonamide-derivatized macroreticular resin was acylated with the symmetric anhydride of a carboxylic acid and catalytic DMAP as follows (this method has proven successful with a number of carboxylic acids and represents an improvement over the PFP ester method initially reported). To a 100-mL round-bottom flask were added 3-(3,4,5-trimethoxyphenyl)propanoic acid (8.6 g, 36 mmol), DCM (60 mL), and DIC (2.8 mL, 18 mmol). After 8 h of stirring, the solution was cooled with an ice bath and the urea precipitate was filtered. The filtrate was added to a 250-mL round-bottom flask fitted with an overhead stirrer along with sulfonamide-derivatized macroreticular resin (10 g, 3.9 mmol), DMAP (44 mg, 0.36 mmol), DIEA (2.1 mL, 12 mmol), and DCM (10 mL). The slurry was stirred for 39 h and filtered, and the resin was washed with THF (200 mL), 5% TFA in THF (to protonate the support-bound acylsulfonamide), THF (200 mL), and MeOH (200 mL). The resin was dried with rotary evaporation and then under high vacuum for 24 h. The acylation of the sulfonamide could be monitored by the ninhydrin test.[3]

For Cleavage. A highly activated acylsulfonamide linker was activated as follows. To a 25-mL round-bottom flask were added resin (500 mg, 0.17 mmol), DMSO (4 mL), DIEA (136 μL, 0.85 mmol), and bromoacetonitrile or iodoacetonitrile (4.0 mmol). After the mixture was stirred for 24 h, the resin was filtered and washed with DMSO (5 × 5 mL) and THF (3 × 5 mL). To the resin were added THF (3 mL) and amine (3 mmol), and the mixture was stirred for 12 h, after which the resin was filtered and washed with DCM (3 × 5 mL).[4] The filtrate and the washes were combined, washed with 1 M HCl (2 × 20 mL), dried with sodium sulfate, and concentrated with rotary evaporation. In order to remove particulates, the unpurified products were chromatographed. Treatment with *limiting* amounts of an amine nucleophile results in complete consumption of the amine to provide pure amides.

REFERENCES FOR SECTION 3.3.11

1. Backes, B. J.; Virgilio, A. A.; Ellman, J. A. *J. Am. Chem. Soc.* **1996,** *118,* 3055–3056.
2. Backes, B. J.; Ellman, J. A. *J. Am. Chem. Soc.* **1994,** *116,* 11171–11172.
3. A positive test for unreacted amine gives an intense blue color and a positive test for unreacted

sulfonamide gives a pale red color. Kaiser, E.; Colescot, R. L.; Bossinger, C. D.; Cook, P. I. *Anal. Biochem.* **1970**, *34*, 595–598.

4. After alkylation, the linker is activated enough for expedient cleavage with a limiting quantity of amine or a pool of amines.

3.3.12   SCAL (Safety-Catch Acid-Labile) Linker[1]

Points of Interest

1. The SCAL linker was prepared in 14 steps in solution; however, it is commercially available.

2. During peptide synthesis the SCAL linker is fully compatible with Fmoc and/or Boc protecting groups.

3. The SCAL linker is activated and cleaved in one pot under acidic, sulfoxide-reducing reaction conditions.

4. The SCAL linker is commercially available from CoshiSoft PeptiSearch; see: http://www.azstarnet.com/~mlebl/linker.htm

Literature Summary

For Cleavage.   After standard Fmoc- and/or Boc-amino acid synthesis, peptide amides were cleaved from resin with 1M Me_3SiBr/thioanisole/TFA for 2 h at 0°C (sulfoxides can also be reduced with $SiCl_4$/TFA/anisole). An 11 amino acid sequence related to human calcitonin was obtained in 95% purity (HPLC).

REFERENCES FOR SECTION 3.3.12

1. Patek, M.; Lebl, M. *Tetrahedron Lett.* **1991**, *32*, 3891–3894.

3.3.13   Oxime Resin.[1,2]

Points of Interest

1. The oxime resin is compatible with Boc strategies, but not Fmoc strategies. The removal of Boc groups from substrates on oxime resin is generally performed with only 25% TFA in DCM due to the partial lability of the oxime ester to TFA.

2. The oxime resin has been used for the synthesis of cyclic peptides and segment condensations as well as the preparation of hydrazides and a variety of amides.[3] Cleavage with amines can be very slow; however, with acetic acid as a catalyst, the cleavage from oxime resin is typically complete within 1 day. To minimize diketopiperazine formation ($t_{1/2}$ = 27 min for oxime support-bound Glu(OBz)-Ser(Bz)), neutralization of the TFA salt that is normally used in solid-phase peptide synthesis was omitted and coupling was accomplished with 3 equiv of the symmetric anhydride and a slight excess of DIEA.

3. The oxime resin was used to demonstrate a support-bound Pictet–Spengler reaction.[4]

4. Both the coupling of the first amino acid to and hydrazinolysis from the oxime resin occurred with little racemization.

5. The oxime resin is commercially available.

Literature Procedures

For Loading onto the Oxime Resin. To 2.00 g of oxime resin (0.95 mmol/g) were added 25 mL of DCM, 1.20 mmol of Boc-Leu·H$_2$O, and 1.20 mmol of DCC. The mixture was shaken for 15 h to yield 2.14 g of Boc-Leu-derivatized oxime resin after washing with DCM, DMF, *i*-PrOH, and DCM and drying *in vacuo*. The substitution level was determined to be 0.27 mmol/g by amino acid analysis. The rest of the synthesis was carried out automatically by a peptide synthesizer programmed to wash with DCM (3×) and DMF (3×), cap with 3 mL of acetic anhydride, 1 mL of DIEA, and 12 mL of DMF (30 min), wash with DMF (2×), *i*-PrOH/DCM (3×), and DCM (6×), prewash with 25% TFA and then deprotect with 25% TFA (30 min), wash with DCM (4×), *i*-PrOH, DCM (2×), *i*-PrOH, and DCM (4×), couple with 3 equiv of symmetric anhydride and 2.2 equiv DIEA in DCM (1 × 2 h), and finally wash with DCM (2×), DMF, *i*-PrOH/DCM (2×), and DCM (2×) for a complete cycle.

For Cleavage. Oxime resin-bound Boc-Glu(OBz)-Leu (0.98 g) was shaken with 0.5 M anhydrous hydrazine in CHCl$_3$/MeOH (2:1) for 10 min, filtered, and washed twice with 10-mL portions of CHCl$_3$ and once with 10 mL of MeOH. To the combined washes was added 20 mL of water, and the organic layer was separated and evaporated *in vacuo* to give an oil. After several MeOH azeotropes, a glassy solid was crystallized from 3 mL of EtOH by the addition of about 4 mL of H$_2$O to yield 114 mg of the protected dipeptide hydrazide (99%). Oxime resin-bound substrates have also been cleaved with a 1.5-fold excess of an amino ester in DCM for 12 h, followed by extraction with 5% aqueous citric acid (3×) and water (3×) to remove the excess amino ester.[2] Cleavage with aniline, but not 4-nitroaniline, in CHCl$_3$ for 24 h with 2% acetic acid catalyst provided the corresponding anilide in 81% isolated yield.[3]

REFERENCES FOR SECTION 3.3.13

1. DeGrado, W. F.; Kaiser, E. T. *J. Org. Chem.* **1980**, *45*, 1295–1300.
2. DeGrado, W. F.; Kaiser, E. T. *J. Org. Chem.* **1982**, *47*, 3258–3261.
3. Voyer, N.; Lavoie, A.; Pinette, M.; Bernier, J. *Tetrahedron Lett.* **1994**, *35*, 355–358.
4. Mohan, R.; Chou, Y.-L.; Morrissey, M. M. *Tetrahedron Lett.* **1996**, *37*, 3963–3966.

3.4   LINKERS FOR ALCOHOLS AND AMINES

TABLE 3.12   Acid-Labile Linkers for Alcohols and Amines

Linker name	Page	Cleavage conditions and comments
Trityl resins	52	1–5% TFA/DCM + 5% *i*-Pr₃SiH: for both alcohols and amines
THP linker	55	PPTS in 1:1 DCE/butanol, 60°C (12 h), or 95:5 TFA/DCM: for alcohols
Ketal and acetal linkers	56	30% TFA/H₂O (3 h) for alcohols, 95% TFA/H₂O (12 h) for diols
Solid-phase CBZ equivalent	58	TFA/DCM (1 h) or Pd(OAc)₂, H₂(g), 45 psi (16 h)

TABLE 3.13   Silyl-Based Linkers for Alcohols

Linker name	Page	Cleavage conditions and comments
Silyl chloride resin	59	TBAF in DCM (5 h)

TABLE 3.14   Other Linkers for Amines and Alcohols

Linker name	Page	Cleavage conditions and comments
Ester linkage for peptide alcohols	60	NH₃/MeOH, pressure bottle (72 h), **or** hydrazine/DMF (24 h): for alcohols
Sulfonate ester linker for alcohols	61	Oligosaccharides were cleaved from resin with iodide, acetate, or azide nucleophiles
p-Nitrophenyl carbonate resin	62	Cyclic cleavage: NEt₃/MeOH, 55–90°C (48 h): for amines
ADCC linker for primary amines	63	H₂NNH₂/DMF, but stable to other bases and acids
REM linker for tertiary amines	64	Quaternization by alkylation, cleavage by elimination

3.4.1   Trityl Resins: General Information [1]

Points of Interest

1. Trityl resins have been used for the immobilization of alcohols and amines. Diols and diamines are effectively monoprotected in the process, allowing the other function to be further elaborated.

2. Cleavage of amines and alcohols from trityl resins is generally achieved with 1–5% TFA in DCM containing 5% triisopropylsilane (TIS). Original studies used much harsher conditions than are necessary for cleavage.

3. Substitution of trityl linkers with alkyl, alkoxide, or chloride functional groups results in substituted trityl linkers with different susceptibilities to acidic cleavage.[1]

4. The addition of silanes such as TIS accelerates the cleavage of carboxy, hydroxy, phenoxy, and thiol functions from trityl resins. The cleavage rates of amino, imino, and amide functions are not appreciably affected by silanes.[2]

5. Because trityl resins are heat sensitive, in addition to acid sensitive, excessive heating can result in the premature loss of product from resin.

Representative Examples

See Novabiochem's 1996 Combinatorial Chemistry Catalog for tables comparing different trityl resins.

3.4.2   Original Studies: Trityl Resins for Symmetrical but Not α- or β-Polyols[2]

Points of Interest

1. Insoluble polymers containing trityl chloride residues were used to block one primary alcohol functional group of several polyhydroxy alcohols. After the remaining hydroxyl groups were protected by benzoylation, the ether linkage between the polymer and the protected alcohol was cleaved off support in acidic medium. The reactions worked best with symmetrical diols.

2. In addition to the usual selectivity of trityl chloride in solution for primary vs secondary or tertiary alcohols, the support-bound reagent should react with only one end of a molecule containing several primary hydroxyl groups which is present in large excess in solution.[2]

3. The trityl chloride polymers could be regenerated in one step without degradation.

4. Both a macroreticular resin and a standard gel-form resin (1% divinylbenzene) were used. The macroreticular resins were less reactive and led to trityl resins of relatively low capacity (0.6–0.9 vs 0.9–2.3 mequiv/g for swellable resin).

5. Trityl chloride resins are commercially available from a number of sources.

Literature Procedures

For Loading onto the Linker.   A mixture of 3.92 g (5.09 mequiv) of trityl resin and 1.3 mL of dry 1,4-butanediol in 40 mL of dry pyridine was stirred for 2 days at room temperature. After addition of 10 mL of benzoyl chloride, the stirring was continued for another 24 h. The resin was then filtered and rinsed repeatedly with 3:1 dioxane–water, dioxane, methyl ethyl ketone, water, methyl ethyl ketone, and methanol. After drying, 83% of the sites were derivatized with 1,4-butanediol monobenzoate by weight increase.

For Cleavage.   (Note: Cleavage of amines and alcohols from trityl resins is generally achieved with 1–5% TFA in DCM containing 5% triisopropylsilane. Original studies used much harsher conditions than are necessary for cleavage.) After a suspension of 4.2 g of 1,4-butanediol monobenzoate trityl resin in 40 mL of carbon tetrachloride was stirred for 2 h, a current of HBr was bubbled through it for 45 min. The only product obtained after cleavage was the alkyl bromide. Cleavage with anhydrous TFA resulted in a mixture of alcohols and trifluoroacetates (which could easily be hydrolyzed during workup). Cleavage with TFA/H_2O afforded the alcohol directly. The mixture was then filtered and the polymer washed with portions of toluene. After the filtrate was washed with aqueous sodium bicarbonate, washed with water, and then dried over magnesium sulfate, the solvent was evaporated to yield a slightly colored oil (0.71 g, 78% yield based on trityl resin) that was homogeneous by TLC and NMR. Other alcohols were examined. Benzylic and allylic diols, as well as triols, gave mixtures of products depending on cleavage conditions.

3.4.3   Original Studies: Trityl and Highly Cross-Linked Trityl Resins[3]

Points of Interest

1. Primary symmetrical diols gave mostly symmetrical diols monoblocked by insoluble polymer-bound trityl groups. Acylation followed by acid cleavage from the polymer gave monoacetates and some recovered diols due to "double binding". The extent of "double binding" could be reduced by using more highly cross-linked trityl resins. Partial site isolation on solid support is a kinetic phenomenon that has been observed by a number of researchers. For an example of the subtle relationship of site isolation to resin cross-linking, resin loading, solvent, and temperature, see Chang and Ford.[4] Merrifield and co-workers have described an efficient intersite reaction within the same bead in a swollen polymer network.[5]

2. The more labile 4,4'-dimethoxytrityl group was also prepared on support. Support-bound diols prepared with this resin were partially acylated upon cleavage with 80% acetic acid at room temperature overnight. Anhydrous 0.3 N HCl in dioxane for 48 h at room temperature was found suitable for the desired cleavage. Under these conditions acetate groups are not affected, although free alcohol groups are chlorinated to a small extent (<1%); more chlorination was observed with acid concentrations greater than 0.5 N.

3. The method compares favorably to classical preparations of monofunctionalized diols in solution.

Literature Procedures

For Loading onto the Linker.   A mixture of 5.00 g (1.42 mmol of Cl/g) of trityl resin and 4.30 g (25 mmol) of dry 1,10-decanediol in 60 mL of dry pyridine was stirred at room temperature for 48 h. The polymer was filtered, washed with 25 mL of dry pyridine (4×) and then 25 mL of dry ether (4×), washed in a Soxhlet extractor for 8 h with ether, and dried under high vacuum. Analysis for chloride showed only trace amounts. The IR spectrum was uninformative prior to acylation. A suspension of 1.25 g of functionalized resin (0.55 mmol of diol/g as determined from cleavage) in 60 mL of dry pyridine containing 6 mL (60 mmol) of acetic anhydride was stirred at 60–80°C for 48 h. The polymer-bound monoacetate was filtered, washed with pyridine (2×), ethanol, water (10×), ethanol (4×), chloroform (2×), and ether (4×), and air-dried. The IR spectrum exhibited a strong absorbance at 1725 cm^{-1}.

For Cleavage. (Note: Cleavage of amines and alcohols from trityl resins is generally achieved with 1–5% TFA in DCM containing 5% triisopropylsilane. Original studies used much harsher conditions than are necessary for cleavage.) A 1.0-g sample of the polymer-bound monoacetate was suspended in 40 mL of 0.3 N anhydrous HCl in dry dioxane, and the mixture was stirred at room temperature for 48 h. The polymer was filtered, washed with ethanol, water (10×), ethanol (2×), and ether (4×), and air-dried. The filtrate was extracted with ether, washed with water, dried over magnesium sulfate, and evaporated to afford 106 mg of a light brown oil. The starting diol was separated from the monoacetate by preparative TLC (eluent: 40% ether, 60% benzene).

REFERENCES FOR SECTIONS 3.4.1 to 3.4.3

1. An excellent discussion of trityl resins including a table that directly compares the loading and cleavage conditions for a range of functional groups with a range of trityl resins can be found in Novabiochem's Combinatorial Chemistry Catalog. Calbiochem-Novabiochem International, San Diego, CA (1-800-228-9622); Calbiochem-Novabiochem AG, Läuflefingen, Switzerland (+41 62 299 19 22).
2. Frechet, J. M. J.; Nuyens, L. J. *Can. J. Chem.* **1976**, *54*, 926–934.
3. Fyles, T. M.; Leznoff, C. C. *Can. J. Chem.* **1976**, *54*, 935–942.
4. Chang, Y. H.; Ford, W. T. *J. Org. Chem.* **1981**, *46*, 5364–5371.
5. Bhargava, K. K.; Sarin, V. K.; Trang, N. L.; Cerami, A.; Merrifield, R. B. *J. Am. Chem. Soc.* **1983**, *105*, 3247–3251.

3.4.4 Tetrahydropyranyl (THP) Linker: For Alcohols[1]

Points of Interest

1. Primary and secondary alcohols are both loaded to the THP linker and cleaved from the THP linker under mild conditions.

2. Loading was consistently between 0.43 and 0.54 mequiv/g (66–95% loaded, Merrifield resin contained 0.77 mequiv/g of chloride) with a range of different primary and secondary alcohols. Tertiary alcohols showed poor loading efficiencies and define a limit to the coupling strategy.

3. The THP linker is stable to a range of reaction conditions including strongly nucleophilic or basic reagents.

4. Carboxylic acids can be loaded onto the THP linker under more forcing conditions.[2]

5. The stereocenter in the support-bound THP does not complicate use with chiral alcohols, because upon cleavage the THP group remains on support. Consequently, purification and analytical evaluation of products after cleavage from support do not involve diastereomeric mixtures.

Literature Summary

For Loading onto the Linker. The dihydropyran-functionalized support was prepared by treating Merrifield resin with 3 equiv. of the sodium salt of 6-(hydroxymethyl)-3,4-dihydro-*2H*-pyran in freshly distilled dimethylacetamide at room temperature for 16 h. The resin was then rinsed with DCM, DMF/water (4×), DMF (3×), and DCM (3×) and dried *in vacuo*. The DHP-derivatized resin is stable indefinitely. Alcohols were loaded onto the DHP-derivatized resin with 2 equiv of PPTS and 5 equiv of the alcohol (0.4 M in DCE) at 80°C for 16 h. Alternatively, the alcohol can be coupled to support with 1 equiv of anhydrous *p*-TsOH and 5 equiv of alcohol (0.4 M in DCE) at 0°C for 16 h. For most alcohols shorter reaction times or lower temperatures can be employed.

For Cleavage. Alcohol-derivatized support (1.0 g, 0.74 mmol) in 20 mL of 1:1 DCE/butanol was treated with PPTS (370 mg, 1.48 mmol). The flask was stoppered and heated at 60°C for 16 h. The solution was isolated by filtration and concentrated *in vacuo*. The resulting material can be separated from the PPTS by either extraction or flash chromatography. Alternatively, alcohols are cleaved from the resin with 95:5 TFA/water; this cleavage cocktail effectively removes standard acid-labile side-chain protecting groups employed in Fmoc-amino acid peptide synthesis.

REFERENCES FOR SECTION 3.4.4

1. Thompson, L. A.; Ellman, J. A. *Tetrahedron Lett.* **1994**, *35* (50), 9333–9336.
2. Liu, G., unpublished results.

3.4.5   Ketal and Acetal Linkers: For Diols and Alcohols[1]

Ketal linkage   Acetal linkage

Points of Interest

1. Both ketal and acetal linkers were used for the preparation of symmetrical HIV protease inhibitors. Analogs of known protease inhibitors were prepared from protected diamino alcohol or diamino diol core units. The cores were extended in both directions by automated solid-phase synthesis.

2. Alcohols and diols were preloaded onto linkage units which were then loaded onto resin via standard amide bond formation.

3. The diamino alcohol derivatives were completely cleaved from resin with 30% aqueous TFA for 3 h. For the diamino diol derivatives, however, the cleavage required much more drastic conditions. The best result was obtained with 95% TFA overnight; however, the diol products contained contaminating compounds still carrying the linker.

4. HPLC analysis revealed sample purities ranging from 30 to 70% for the mono-ol series and from 20 to 50% for the diol series. In general, purities of compounds in the mono-ol series were better than those of the corresponding compounds in the diol series, probably due to the unexpected long reaction time required for cleavage of the diol products.

5. The ketal linker has been used for the support-bound monoprotection of diketones.[2]

Literature Procedures

For Loading onto the Resin. The diamino diol or diamino alcohol cores (prederivatized with the linkers in solution) (5.0 mmol, 1.5 equiv relative to the resin) were dissolved in 10 mL of NMP and 10 mL of DCM. To the solution were added DIC (0.79 mL, 5.0 mmol) and 0.68 g of HOBt. After 30–60 min of stirring, the solution was mixed with 5.0 g of NovaBead MBHA resin (0.65 mmol/g). (MBHA resin was used to minimize cleavage of the linker from resin; however, for several compounds in the diol series, a (M + 97) peak derived from the expected product still coupled to the linker was observed by mass spectrometry.) The mixture was shaken overnight and then filtered. The resin was washed with NMP (5×), NMP–DCM (5×), and DCM (5×) and vacuum-dried. By weight gain the loading level was 0.5–0.55 mmol/g; a more quantitative value could be obtained by Fmoc quantitation.

For Cleavage. After bidirectional automated solid-phase synthesis the products were cleaved from resin as follows. The resin containing molecules with the diamino diol core were cleaved by treating the resins with 95% aqueous TFA overnight, and molecules derived from the diamino alcohol core were cleaved by treating the resin with 30% TFA in DCM for 3 h.

REFERENCES FOR SECTION 3.4.5

1. Wang, G. T.; Li, S.; Wideburg, N.; Krafft, G. A.; Kempf, D. J. *J. Med. Chem.* **1995**, *38*, 2995–3002.
2. Xu, Z. H.; McArthur, C. R.; Leznoff, C. C. *Can. J. Chem.* **1983**, *61*, 1405–1409.

3.4.6   Linkers for Amines: A Solid-Phase CBZ Chloride Equivalent [1]

X = Cl, imidazole

Points of Interest

1. Treatment of TentaGel-S resin (containing the Wang linker) with phosgene produced a support-bound chloroformate carbonyl IR stretch at 1776 cm^{-1}. Other commercially available Wang resins that were examined did not produce the chloroformate under these conditions; however, the other Wang resins could still be derivatized with carbonyldiimidazole.

2. Amino-terminal support-bound peptide carbamates could be cleaved with acid or by hydrogenation.

3. Support-bound acylimidazolines were treated with benzylmagnesium chloride to form the corresponding ester in 70% yield.

4. Burdick and coworkers have also prepared a support-bound chloroformate that was further derivatized with 1,4-phenylenediamine for the solid-phase synthesis of peptide p-nitroanilides.[2]

5. Acylimidazole-modified Wang resin was employed for the Wittig olefination of support-bound amino aldehydes in the solid-phase synthesis of olefin and hydroxyethylene peptidomimetics.[3]

6. The support-bound imidazolide carbamate is commercially available from Advanced ChemTech.

Literature Summary

For Loading onto the Support-Bound Chloroformate.   A solution of phosgene in toluene (620 μL, 1.93 M, 1.20 mmol) was added to a stirred mixture of specifically Tentagel-S-OH (Rapp Polymer, 376 mg, 0.32 mmol/g, 120 μmol) in 5 mL THF at room temperature under N$_2$. After 2 h, the resin was filtered, washed with 2 × 10 mL of THF, ether, THF, and ether, and dried on a vacuum pump to afford 400 mg of the support-bound chloroformate. To a portion of the derivatized resin (325 mg) in 2 mL of THF was added glycine N-carboxyanhydride (13 mg, 128 μmol), followed by the addition of a solution of N-methylmorpholine (22 μL, 200 μmol) in 100 μL of THF via a syringe pump over 2 h. After an additional 2 h, the resin was filtered, washed twice each with THF, ether, DCM, and ether, and dried on a vacuum pump to afford 347 mg of resin. Support-bound N-carboxyanhydride (250 mg, 80 μmol) was resuspended in 5 mL of THF, and glycine ethyl ester hydrochloride (112 mg, 800 μmol) was added, followed by N-methylmorpholine (88 μL, 800 μmol). After 16 h, the resin was filtered, washed twice each with MeOH, DCM, MeOH, and ether, and dried on a vacuum pump to give 261 mg of derivatized resin.

For Loading onto the Support-Bound Imidazolide Carbamate.   To Wang resin (Fluka, 0.6–0.8 mmol/g, 1.00 g) in 15 mL THF under N$_2$ was added

1,1'-carbonyldiimidazole (567 mg, 3.50 mmol), and the resulting slurry was stirred for 2 h. The resin was filtered, washed twice each with THF, ether, THF, and ether, and dried on a vacuum pump to give 1.57 g of derivatized resin. The resin-bound imidazolide carbamate (500 mg, 0.35 mmol) was resuspended in 5 mL of THF and 5 mL of N-methylpyrrolidinone. Phe-Val-Phe-OMe hydro-chloride (807 mg, 1.75 mmol) was added followed by 1 mL of N-methyl-morpholine. The stirred mixture was heated under N_2 in a 60°C oil bath for 4 h. The resin was filtered, washed twice each with DCM, MeOH, THF, ether, THF, and ether, and dried on a vacuum pump to give 607 mg of resin.

For Cleavage with TFA. Derivatized resin (50 mg, 35 μmol) was treated with 1 mL DCM and 1 mL of TFA at room temperature. After the resulting slurry was stirred for 3 h, 1 mL of MeOH was added, and the resin was filtered and washed with 2 × 2 mL of MeOH, DCM, and MeOH. The filtrate was concentrated *in vacuo* to give 13 mg (24 μmol, 85% yield at 0.8 loading) of a white solid, which was identical with authentic Phe-Val-Phe-OMe by IR, MS, and HPLC.

For Cleavage by Hydrogenation. The same resin (50 mg, 35 μmol) was added to a solution of $Pd(OAc)_2$ (31 mg, 140 μmol) in 10 mL of DMF in a small Parr bottle. The resulting slurry was placed on a Parr shaker under 45 psi of H_2 for 16 h. Then the catalyst was filtered through Supercell and washed with 2 × 5 mL of MeOH, DMF, and MeOH, and the filtrate was concentrated *in vacuo* to give 20 mg. MS and HPLC showed Phe-Val-OMe as the major product (87% yield).

REFERENCES FOR SECTION 3.4.6

1. Hauske, J. R.; Dorff, P. *Tetrahedron Lett.* **1995**, *36*, 1589–1592.
2. Burdick, D. J.; Struble, M. E.; Burnier, J. P. *Tetrahedron Lett.* **1993**, *34*, 2589–2592.
3. Rotella, D. P. *J. Am. Chem. Soc.* **1996**, *118*, 12246–12247.

3.4.7 Silyl Chloride Linkers for Alcohols [1,2]

R = Me, Ph, *i*-Pr

Points of Interest

1. A competition experiment demonstrated that the polymer preferentially silylates sterically less hindered alcohols. For example, 3β-cholestanol and tes-tosterone were easily separated with the silyl chloride resin.

2. The support-bound silyl linker was used to convert nonane-1,9-diol to the fall army worm moth sex pheromone.[1]

3. Danishefsky and coworkers have found that the isopropylsilyl linker performed much better for solid-phase oligosaccharide synthesis with glycals.[2]

4. The silyl polymer is reasonably stable in air and was stored in the dry form for up to 1 month.

Literature Procedures

For Loading onto the Linker. To 1.0 g of the phenylsilyl chloride resin (loading not determined) in DCM (20 mL) were added octan-1-ol (1.3 g, 10 mmol) and DIEA (1.7 mL, 10 mmol). The mixture was stirred under a nitrogen atmosphere at room temperature for 24 h. The polymer was filtered and continuously washed with diethyl ether in a Soxhlet extractor for 4 h. Silylated polymer was dried at 60–70°C under reduced pressure.

For Cleavage. The polymer was treated with 2 mmol of TBAF in DCM (15 mL) for 5 h, followed by treatment with water.

REFERENCES FOR SECTION 3.4.7

1. Chan, T. H.; Huang, W. Q. *J. Chem. Soc., Chem. Commun.* **1985**, 909–911.
2. Randolph, J. T.; McClure, K. F.; Danishefsky, S. J. *J. Am. Chem. Soc.* **1995**, *117*, 5712–5719.

3.4.8 Linker for Preparation of C-Terminal Peptide Alcohols (and Peptide Aldehydes) [1]

Points of Interest

1. Protected amino alcohols are esterified with succinic anhydride. After an extractive workup the hemisuccinates are directly coupled to benzhydrylamine resin (BHA) under standard DIC/HOBt conditions.
2. After peptide synthesis, the products are cleaved from support via ammonolysis or hydrazinolysis.
3. Modified polystyrene resins have been described for the hydrolysis of support-bound esters to carboxylic acids.[2]
4. C-terminal peptide aldehydes,[3] as well as olefin and hydroxyethylene peptidomimetics derived from amino aldehydes,[4] have been prepared on solid support.

Literature Summary

After standard methods for peptide synthesis, the products were liberated from resin by treatment either with ammonia/MeOH in a pressure bottle (72–96 h) or with an excess of hydrazine in DMF (24 h) in 14–85% yield based on loading. For ammonolysis: ammonia (40 mL) was condensed into a chilled (−40°C) pressure bottle containing the peptide resin (1.5 g) and methanol (60 mL). The vessel was sealed and the solution was stirred (72–96 h). After the pressure bottle was chilled ($-40°C$), the vessel was opened to allow the ammonia to evaporate. The methanolic solution was then filtered and the resin was washed with methanol. The filtrate was concentrated *in vacuo* and the resulting crude residue was dissolved in acetic acid, lyophilized, purified by HPLC, and relyophilized. For hydrazinolysis: the peptide resin (1.5 g) was treated with hydrazine (4.0 mL) in DMF (25 mL). After 24 h of shaking, the mixture was filtered and the resin was washed with DMF (2 × 25 mL). The combined filtrates were then concentrated under high vacuum (0.1 Torr, bath temperature 45°C) to constant weight. The product was then precipitated with ether to afford a solid that was dissolved in acetic acid/water (6 mL, 5 : 1) and purified by HPLC.

REFERENCES FOR SECTION 3.4.8

1. Swistok, J.; Tilley, J. W.; Danho, W.; Wagner, R.; Mulkerins, K. *Tetrahedron Lett.* **1989**, *30*, 5045–5048.
2. Tilak, M. A.; Hollinden, C. S. *Tetrahedron Lett.* **1968**, *11*, 1297–1300.
3. (a) Murphy, A. M.; Dagnino, R.; Vallar, P. L.; Trippe, A. J.; Sherman, S. L.; Lumpkin, R. H.; Tamura, S. Y.; Webb, T. R. *J. Am. Chem. Soc.* **1992**, *114*, 3156–3157. (b) Fehrentz, J. A.; Paris, M.; Heitz, A.; Velek, J.; Liu, C. F.; Winternitz, F.; Martinez, J. *Tetrahedron Lett.* **1995**, *36*, 7871–7874.
4. Rotella, D. P. *J. Am. Chem. Soc.* **1996**, *118*, 12246–12247.

3.4.9 Sulfonate Ester Linkage for Oligosaccharide Synthesis [1]

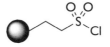

0.65 mmol/g

Points of Interest

1. The sulfonyl chloride resin was prepared in three steps (70% yield) from (chloromethyl)polystyrene (Merrifield resin).

2. Diastereomerically pure disaccharides were prepared by solid-phase synthesis using both iodo acetates or trichloroacetimidates. As a more demanding test of the methodology, a trisaccharide was prepared in 67% isolated yield with only trace amounts ($≤5–10\%$ by NMR) of diastereomers.

3. Nucleophilic cleavage from resin with NaI afforded iodo saccharides which could then be reduced to the corresponding saccharides (92% yield) by treatment with Bu_3SnH (C_6H_6, AIBN, 80°C). Alternatively, treatment with NaOAc or NaN_3 in DMF afforded the corresponding 6-acetoxy or 6-azido sugars.

Literature Procedures

For Loading onto Sulfonyl Chloride Resin. Sulfonyl chloride resin (500 mg, 0.325 mmol) was suspended in 4 mL of DCM. Methyl 4-O-acetyl-2,3-di-

O-benzyl-α-D-glucopyranoside (541 mg, 1.30 mmol) and 1 mL of Et₃N were added, and the mixture was stirred overnight at room temperature. The resin was then collected on a coarse-fritted funnel, washed (2 × 25 mL each) with DCM, THF, MeOH, THF/H₂O (2:1), THF/H₂O (2:1), THF, MeOH, DCM, and Et₂O, and dried overnight under vacuum to yield 619 mg of derivatized resin (0.31 mmol, 96% by mass balance).

For Cleavage.   Glucoside resin (49 mg, 0.025 mmol) was suspended in 2 mL of 2-butanone. NaI (500 mg, 3.31 mmol) was added, and the mixture was heated to 65°C and stirred overnight (20 h). The cooled resin was collected on a coarse-fritted funnel and washed (2 × 10 mL each) with H₂O, THF/H₂O (2:1), H₂O, H₂O, THF/H₂O (2:1), and THF. The filtrate was diluted with 30 mL of EtOAc, 10 mL of H₂O, and 10 mL of brine. The phases were separated, the aqueous phase was extracted with 3 × 25 mL of EtOAc, and the combined organics were dried over MgSO₄ and concentrated. The residue was purified by flash chromatography, eluting with a gradient system of 9:1 to 1:1 hexanes–EtOAc, to yield 20 mg (0.025 mmol, 100%) of the known glucoside as a white solid based on the calculated loading.

REFERENCES FOR SECTION 3.4.9

1. Hunt, J. A.; Roush, W. R. *J. Am. Chem. Soc.* **1996**, *118*, 9998–9999.

3.4.10   Linkers for Amines: A Support-Bound *p*-Nitrophenyl Carbonate [1]

Points of Interest

1. The nitrophenyl carbamate linker was initially reported by Leznoff for the synthesis of unsymmetrical diamines.[2]

2. The nitrophenyl carbamate linker was formed on support in nearly quantitative yield by gel-phase ¹³C NMR (elemental analysis for nitrogen gave a loading capacity of 0.84 mmol/g). The support-bound carbonate can be stored for at least 6 months without loss of activity.

3. Zwitterionic amino acids were loaded onto the nitrophenyl carbamate linker on large scale. The zwitterionic amino acids were dissolved with light heating in DMF using 2.5 equiv of N,O-bis(trimethylsilyl)acetamide (BSA) for solubility.

4. Hydantoins were cleaved from the derivatized resin with triethylamine in methanol. The base-catalyzed cyclization can be performed in acetonitrile instead of methanol but lower yields and product purities were generally ob-

served. Attempts to use other protic solvents such as isopropyl and ethyl alcohol during the cyclic cleavage step were unsuccessful.

5. From a library of 800 individual hydantoins, a random analytic sampling (15%) by FD MS showed that in 90% of the cases, desired product was obtained. Regardless of the mass balance, the products were obtained in 67–99% purity by HPLC presumably because of the cyclic cleavage (see p 158).

Literature Summary

p-Nitrophenyl chloroformate (1.31 g, 6.50 mmol, 2 equiv) was added in one portion to a stirring slurry of 3.26 g of 1% cross-linked (hydroxymethyl)polystyrene (1 mmol/g) and N-methylmorpholine (659 mg, 6.50 mmol, 2 equiv) in DCM at 0°C. The reaction mixture was warmed to room temperature overnight, filtered, and then washed with DCM. Drying overnight in a vacuum oven afforded 3.28 g of derivatized resin. ^{13}C NMR (CDCl$_3$) δ 70.96 (broad s, P–CH$_2$O–). IR (KBr) 1761 cm^{-1}. Gel-phase ^{13}C NMR spectra were obtained using the procedure of Giralt.[3] Selected amino acids (4 equiv) were dissolved with light heating in DMF using N,O-bis(trimethylsilyl)acetamide (BSA; 10 equiv) and then coupled with the support-bound activated carbonate with DMAP (2 equiv). Rinses between steps were not described. Amide formation was then carried out overnight with excess primary amine (4 equiv), HOBt–H$_2$O (4 equiv), and DCC (4 equiv). Treatment of the support-bound amide with excess triethylamine (14 equiv) in methanol for 48 h between 55 and 90°C afforded hydantoins.

REFERENCES FOR SECTION 3.4.10

1. Dressman, B. A.; Spangle, L. A.; Kaldor, S. W. *Tetrahedron Lett.* **1996**, *37*, 937–940.
2. (a) Dixit, D. M.; Leznoff, C. C. *Isr. J. Chem.* **1978**, *17*, 248–252. (b) Dixit, D. M.; Leznoff, C. C. *J. Chem. Soc., Chem. Commun.* **1977**, 798–799.
3. Giralt, E.; Rizo, J.; Pedroso, E. *Tetrahedron* **1984**, *40*, 4141–4152.

3.4.11   An ADCC Linker for Primary Amines[1]

Points of Interest

1. The ADCC linker was developed for the attachment of primary amines to solid support. The ADCC linker is stable to acidic conditions (e.g., TFA), basic conditions (e.g., piperidine, DBU), yet completely labile to 2% hydrazine in DMF.

2. The ADCC linker molecule is based on a protecting group for amines derived from acetyldimedone.[2]

3. A traceless linker for the preparation of phthalhydrazides using the Ing-Manske procedure for the cleavage of phthalimides by hydrazines has also been described.[3]

Literature Summary

For Loading onto Resin.   The enamide was formed in DMF between the linker molecule (derived from acetyldimedone) and L-phenylalanine methyl ester. The enamide (2.5-fold excess relative to support loading) was attached with 1,1,3,3-tetramethyl-2-(2-oxo-1(2*H*)pyridyluronium tetrafluoroborate (TPTU) to (aminomethyl)polystyrene. Both the ninhydrin test and FT-IR of the support indicated formation of the amide bond.

For Cleavage.   As a model system, the L-phenylalanine methyl ester was cleaved from support employing a 2% hydrazine hydrate solution in DMF for 5 min under Ar. The support was filtered and washed several times with DMF. Complete recovery of the L-phenylalanine methyl ester was observed following evaporation of the combined DMF solutions. Signal bands corresponding to an ester function were not detected in FT-IR of the support after cleavage.

REFERENCES FOR SECTION 3.4.11

1. Bannwarth, W.; Huebscher, J.; Barner, R. *Bioorg. Med. Chem. Lett.* **1996**, *6*, 1525–1528.
2. Bycroft, B. W.; Chean, W. C.; Chhabra, S. R.; Hone, N. D. *J. Chem. Soc., Chem. Commun.* **1993**, 778.
3. Nielsen, J.; Rasmussen, P. H. *Tetrahedron Lett.* **1996**, *37*, 3351–3354.

3.4.12   REM Linker: For the Preparation of Tertiary Amines[1]

Points of Interest

1. Tertiary amines are liberated from support via Hoffmann elimination.

2. The REM linker is *Re*generated after cleavage of the product and is functionalized via a *M*ichael reaction.

3. The purity of the products is good (>90% by gas chromatography), presumably because only the desired quaternized amines are susceptible to the cleavage conditions. The method is compatible with esters, anilines, or Boc-protected amines, although the product from the quaternization of a bisbenzylic amine on support was obtained in very low yields (8%).

Literature Summary

(Hydroxymethyl)polystyrene resin (1 g, 0.58 mmol) was added to a 10-mL polypropylene tube. DCM (7 mL) and DIEA (866 μL, 5 mmol) were added, followed by acryloyl chloride (404 μL, 5 mmol). The vessel was then placed on a tube rotator for 4 h at room temperature. The resin was washed with DCM (3×) and MeOH (2×) and dried *in vacuo*. A portion of the acylated resin (0.3 g, 0.17 mmol) was swollen with a mixture of DMF (4 mL) and a secondary amine (3 mmol). The tube was agitated on the rotator for 18 h at room temperature to give a support-bound tertiary amine (note: when Michael addition was performed with a primary amine, a support-bound tertiary amine was obtained by reductive alkylation). The resin was washed with DMF (3×), DCM (3×), and MeOH (2×) and dried *in vacuo*. The resin was suspended in a solution of an alkylating agent (1.5 mmol) in DMF (4 mL) and was agitated on the rotator for 18 h at room temperature. Following quaternization, the resin was washed with DMF (3×), DCM (3×), and MeOH (2×) and dried *in vacuo*. A suspension of the resin in DMF (4 mL) containing DIEA (106 μL, 0.6 mmol) was agitated on the rotator for 18 h at room temperature. The resin was washed with DMF (3×), DCM (3×), and MeOH (2×) and the filtrate was evaporated. The resulting white solid was distributed between ethyl acetate and 5% aqueous sodium carbonate. The organic layer was removed and the aqueous layer was washed with ethyl acetate (2×). The combined organic layers were dried and evaporated. Trace baseline material was removed with silica gel, employing 40% ether in heptane as the eluent.

REFERENCES FOR SECTION 3.4.12

1. Morphy, J. R.; Rankovic, Z.; Rees, D. C. *Tetrahedron Lett.* **1996**, *37*, 3209–3212.

3.5   LINKERS FOR HYDROCARBONS AND OTHER FUNCTIONAL GROUPS

TABLE 3.15   Traceless and Other Linkers for Hydrocarbons

Linker name	Page	Cleavage conditions and comments
Traceless silyl linker	66	HF (*Caution!*), stable to TFA
Aryl silyl linker	67	TFA for electron-rich substrates and CsF in DMF/H$_2$O at 110°C for electron-poor substrates
Han's silyl linkers	68	3 equiv of ICl in DCM (instantaneously), **or** 6 equiv of Br$_2$/3 equiv of pyridine in DCM, 0°C (2 h): protiodesilylation was substrate dependent
Silyl ether linkage	69	TBAF in DMF at 60°C for Si–C cleavage
Triazine linkage	70	Neat MeI at 110°C for 24 h
Reissert linkage	71	1 M KOH in H$_2$O/THF (1:2)

TABLE 3.16   Linkers for Sulfur Attachment to Resin

Linker name	Page	Cleavage conditions and comments
Thioester linker	73	HF (*Caution!*)
Disulfide linker	74	DTT for thiolysis, or TCEP for reduction

3.5.1   Traceless Silyl Linkers: For the Preparation of 1,4-Benzodiazepines [1]

Points of Interest

1. The traceless silyl linker was used for the preparation of 1,4-benzodiaze-pine derivatives.

2. The silylstannane linker was attached to aminomethyl resin as the acti-vated cyanomethyl ester to avoid protiodestannylation observed in the presence of the free acid or HOBt.

3. Gram quantities of the silylstannane linker were prepared via a conver-gent multistep synthesis in solution.

4. Electron-poor substrates are more difficult to cleave. A more labile linker substituting germanium for silicon has been prepared and successfully used for the preparation of 1,4-benzodiazepine derivatives. 1,4-Benzodiazepines were cleaved from the germanium linker with TFA/Me$_2$S/H$_2$O (85:10:5).[2]

5. In addition to protiodesilylation, both the silyl and germanium linkers were efficiently cleaved with other electrophiles such as bromine.

Literature Procedures

For Loading onto the Resin.   In an oven-dried Schlenk flask under nitro-gen the (cyanomethyl)silylstannane (3 equiv relative to resin) was dissolved in minimal NMP. Hunig's base (4.5 equiv), aminomethyl resin (1 equiv), and DMAP (3 equiv) were added and the solution was mixed at 70°C for 24 h at which point the ninhydrin test indicated that the reaction had gone to >85% completion. The solid-phase synthesis of 1,4-benzodiazepines proceeded as pre-viously described.[3]

For Cleavage and Isolation of 1,4-Benzodiazepines.   For 1,4-benzodiaze-pines incorporating side-chain protecting groups, the resin was treated with 15 mL of 95:5:10 trifluoroacetic acid/water/dimethyl sulfide for 30 min. No cleavage of the silyl support-bound benzodiazepine (<2%) was observed under these conditions. The cleavage solution was removed by filtration cannula, and the resin was rinsed with DCM (5×) and MeOH (5×). The resin was dried and then transferred to a perfluorinated plastic reaction vessel for HF cleavage. (*Caution:* Hydrogen fluoride gas is **extremely** toxic and reaction with this gas should only be performed with the proper equipment and training. There are two excellent reviews[4] that address safety issues with respect to the handling and use of this reagent.) Anhydrous hydrogen fluoride gas (Matheson) was condensed in the reaction vessel and allowed to react (under slight pressure since HF boils at 19.5°C) for 12 h, at which point the hydrogen fluoride was removed with gentle nitrogen flow. The vapor was removed with two sequential KOH traps. The resin was rinsed with 20 mL of 1:4 MeOH/DCM (5×). Concentration of the combined filtrates then provided the unpurified products. The 1,4-benzodiazepines were purified by silica gel chromatography and fully characterized (50–68% yield).

3.5.2 Protiodetachable Arylsilane Linker: For the Preparation of Biaryls[5]

Points of Interest

1. The arylsilyl linkers were prepared in only a few steps in solution.
2. Suzuki and Grignard reactions were performed on different silyl-bound substrates.
3. Electron-poor substrates were more difficult to cleave by protiodesilylation. Because protiodesilylation is substrate dependent, a current limitation of all silyl–carbon linkers is the lack of mild, general cleavage conditions.

Literature Procedures

For Loading onto Resin. A solution of 4-((((4-bromophenyl)dimethyl-silyl)methyl)oxy)benzyl alcohol (12.0 g, 34.1 mmol) in dry THF (50 mL) was added to a suspension of sodium hydride (95%, 1.21 g, 50.4 mmol) in dry THF (5 mL) and stirred at 45°C until gas evolution ceased. Merrifield resin (1.4 mmol of Cl⁻/g) (12.0 g, 16.8 mmol) was added along with additional dry THF (25 mL), and the reaction mixture was stirred at 65–70°C for 72 h. The reaction mixture was cooled to room temperature and quenched with MeOH

(5 mL). The resin was filtered and washed with THF, MeOH, MeOH/H$_2$O
(1:1), MeOH, DCM, and finally MeOH. The product was dried *in vacuo* for
24 h to yield 14.3 g of the silicon-linked polymer.

For Cleavage.   Biaryls prepared by a support-bound Suzuki coupling reaction were removed from the silyl resin with neat TFA at room temperature;
however, for electron-poor compounds (e.g., those bearing a cyano group)
cleavage required treatment with CsF in 4:1 DMF/water at 110°C.

3.5.3   Silicon-Directed Ipso Substitution of Polymer-Bound Arylsilanes[6]

X = H, Br, and/or OMe

Points of Interest

1. Han's arylsilyl linker was prepared via four high-yielding steps in
solution.
2. The arylsilyl linker was attached to commercially available Wang resin
by a straightforward alkylation.
3. The arylsilyl linker is very stable to bases or fluorides. A range of biaryls
were prepared via a Suzuki cross-coupling on support.[7]
4. Electrophilic cleavage [iododesilylation, bromodesilylation, and protio-
desilylation] was performed on different modified resins. Yields were high for
iodonation and bromination, but low for protiodesilylation.

Literature Summary

For Loading onto the Resin.   Wang resin was treated with chloromethoxy-
silyl linkers in DMF at 40–50°C in the presence of Hunig's base under N$_2$(g)
overnight. The reaction was monitored by the disappearance of the hydroxyl
absorption band at 3500 cm^{-1} in the IR spectrum.

For Cleavage.   Following the Suzuki coupling, biaryl products were
cleaved from resin under different conditions. Iodination with ICl (3 equiv) in
DCM was nearly instantaneous. Greater than 90% of the product was released
from the solid support within 10 min. Longer reaction times (2–24 h) were
avoided due to the presence of various silicon species. Bromination with bromine (6 equiv)/pyridine (3 equiv) in DCM was usually complete after 2 h at 0°C.
Conversely, cleavage by protiodesilylation provided variable yields and was
substrate dependent.

3.5.4 A Silyl Ether Linker for Solid-Phase Organic Synthesis[8]

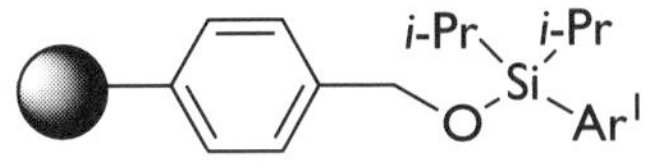

i-Pr　*i*-Pr
Si
O　　Ar[1]

Points of Interest

1. Mild fluoride-mediated desilylation of the resin-bound intermediates produces aromatic compounds in generally high yield and purity.

2. The silicon–oxygen bond could be cleaved by protio-ipsodesilylation conditions (TFA); alternatively, the silicon–carbon bond could be cleaved by fluoride-induced hydrodesilylation of the siloxane (TBAF in DMF at 60°C).

Representative Examples

Ortholithiation, carbonyl addition, oxidation, alkylation, and condensation reactions were all performed to demonstrate the versatility of the support-bound silyl ether linker.

Literature Procedures

For Loading onto the Resin. To a suspension of (hydroxymethyl)polystyrene resin (Bachem California, 6.00 g, 1.0 mequiv/g) and imidazole (2.45 g, 36 mmol) in DMF (40 mL) was added an arylchlorosilane (6.92 g, 24.1 mmol). The resulting mixture was agitated at room temperature for 45 h. The resin was collected by filtration and washed successively with DMF (3 × 30 mL), THF (3 × 20 mL), and DCM (3 × 30 mL). The product was dried to give the derivatized resin (7.10 g) as an off-white solid. The resin was resubjected to the above reaction conditions in order to maximize the loading.

For Cleavage. A suspension of derivatized resin (0.209 g) in 2 mL of DMF was treated with 1 mL of TBAF in THF (1 M, 1 mmol), and the mixture was heated at 65°C with agitation for 1 h. The mixture was allowed to cool to room temperature and diluted with H_2O (10 mL). The mixture was filtered, washing with Et_2O (3 × 10 mL). The organic phase was washed with H_2O (10 mL) and then brine (10 mL), dried, and concentrated. The residue was dissolved in $CHCl_3$ and filtered through a plug of basic Al_2O_3. The filtrate was concentrated and then this process was repeated a second time to give desired aryl product in 57% yield.

REFERENCES FOR SECTIONS 3.5.1 TO 3.5.4

1. Plunkett, M. J.; Ellman, J. A. *J. Org. Chem.* **1995**, *60*, 6006–6007.
2. Plunkett, M. J.; Ellman, J. A. *J. Org. Chem.* **1997**, *62*, 2885–2893.
3. Plunkett, M. J.; Ellman, J. A. *J. Am. Chem. Soc.* **1995**, *117*, 3306–3307.
4. (a) Skakihara, S. In *Chemistry and Biochemistry of Amino Acids, Peptides, and Proteins;* Weinstar, B., Ed.; New York, **1971**; pp 51–85. (b) Lenard, J. *Chem. Rev.* **1969**, *69*, 625–636.

5. Chenera, B.; Finkelstein, J. A.; Veber, D. F. *J. Am. Chem. Soc.* **1995**, *117*, 11999–12000.
6. Han, Y.; Walker, S. D.; Young, R. N. *Tetrahedron Lett.* **1996**, *37*, 2703–2706.
7. Frenette, R.; Friesen, R. *Tetrahedron Lett.* **1994**, *35*, 9177–9179.
8. Boehm, T. L.; Showalter, H. D. H. *J. Org. Chem.* **1996**, *61*, 6498–6499.

3.5.5 Phosphonium Salts as Traceless Supports for Solid-Phase Synthesis [1]

Points of Interest

1. Conditions allowing cleavage to products lacking polar functionality by either Wittig or hydrolytic pathways were developed.

2. The phosphonium group was stable to a range of reaction conditions including reduction of a nitro group with sodium dithionite (followed by HBr treatment to restore the bromide counterion) and acylation of the resulting aniline.

3. Acylation of the aniline to the anilide was monitored by the lack of a yellow Schiff base formation on treatment of a few beads with methanolic 4-nitrobenzaldehyde.

4. Following Wittig cleavage, excess aldehyde was removed with (aminomethyl)polystyrene or Girard's Reagent T.

5. The phosphonium salt was readily prepared from commercially available polymer-bound triphenylphosphine by the method of Ford.[2]

Literature Procedures

For Cleavage by Wittig Reaction. A suspension of polymer-bound (2-((4-methoxybenzoyl)amino)benzyl)triphenylphosphonium bromide (500 mg) in dry methanol (15 mL) was treated with 2.0 M sodium methoxide (0.9 mL, 1.8 mmol) immediately followed by methyl 4-formylbenzoate (295 mg, 1.8 mmol). The mixture was heated under reflux for 2 h, cooled, treated with glacial acetic acid (0.5 mL) and (carboxymethyl)trimethylammonium chloride hydrazide (452 mg, 2.7 mmol), and then stirred at room temperature overnight. Alternatively, excess aldehyde could be removed with aminomethyl resin with acetic acid in MeOH/dioxane for 18 h. The mixture with Girard's Reagent T was filtered through Kieselguhr and washed well with DCM. The filtrate was washed with water (3×) and brine, dried with sodium sulfate, and evaporated *in vacuo* to leave a 3:1 mixture of (*E*)- and (*Z*)-stilbene (82% yield).

For Cleavage by Hydrolysis. A suspension of polymer-bound (2-((4-methoxybenzoyl)amino)benzyl)triphenylphosphonium bromide (500 mg) in methanol (15 mL) was treated with 2.0 M sodium methoxide (0.9 mL, 1.8 mmol) and heated under reflux for 4.5 h. The mixture was cooled, filtered through Kieselguhr, and washed well with DCM. The filtrate was washed with water (3×) and brine, dried with sodium sulfate, and evaporated *in vacuo* to leave traceless methyl derivative (81% yield).

For Cleavage by Intramolecular Wittig Reaction. A suspension of polymer-bound (2-((4-methoxybenzoyl)amino)benzyl)triphenylphosphonium bromide (500 mg) in toluene (25 mL) and DMF (5 mL) was distilled until approximately 5 mL of distillate was collected. Potassium *tert*-butoxide (134 mg, 1.2 mmol) was added and the mixture was heated under reflux for 45 min. The mixture was cooled, acidified with 2 N HCl, filtered through Kieselguhr, and washed well with DCM. The filtrate was washed with water (3×) and brine, dried with sodium sulfate, and evaporated *in vacuo* to provide the indole derivative (78% yield).

REFERENCES FOR SECTION 3.5.5

1. Hughes, I. *Tetrahedron Lett.* **1996**, *37*, 7595–7598.
2. Bernard, M.; Ford, W. T. *J. Org. Chem.* **1983**, *48*, 326.

3.5.6 A Novel 3-Propyl-3-(benzyl-supported)triazine Linkage for the Solid-Phase Synthesis of Phenylacetylene Oligomers[1]

Points of Interest

1. The triazene linkage is obtained by reaction of arenediazonium tetrafluoroborate salts with an *n*-propylamino-modified Merrifield resin.

2. Resin-bound (trimethylsilyl)acetylene was deprotected with TBAF followed by the coupling of an aryl iodide with a palladium(0) catalyst in an iterative manner. Phenylacetylene oligomers were liberated from the 1-aryl-3-propyl-3-(benzyl-supported)triazene group by reaction with iodomethane, producing an aryl iodide.

Literature Procedures

For the Preparation of Propylaminomethylated Polystyrene Resin. A suspension of (chloromethyl)polystyrene–1% divinylbenzene copolymer beads (20.0 g, 0.70 mequiv of chloride) and *n*-propylamine (100 mL, 1.22 mol) was degassed and heated at 70°C for 3 days in a sealed tube and agitated periodically. The polymer was transferred to a coarse sintered glass filter using DCM and washed with DCM (400 mL). The resin was thoroughly washed via a standard protocol.[2] The resin was then placed in a 1-L round-bottom flask and fitted with a magnetic stirrer and reflux condenser at 70°C. The resin was stirred slowly with dioxane/2 N NaOH (1/1 v/v, 400 mL) for 30 min, and the solvent was removed by aspiration through a coarse sintered glass filter. This was repeated once more with dioxane/2 N NaOH, twice each with dioxane/H$_2$O, DMF, and MeOH, and finally benzene. The resin was then rinsed with hot MeOH, hot benzene, hot MeOH, hot DCM, and MeOH and dried *in vacuo* to a constant mass to give 19.58 g of (propylaminomethyl)polystyrene resin.

For the Preparation of Triazine Polystyrene Resin. To a chilled (0°C) suspension of (propylaminomethyl)polystyrene–1% divinylbenzene resin (4.32 g, 0.522 mequiv/g of nitrogen), finely ground potassium carbonate (590 mg, 4.5 mmol), and DMF (50 mL) was added 3-bromo-5-(hexyloxy)benzenediazonium tetrafluoroborate (1.0 g, 2.69 mmol) in portions over 1 h. After each addition, an aliquot of the DMF supernatant was diluted in diethylamine and analyzed by GC. After diethyltriazene was detected, the additions were ceased and the suspension was transferred to a fritted filter using DMF, washed sequentially with 120 mL each of MeOH, H$_2$O, MeOH, THF, and MeOH, and dried *in vacuo* to a constant mass to give 4.85 g (0.389 mequiv/g, 84%) of light yellow triazene resin.

For Cleavage. A suspension of the derivatized triazene resin (610 mg, 0.389 mmol) and iodomethane (6.1 mmol) was degassed and heated at 110°C in a sealed tube for 24 h. After the iodomethane was removed *in vacuo*, the product was extracted from the resin using hot DCM. The resulting solution was cooled and filtered through a plug of silica gel in hexanes to give the expected iodobenzene derivative in 98% yield.

REFERENCES FOR SECTION 3.5.6

1. (a) Nelson, J. C.; Young, J. K.; Moore, J. S. *J. Org. Chem.* **1996**, *61*, 8160–8168. (b) Young, J. K.; Nelson, J. C.; Moore, J. S. *J. Am. Chem. Soc.* **1996**, *116*, 10841–10842.
2. Patterson, J. A. *Biochemical Aspects of Reactions on Solid Supports*; Academic Press: New York, 1971.

3.5.7   Reissert-Based "Traceless" Amide Linkage to Support[1]

Points of Interest

1. Support-bound Reissert complexes were used to prepare isoquinoline–isoxazoline heterocycles.

Literature Procedures

For Loading onto the Resin.   Benzoic acid-functionalized polystyrene–2% divinylbenzene resin[2] (see p 12) was converted to polymer-bound benzoyl chloride by treatment with $SOCl_2$ (DMF, reflux, 2–3 h).[3] Reissert formation was accomplished by treating the DCM-swollen, support-bound acyl chloride with 5 equiv of isoquinoline and TMSCN at room temperature for 48 h. FT-IR analysis showed loss of the acyl chloride stretch at 1760 cm^{-1} and formation of an amide stretch at 1680 cm^{-1} (a very weak cyano absorption is found between 2337 and 2367 cm^{-1}).

For Cleavage from Resin.   Following solid-phase synthesis, the resin was swollen in THF and treated with aqueous KOH (1 M; 2:1 THF–H_2O) to effect Reissert hydrolysis. Subsequent filtration and ether wash followed by aqueous workup provided isoquinolines.

REFERENCES FOR SECTION 3.5.7

1. Lorsbach, B. A.; Miller, R. B.; Kurth, M. J. *J. Org. Chem.* **1996**, *61*, 8716–8717.
2. Flyes, T. M.; Leznoff, C. C. *Can. J. Chem.* **1976**, *54*, 935–942.
3. Flyes, T. M.; Leznoff, C. C.; Weatherston, J. *Can. J. Chem.* **1978**, *56*, 1031–1041.

3.5.8   Thioester Linker for the Preparation of Thioacids[1]

Points of Interest

1. Peptide α-thioacids are used for the chemical ligation of unprotected peptide segments.

2. The thioacid-producing linkers were prepared in solution as previously described, but without the inconvenience of using hydrogen sulfide gas.[2]

3. Polymer-supported EDCI (see p 266) has been used for the synthesis of the related primary and aryl thiol esters in solution.[3]

4. Sulfonamides have been prepared and cleaved from the Rink amide resin.[4]

Literature Summary

Boc-amino acids were preloaded onto the linker in solution. The appropriate amino acyl form of this linker was coupled to the resin via the terminal carboxyl group, and the desired peptide was then assembled in a stepwise fashion from the C-terminal amino acid which was incorporated into the linker prior to coupling to the resin. Acidolytic cleavage with HF generated a peptide with a C-terminal $^\alpha$COSH according to known procedure. (**CAUTION:** Hydrogen fluoride gas is **extremely** toxic and reaction with this gas should only be performed with the proper equipment and training. There are two excellent reviews[5] that address safety issues with respect to the handling and use of this reagent.)

REFERENCES FOR SECTION 3.5.8

1. Canne, L. E.; Walker, S. M.; Kent, S. B. H. *Tetrahedron Lett.* **1995,** *36,* 1217–1220.
2. Yamashiro, D.; Li, C. H. *Int. J. Pept. Protein Res.* **1988,** *31,* 322–334.
3. Adamczyk, M.; Rishpaugh, J. R. *Tetrahedron Lett.* **1996,** *37,* 4305–4308.
4. Beaver, K. A.; Siegmund, A. C.; Spear, K. L. *Tetrahedron Lett.* **1996,** *37,* 1145–1148.
5. (a) Skakihara, S. In *Chemistry and Biochemistry of Amino Acids, Peptides, and Proteins;* Weinstar, B., Ed.; New York, **1971;** pp 51–85. (b) Lenard, J. *Chem. Rev.* **1969,** *69,* 625–636.

3.5.9   Disulfide Linker for Solid-Phase Synthesis [1,2]

Points of Interest

1. A reversible linkage through the use of disulfide bonds was originally developed for solid-phase peptide synthesis. The disulfide linkage was employed for the solid-phase synthesis of β-turn mimetics (see p 156).[3]

2. Disulfide cleavage can be performed under very mild conditions and in high yield by reduction with thiols or trialkylphosphines. Quantitative reduction occurs with dithiothreitol (DTT) without employing excess reagent, due to the thermodynamic stability of the six-membered cycle in the reduced form.

3. Mixed disulfides were initially prepared in solution, purified, and char-

acterized before being bound to the support by a standard amide bond. Alternatively, a stepwise procedure was developed for the introduction of the disulfide on support. The mercapto acid, bound to the support by its carboxyl function, was reacted with different sources of mercapto amines.

4. Estimation of the amino groups was achieved by a rapid, nondestructive method involving treatment of amino groups bound to the support (0.2–1.5 mequiv) by an excess (10–100×) of pyridine hydrochloride in DCM (0.3 M), selective precipitation of pyridine hydrochloride by 10–fold dilution with anhydrous n-hexane, and spectrophotometric determination of pyridine in the supernatant ($\epsilon_{252nm} = 2180$).[4]

Literature Procedures

For Loading the Disulfides onto Resin. Protected, bifunctional, mixed disulfide handles were originally prepared in solution and loaded onto resin via a standard amide bond. The disulfide handles were loaded onto 0.15 mmol of Expansin resin (0.468 mmol of NH_3Cl/g) with TBTU activation of the free acid. Estimation of the remaining free amino groups was determined by the pyridine hydrochloride assay described above. Alternatively, the disulfides were prepared on support as follows. Expansin resin (3.0 g, 0.59 mmol of NH_2/g) was washed with DMF (4×), 10% DIEA in DMF (2×), and DMF (4×) and reacted for 2 h with 4 equiv each of S-acetyl-2-mercapto-2-isobutyric acid, HOBt, and TBTU and 8 equiv of DIEA (pH 8) in 20 mL of DMF. At the end of the reaction the ninhydrin test was fully negative and the SAMI (S-acetyl-2-mercapto-2-isobutyric acid) resin was washed with DMF (4×), capped for 30 min with 20 mL of Ac_2O/pyridine/DMF (2:2:20), and washed with DMF (4×). S-Deacetylation was achieved by three successive 30-min treatments with 0.1 N NaOH, and the S-deprotected polymer was washed twice with water. Disulfide formation was obtained by 2-h reaction of the 2-mercapto-2-isobutyl resin with 4.95 g (22 mmol, 12.3 equiv) of cystamine dihydrochloride and 0.8 g of NaOH (20 mmol, 11.2 equiv) in 40 mL of water. Three 1-mL portions of 30% H_2O_2 were added at $t = 0$, 30, and 60 min. Lastly, the handle resin was washed with water (3×), neutralized with 22 mL of 0.1 N HCl, washed with water (4×), DMF, and diethyl ether, and dried *in vacuo*. A loading of 96% of the theoretical was observed after 48 h of hydrolysis (110°C) in 6 M HCl containing 5% DMSO (0.501 mmol/g).

For Solid-Phase Synthesis on Disulfide Linkers. Symmetrical anhydrides were used for Boc-amino acid solid-phase synthesis on disulfide handles. Alternatively, Fmoc-amino acids were activated with equimolar amounts of TBTU and HOBt diluted to 0.3 M with 8% DIEA in DMF according to the manufacturer-specified protocols (Milligen 9050 Pepsynthesizer, Millipore-UK).

For Cleavage. At the end of the stepwise synthesis an aliquot of the resin was submitted to acid hydrolysis for amino acid analysis. After thiolysis with dithiothreitol (DTT) or reduction with tris(2-carboxyethyl)phosphine (TCEP), reanalysis of the amino acid content was used to determine the thiolysis percent conversion (95% conversion). For DTT cleavage: under nitrogen pressure, 500 mg of deprotected peptidyl resin (0.1–0.25 mmol peptide) was washed with DMF (3×), 10% DIEA in DMF (2×), DMF (3×), and water (3×) and then reacted for 2 h at room temperature in 20 mL of degassed 0.1 M borate buffer (pH 8.9) with 1–1.5 mol of DTT per mole of peptide in the polymer (the final pH was readjusted with diluted aqueous NaOH). The reactor content was then filtered out, and the resin was washed with 10 mM HCl (3×), MeOH, and diethyl ether, dried *in vacuo*, and hydrolyzed for amino acid analysis. A similar protocol was used with 1–1.5 mol of TCEP·HCl per mole of peptide on the polymer in 20 mL of degassed 0.05 M sodium acetate buffer (pH 4.5).

REFERENCES FOR SECTION 3.5.9

1. Méry, J.; Brugidou, J.; Derancourt, J. *Pept. Res.* **1992**, *5*, 233–240.
2. Méry, J.; Granier, C.; Juin, M.; Brugidou, J. *Int. J. Pept. Protein Res.* **1993**, *42*, 44–52.
3. Virgilio, A. A.; Schürer, S. C.; Ellman, J. A. *Tetrahedron Lett.* **1996**, *37*, 6961–6964.
4. Dorman, L. C. *Tetrahedron Lett.* **1969**, *28*, 2319–2321.

3.5.10   Linkages for Other Functional Groups

1. C-terminal peptide aldehydes,[1] as well as olefin and hydroxyethylene peptidomimetics derived from amino aldehydes,[2] have been prepared on solid support. Aldehydes have also been synthesized via reductive cleavage of support-bound Wienrib amides.[3]

2. Transesterification of support-bound esters has been performed under a variety of conditions.[4,5]

3. Hydroxylamines were attached to Wang resin using Mitsunobu loading conditions in the solid-phase synthesis of hydroxamic acids.[6]

REFERENCES FOR SECTION 3.5.10

1. Murphy, A. M.; Dagnino, R.; Vallar, P. L.; Trippe, A. J.; Sherman, S. L.; Lumpkin, R. H.; Tamura, S. Y.; Webb, T. R. *J. Am. Chem. Soc.* **1992**, *114*, 3156–3157.
2. Rotella, D. P. *J. Am. Chem. Soc.* **1996**, *118*, 12246–12247.
3. Fehrentz, J. A.; Paris, M.; Heitz, A.; Velek, J.; Liu, C.-F.; Winternitz, F.; Martinez, J. *Tetrahedron Lett.* **1995**, *36*, 7871–7874.
4. Frenette, R.; Friesen, R. W. *Tetrahedron Lett.* **1994**, *35*, 9177–9180.
5. Seebach, D.; Thaler, A.; Blaser, D.; Ko, S. Y. *Helv. Chim. Acta* **1991**, *74*, 1102–1119.
6. Floyd, C. D.; Lewis, C. N.; Patel, S. R.; Wittaker, M. *Tetrahedron Lett.* **1996**, *37*, 8045–8048.

4

COMBINATORIAL SOLID-PHASE SYNTHESIS

The majority of combinatorial studies reported in the literature employ solid-phase synthesis. The primary advantages of solid-phase synthesis are that a large excess of reagents may be employed to drive reactions to completion and that the excess of reagents may be removed by filtration, thus eliminating the tedious workup often associated with organic synthesis in solution. Since general and high-yielding reactions can be performed on solid support, solid-phase synthesis is particularly well suited to library generation involving multistep transformations. Literature procedures and references for condensations, carbon–carbon bond formation, Mitsunobu and substitution reactions, and oxidations and reductions on solid support are described. Examples of specific reactions on solid support ranging from amide bond formation to the palladium-catalyzed amination of support-bound aryl halides are provided. The final section covers the solid-phase synthesis of heterocyclic compounds.

4.1 CONDENSATIONS TO PREPARE AMIDES, ESTERS, UREAS, IMINES, AND PHOSPHORUS COMPOUNDS

Condensation reactions to form amides on solid support are the basis of solid-phase peptide synthesis. In addition to the repetitive formation of amide bonds during chain elongation, esterification reactions are often employed to load the first amino acid onto the resin. A variety of different methods and reagents for these fundamental transformations have been developed; a more detailed treatment can be found in reviews and books on amide bond formation.[1-4]

Condensation reactions have also found wide use in combinatorial organic synthesis because they potentially can be extremely high-yielding and general solid-phase transformations. However, the efficiency of condensation reactions (even amide bond formation) is dependent on the specific electronic and steric environment of the reactants. Accordingly, a range of reactants has been developed specifically for condensation reactions. The following general observations can be used as a starting point for the investigation of challenging amide bond formations. For electron-poor nucleophiles (e.g., 2-aminobenzophenones), activated acylating agents with hard leaving groups such as protected amino acid fluorides and symmetric anhydrides are often efficient. Alternatively, for sterically hindered nucleophiles (e.g., certain secondary amines), HATU/HOAt or PyBrop/DIPEA (with heat) have been successfully employed. Challenging condensations are common in combinatorial synthesis where diverse building blocks, such as electron-poor amines or branched carboxylic acids, are often incorporated into libraries. Thus, despite their fundamental nature, determining general reaction conditions for condensations can still require significant optimization. In addition to the traditional solid-phase condensations employed in peptide and oligonucleotide synthesis, protocols for solid-phase condensations such as imine and urea formation have been developed to generate combinatorial libraries. To assist in this process, examples of various solid-phase condensation procedures are compiled and referenced.

REFERENCES FOR SECTION 4.1

1. Bodanszky, M.; Klausner, Y. S.; Ondetti, M. A. *Peptide Synthesis*, 2nd ed.; Wiley: New York, 1976.
2. Haslam, E. *Tetrahedron* **1980**, *36*, 2409.
3. Fields, G. B.; Noble, R. L. *Int J. Pept. Protein Res.* **1990**, *35*, 161–214.
4. Hudson, D. *J. Org. Chem.* **1988**, *53*, 617–624.

4.1.1 Amide Bond Formation

4.1.1.1   General Points of Interest for Amide Bond Formation on Solid Support [1]

DIC   HOBt   PyBOP   HBTU

1. Carbodiimides are perhaps the most common reagents used for preparing amide bonds on solid support.[2] Diisopropylcarbodiimide (DIC) has the advantage of forming a more soluble urea. Hydroxybenzotriazole (HOBt) is commonly added to inhibit side reactions and reduce racemization.[3]

2. Optimized carbodiimide coupling protocols have been reported for Boc and Fmoc solid-phase peptide synthesis.[2,4]

3. Other popular activating agents include benzotriazolyloxytris(dimethylamino)phosphonium hexafluorophosphate (PyBOP)[5] and 2-(1H-benzotriazol-1-yl)-1,1,3,3-tetramethyluronium hexafluorophosphate (HBTU),[6] both of which require basic conditions. Although more expensive, the phosphonium and uronium activating agents react with faster rates than the standard carbodiimides.

4.1.1.2 A Comparison of Different Coupling Procedures with Methodological Implications[7]

OAIU   Symmetric anhydride   Acyloxyphosphonium salt   HOBt active ester

Points of Interest

1. Competition experiments were used to assess the relative efficiencies of a wide range of coupling reagents.

2. Initial experiments showed that diisopropylcarbodiimide (DIC) and dicyclohexylcarbodiimide (DCC) were equally effective in the formation of symmetrical anhydrides. When symmetric anhydrides are prepared in DCM, often the ureas are removed by filtration and the symmetric anhydrides are isolated and redissolved in DMF. However, the isolation of the anhydrides was unnecessary when the proportion of DMF present in the activation solvent mixture did not exceed 25% in DCM.

3. Under optimized conditions, both BOP/HOBt and DIC/HOBt were superior reagents for peptide bond formation in competition experiments relative to the analogous preformed symmetrical anhydrides. This is in contrast to previous descriptions that suggested the symmetrical anhydrides provide an order of magnitude increased activity and efficiency over the direct carbodiimide method.

4. Both O-acylisoureas (OAIU) and symmetric anhydrides have been proposed as intermediates following carbodiimide activation (without the addition of HOBt) depending on the exact reaction conditions. The following optimal conditions were developed for acylation of support-bound leucine with Fmoc-amino acids: an exactly 2-min preactivation of the Fmoc-amino acid with 1 equiv of DIC in 1:1 DCM/DMF (in which no precipitation was observed even for reaction times longer than 2 h). OAIU was implicated under these reaction conditions based on kinetic data relative to the corresponding symmetric anhydrides.

5. In addition to being more activated under optimal conditions, OAIU are unquestionably far simpler to form than the corresponding symmetric anhy-

drides that can involve a more lengthy procedure for preactivation, filtration, evaporation, redissolution, and addition processes.

Literature Procedures for Solid-Phase Peptide Bond Formation

With DIC. Stock solutions of both the Fmoc-amino acid and DIC (1 equiv) in 1:1 DCM/DMF (0.4 M) were combined with rapid mixing for exactly 2 min to preactivate the amino acid derivative prior to acylation of support-bound leucine.

With DIC/HOBt. Activation was implemented by simply mixing the stable stored DMF solutions of the Fmoc-amino acids dissolved in the presence of HOBt with an equal volume of an identically concentrated solution of DIC in DCM. Fmoc-amino acids are far more soluble in DMF than they are in DCM, and their solutions are stable (in the absence of amines) for at least 7 days.

With BOP/HOBt. Optimal coupling conditions included 1 equiv of BOP, 1 equiv of HOBt, 1.5 equiv of NMM, and 1 equiv of Fmoc-amino acid in DMF (0.2 M solution); 10 min of preactivation was allowed. Peptide bond formation on solid support was mostly complete after 10 min.

4.1.1.3   Amide Bond Formation with Amino Acid Halides

Protected amino acid fluoride   TFFH

Points of Interest

1. Peptide synthesis with amino acid halides has been reviewed.[8]

2. Protected amino acid chlorides have found limited use due to the instability of protected amino acids with functional side chains; for example, Fmoc- or Boc-aspartic acid chlorides containing *tert*-butyl ester side chains simply are too unstable to isolate. However, most of the analogous protected amino acid fluorides are stable crystalline compounds and practical acylating agents. Boc-amino acid fluorides and (benzyloxycarbonyl)amino acid fluorides,[9] as well as Fmoc-amino acid fluorides,[10] have been employed for peptide synthesis.

3. Protected amino acid fluorides can be prepared *in situ* with tetramethylfluoroformamidinium hexafluorophosphate (TFFH)[11] or (diethylamino)sulfur trifluoride (DAST). TFFH is a promising coupling reagent because it is inexpensive, easily handled, and nonhygroscopic.

4. Fmoc-amino acid fluorides were the reagent of choice for performing general and high-yielding acylations of poorly nucleophilic support-bound 2-aminobenzophenones (see p 145).[12]

4.1.1.4   Azabenzotriazole-Based Coupling Reagents HOAt and HATU for Solid-Phase Peptide Synthesis

HOAt   HATU

Points of Interest

1. Hydroxybenzotriazole (HOBt) is commonly added to inhibit side reactions and reduce racemization.[13] 1-Hydroxy-7-azabenzotriazole (HOAt), a more efficient additive, has recently been developed.[14,15]

2. HOAt and its corresponding uronium (HATU) and phosphonium salts were shown to be superior to their benzotriazole analogs in solid-phase peptide synthesis involving hindered amino acids when coupling times were shortened and excesses of reagents were reduced in order to emphasize the differences.[16]

3. HATU was found to be the optimal reagent for the acylation of support-bound secondary amines in the solid-phase synthesis of β-turn mimetics (see p 154).[17]

4. Sterically hindered support-bound secondary amines have also been acylated by using THF as the solvent and PyBrop as the activating agent in the presence of DIEA at 50°C in the solid-phase synthesis of diketopiperazines (see p 168).[18]

5. HATU and HOAt have been used for inverse amide bond formation between a support-bound carboxylic acid and an amino ester in solution.[19]

4.1.1.5   Alternative Procedures for Amide Bond Formation during Cleavage from Resin

1. A range of support-bound active esters and reagents have been employed for amide bond formation; for examples of support-bound DMAP,[20,21] EDC,[22] and nitrophenyl esters,[23] see the section on support-bound reagents (p 262).

2. Inverse amide bond formation by activation of a support-bound amino acid (N-attached to 2-chlorotrityl resin) has been reported. KOTMS in THF was employed for mild methyl ester hydrolysis on solid support. Effective resin swelling permitted complete ester cleavage, whereas aqueous-base conditions generally did not.[24]

3. A highly activated acylsulfonamide linkage can be cleaved with limiting quantities of amine nucleophiles (see p 267).[25,26]

4. Amides have been prepared by aluminum chloride cleavage of unactivated resin-bound esters under ambient conditions.[27]

5. Hydroxamic acids and sulfonamides were prepared by solid-phase synthesis on Wang resin.[28]

6. The Kaiser oxime resin can be treated with amine nucleophiles to form amides (see p 50).

REFERENCES FOR SECTION 4.1.1

1. Fields, G. B.; Noble, R. L. *Int J. Pept. Protein Res.* **1990,** *35,* 161–214.
2. Merrifield, R. B.; Singer, J.; Chait, B. T. *Anal. Biochem.* **1988,** *174,* 399–414.
3. König, W.; Geiger, R. *Ber. Dtsch. Chem. Ges.* **1970,** *103,* 788.
4. Hudson, D. *J. Org. Chem.* **1988,** *53,* 617–624.
5. Coste, J.; Le-Nguyen, D.; Castro, B. *Tetrahedron Lett.* **1990,** *31,* 205–208.
6. Knorr, R.; Trzeclak, A.; Bannwarth, W.; Gillessen, D. *Tetrahedron Lett.* **1989,** *30,* 1927–1930.
7. Hudson, D. *J. Org. Chem.* **1988,** *53,* 617–624.
8. Carpino, L. A.; Beyermann, M.; Wenschuh, H.; Bienert, M. *Acc. Chem. Res.* **1996,** *29,* 268–274.
9. Carpino, L. A.; Mansour, E.-S. M. E.; Sadat-Aalaee, D. *J. Org. Chem.* **1991,** *56,* 2611–2614.
10. Carpino, L. A.; Sadat-Aalaee, D.; Chao, H. G.; DeSelms, R. H. *J. Am. Chem. Soc.* **1990,** *112,* 9651–9652.
11. Carpino, L. A.; El-Faham, A. *J. Am. Chem. Soc.* **1995,** *117,* 5401–5402.
12. Bunin, B. A.; Ellman, J. A. *J. Am. Chem. Soc.* **1992,** *114,* 10997–10998.
13. König, W.; Geiger, R. *Ber. Dtsch. Chem. Ges.* **1970,** *103,* 788.
14. Carpino, L. A. *J. Am. Chem. Soc.* **1993,** *115,* 4397–4398.
15. Carpino, L. A.; El-Faham, A.; Minor, C. A.; Albericio, F. *J. Chem. Soc., Chem. Commun.* **1994,** 201–203.
16. Carpino, L. A.; El-Faham, A.; Minor, C. A.; Albericio, F. *J. Chem. Soc., Chem. Commun.* **1994,** 201–203.
17. Virgilio, A. A.; Ellman, J. A. *J. Am. Chem. Soc.* **1994,** *116,* 11580–11581.
18. Scott, B. O.; Siegmund, A. C.; Marlowe, C. K.; Pei, Y.; Spear, K. L. *Mol. Diversity,* **1995,** *1,* 125–134.
19. Rotell, D. P. *J. Am. Chem. Soc.* **1996,** *118,* 12246–12247.
20. Shai, Y.; Jacobson, K. A.; Patchornik, A. *J. Am. Chem. Soc.* **1985,** *107,* 4249–4252.
21. Patchornik, A. *CHEMTECH* **1987,** 58–63.
22. Desai, M. C.; Stramiello, L. M. S. *Tetrahedron Lett.* **1993,** *34,* 7685–7688.
23. Cohen, B. J.; Karoly-Hafeli, H.; Patchornik, A. *J. Org. Chem.* **1984,** *49,* 922–924.
24. Hoekstra, W. J.; Maryanoff, B. E.; Andrade-Gordon, P.; Cohen, J. H.; Costanzo, M. J.; Damiano, B. P.; Haertlein, B. J.; Harris, B. D.; Kauffman, J. A.; Keane, P. M.; McComsey, D. F.; Villani, F. J.; Yabut, S. C. *Bioorg. Med. Chem. Lett.* **1996,** *6,* 2371–2376.
25. Backes, B. J.; Virgilio, A. A.; Ellman, J. A. *J. Am. Chem. Soc.* **1996,** *118,* 3055–3056.
26. Backes, B. J.; Ellman, J. A. *J. Am. Chem. Soc.* **1994,** *116,* 11171–11172.
27. Barn, D. R.; Morphy, J. R.; Rees, D. C. *Tetrahedron Lett.* **1996,** *37,* 3213–3216.
28. Floyd, C. D.; Lewis, C. N.; Patel, S. R.; Wittaker, M. *Tetrahedron Lett.* **1996,** *37,* 8045–8048.

4.1.2   Esterification Reactions on Solid Support: First Amino Acid Loading by Ester Bond Formation[1]

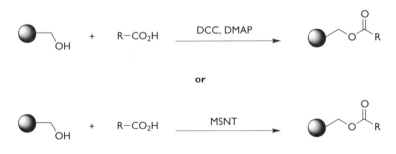

or

Merrifield resin

Points of Interest

1. Many resins are commercially available with the first amino acid pre-loaded.

2. Ester bond formation between a support-bound alcohol and a carboxylic acid can be performed under standard conditions with DCC/DMAP. Potential problems involve partial racemization and deprotection of Fmoc protecting groups in the presence of DMAP. With carboxylic acids other than Fmoc-amino acids, DCC/DMAP esterifications may or may not be problematic.

3. In an investigation into ester formation, in which the alcohol or carboxylic acid component was immobilized on a solid support, MSNT (1-(mesitylene-2-sulfonyl)-3-nitro-1H-1,2,4-triazole) was found to give the best results by far of all the reagents tested (DCC, DCC/HOBt, DCC/DMAP, Ph$_3$P/DEAD).[2]

4. 1,1'-Carbonyldiimidazole has also been employed for *in situ* ester loading.

5. BOP at $-20°$C in DCM has been reported as an efficient reagent for ester and thioester formation in solution.[3]

6. Ester hydrolysis has been reported under nonaqueous conditions for solid-phase synthesis.[4]

7. Alkylation of resin-bound halides with carboxylic acids in the presence of DIEA or as the corresponding cesium salts is often used for ester bond formation (see p 129). Reactions between a range of support-bound alkyl halides (including support-bound benzyl chlorides, alkyl bromides, chlorotrityl chlorides, etc.) and carboxylate anions are quite common in solid-phase peptide and combinatorial synthesis.[5-8]

REFERENCES FOR SECTION 4.1.2

1. Fields, G. B.; Noble, R. L. *Int. J. Pept. Protein Res.* **1990**, *35*, 161–214.
2. *Novabiochem 97/98 Catalog & Peptide Synthesis Handbook*, p 210.
3. Kim, M. H.; Patel, D. V. *Tetrahedron Lett.* **1994**, *35*, 5603–5606.
4. Hoekstra, W. J.; Maryanoff, B. E.; Andrade-Gordon, P.; Cohen, J. H.; Costanzo, M. J.; Damiano, B. P.; Haertlein, B. J.; Harris, B. D.; Kauffman, J. A.; Keane, P. M.; McComsey, D. F.; Villani, F. J.; Yabut, S. C. *Bioorg. Med. Chem. Lett.* **1996**, *6*, 2371–2376.
5. Baleux, F.; Calas, B.; Mery, J. *Int. J. Pept. Protein Res.* **1986**, *28*, 22–28.
6. Barlos, K.; Chatzi, O.; Gatos, D.; Stavropoulos, G. *Int. J. Pept. Protein Res.* **1991**, *37*, 513–520.
7. Kurth, M. J.; Ahlberg Randall, L. A.; Chen, C.; Melander, C.; Miller, R. B.; McAlister, K.; Reitz, G.; Kang, R.; Nakatsu, T.; Green, C. *J. Org. Chem.* **1994**, *59*, 5862–5864.
8. Hanessian, S.; Yang, R. Y. *Tetrahedron Lett.* **1996**, *37*, 5835–5838.

4.1.3   Imine Formation on Solid Support

4.1.3.1   Solid-Phase Imine Formation with Trimethyl Orthoformate[1]

Rink or Sasrin
resin

Points of Interest

1. Trimethyl orthoformate ($(MeO)_3CH$) has been employed as a dehydrating agent with either a DCM or THF cosolvent to improve reagent solubility.[2,3]

2. ^{13}C-labeled building blocks were used to facilitate monitoring imine formation on various solid supports. Although the reactions are typically run for 1–2 h, most unhindered imines are formed in less than 10 min.

3. The imine-forming reaction in the cases when the aldehyde component is bound to the solid support and the amine is present in solution is slower than for the reaction with the opposite orientation (amine immobilized/aldehyde in solution). In most cases, a second treatment could be used to drive the imine formation to completion.

4. When alkyl amines or esters of amino acids were treated with an excess of aldehyde and sodium cyanoborohydride in trimethyl orthoformate, only the monoalkylated products were observed by HPLC.

Representative Examples

Electron-poor as well as electron-rich aldehydes have been used in imine formation. However, electron-rich aldehydes react somewhat more sluggishly than do other aldehydes. Aliphatic aldehydes and ketones usually give nonisolable imines due to imine–enamine tautomerization, although these imines are viable intermediates for *in situ* reduction and cycloaddition reactions.

Literature Procedure[4]

The appropriate support-bound amine (0.3 g, 0.165 mmol) was suspended in a mixture of DCM (1.5 mL) and trimethyl orthoformate (1.5 mL), and the aldehyde (2.3 mmol) was added. After 3 h of agitation, the resin was rinsed with DCM, MeOH, and Et_2O and dried under reduced pressure.

4.1.3.2   Other Examples of Imine Formation on Solid Support

1. Imine transfer between excess imine in solution and a support-bound amino ester was used in the solid-phase synthesis of 1,4-benzodiazepines (see p 148).[5]

2. Imine formation followed by reductive amination of a support-bound aldehyde was used in the preparation of 1,4-benzodiazepine-2,5-diones (see p 150).[6] Amino esters are loaded onto a support-bound aldehyde via a reductive amination. The imines must be reduced immediately upon formation to avoid racemization. Conversely, racemic products can easily be prepared if desired without purchasing L- and D-amino acids.

3. Reductive alkylation involving imine formation was used in the synthesis of a piperazinedione combinatorial library.[7] Sodium triacetoxyborohydride was employed for reductive alkylation to avoid the pH sensitivity and potential reagent toxicity associated with the well-established procedure for the reductive alkylation of peptide resins using sodium cyanoborohydride.[8] The reductive amination procedure included brief sonication and was performed twice to ensure high conversion (see p 167).

4. Benzophenone imine formation has been used to temporarily increase the acidity of the proton on the α-carbon for alkylation in solid-phase unnatural peptide synthesis (see p 118).[9]

5. Imine formation was used in the reductive alkylation of Sieber's Xan linker[10] and the Rink amide linker[11] (see p 135). A two-step procedure was necessary for efficient monoalkylation. It was crucial that the solvent mixture for the reduction contained MeOH. Negligible reduction was observed in straight DMF and the substitution of *i*-PrOH for MeOH also resulted in a poor reduction. Stable imines were not obtained with the PAL linker,[12] prohibiting reductive alkylation of the PAL linker.

6. Support-bound salicylimines were prepared in the preparation of chiral catalysts for the enantioselective cyanation of *meso*-epoxides.[13,14]

REFERENCES FOR SECTION 4.1.3

1. Look, G. C.; Murphy, M. M.; Campbell, D. A.; Gallop, M. A. *Tetrahedron Lett.* **1995**, *36*, 2937–2940.
2. Ruhland, B.; Bhandari, A.; Gordon, E. M.; Gallop, M. A. *J. Am. Chem. Soc.* **1996**, *118*, 253–254.
3. Bunin, B. A. Thesis, University of California at Berkeley, 1996.
4. Ruhland, B.; Bhandari, A.; Gordon, E. M.; Gallop, M. A. *J. Am. Chem. Soc.* **1996**, *118*, 253–254.
5. DeWitt, S. H.; Kiely, J. S.; Stankovic, C. J.; Schroeder, M. C.; Cody, D. M. R.; Pavia, M. R. *Proc. Natl. Acad. Sci. U.S.A.* **1993**, *90*, 6909–6913.
6. Boojamra, C. G.; Burrow, K. M.; Ellman, J. A. *J. Org. Chem.* **1995**, *60*, 5742–5743.
7. Gordon, D. W.; Steele, J. *Bioorg. Med. Chem. Lett.* **1995**, *5*, 47–50.
8. (a) Coy, D. H.; Hocart, S. J.; Sasaki, Y. *Tetrahedron* **1988**, *44*, 835. (b) Hocart, S. J.; Nekola, M. V.; Coy, D. H. *J. Med. Chem.* **1987**, *30*, 739. (c) Tourwé, D.; Piron, J.; Defreyn, P.; Van Binst, G. *Tetrahedron Lett.* **1993**, *34*, 5499.
9. O'Donnell, M. J.; Zhou, C.; Scott, W. L. *J. Am. Chem. Soc.* **1996**, *118*, 6070–6071.
10. Chan, W. C.; Mellor, S. L. *J. Chem. Soc., Chem. Commun.* **1995**, 1475–1477.
11. Rink, H. *Tetrahedron Lett.* **1987**, *28*, 3787–3790.
12. Albericio, F.; Kneib-Cordonier, N.; Biancalana, S.; Gera, L.; Masada, R. I.; Hudson, D.; Barany, G. *J. Org. Chem.* **1990**, *55*, 3730–3743.
13. Cole, B. M.; Shimizu, K. D.; Krueger, C. A.; Harrity, J. P. A.; Snapper, M. L.; Hoveyda, A. H. *Angew. Chem., Int. Ed. Engl.* **1996**, *35*, 1668–1671.
14. Bunin, B. A. Thesis, University of California at Berkeley, 1996.

4.1.4 Urea Formation on Solid Support: A Strategy for Combinatorial Solid-Phase Synthesis of Urea-Linked Diamine Libraries [1]

Points of Interest

1. Preparing bisureas from p-nitrophenyl carbamate and amines provides an alternative to the standard method of urea formation from isocyanates and amines.[2,3]

2. Unprotected diamines and a p-nitrophenyl carbamate intermediate (see p 62) were used for the generation of ureas.

3. The solid-phase methodology for urea formation was initially demonstrated with glutamic acid allyl ester derivatives.[4]

4. An iterative synthesis sequence was performed in only 45 min.

5. Repeating urea units were assembled in the solid-phase synthesis of unnatural biopolymers (see Appendix 3, p 303).[5–7]

Representative Examples

Urea products obtained after 8 steps are of high chemical purity (≥89% pure by HPLC). Monofunctional and bifunctional amines, as well as anilines, were incorporated into ureas.

Literature Summary

Piperidine/DMF (1:1, 20 min) was employed for standard Fmoc deprotection from resin-bound Rink amide linker (50 mg). Fmoc-protected 4-(aminomethyl)benzoic acid was coupled to resin using EDC/DMAP in THF/DCM. Piperidine/DMF (1:1, 20 min) was employed for removal of the second Fmoc group. The resin was washed with DMF (4×) followed by THF/DCM (4×). Then a mixture of p-nitrophenyl chloroformate (0.5 M) and DIPEA (0.5 M) in 1 mL of the THF/DCM (1:1) was added and the resulting slurry was mixed for 30 min. The resin was then washed with THF/DCM (2×) followed by the addition of the desired diamine (0.5 M) and triethylamine (0.5 M) in 1 mL of DMF. After 15 min of mixing, the resin was washed with DMF (4×) and THF/DCM (4×). Then a mixture of p-nitrophenyl chloroformate (0.5 M) and diisopropylamine (0.5 M) in 1 mL of THF/DCM (1:1) was added and the reaction mixture was mixed for 30 min. The resin was washed with THF/DCM (2×) followed by the addition of the desired amine (0.5 M) and triethylamine (0.5 M) in 1 mL of DMF. After 15 min of mixing, the resin was washed with DMF (4×), THF/DCM (4×), and glacial acetic acid (4×). The resin was then treated with TFA/H$_2$O (9:1) for 2 × 20 min. The solution was collected, concentrated down to an oil, dissolved in glacial acetic acid, and lyophilized to yield the desired trimer.

REFERENCES FOR SECTION 4.1.4

1. Hutchins, S. M.; Chapman, K. T. *Tetrahedron Lett.* **1995**, *36*, 2583–2586.
2. Kick, E. K.; Ellman, J. A. *J. Med. Chem.* **1995**, *38*, 1427–1430.
3. Meyers, H. V.; Dilley, G. J.; Durgin, T. L.; Powers, T. S.; Winssinger, N. A.; Zhu, W.; Pavia, M. R. *Mol. Diversity* **1995**, *1*, 13–20.
4. Hutchins, S. M.; Chapman, K. T. *Tetrahedron Lett.* **1994**, *35*, 4055–4058.
5. Burgess, K.; Linthicum, D. S.; Shin, H. W. *Angew. Chem., Int. Ed. Engl.* **1995**, *34*, 907–909.
6. Hutchins, S. M.; Chapman, K. T. *Tetrahedron Lett.* **1995**, *36*, 2583–2586.
7. (a) Kim, J.-M.; Bi, Y.; Paikoff, S. J.; Schultz, P. G. *Tetrahedron Lett.* **1996**, *37*, 5305–5308. (b) Kim, J.-M.; Wilson, T. E.; Norman, T. C.; Schultz, P. G. *Tetrahedron Lett.* **1996**, *37*, 5309–5312.

4.1.5   Condensations with Phosphorus Compounds on Solid Support

4.1.5.1   The Preparation of Phosphodiesters in Solid-Phase Oligonucleotide Synthesis[1]

B = base, G,C,A,T

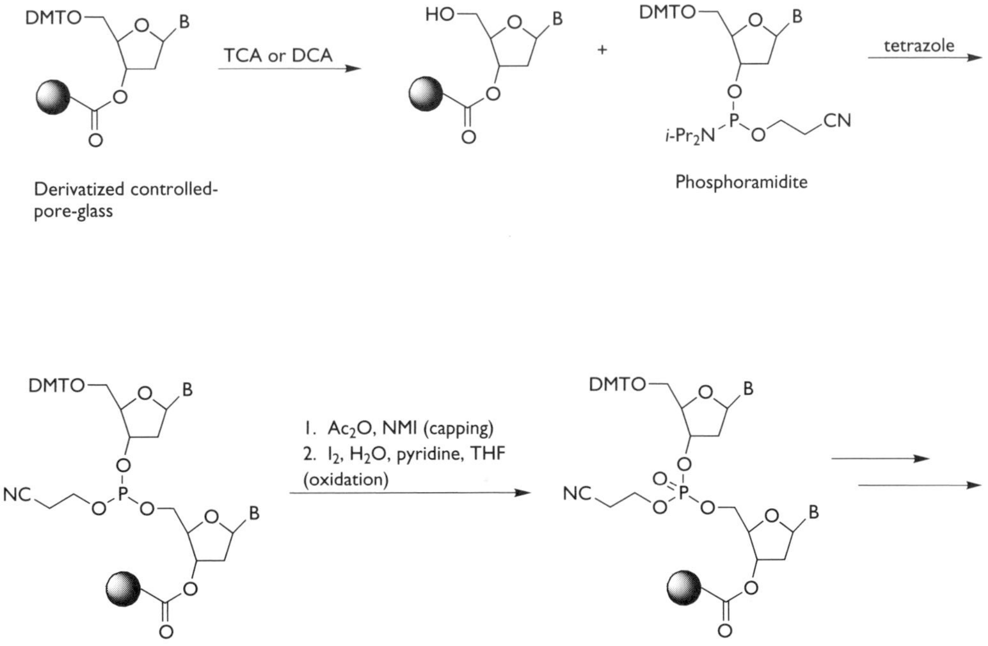

Derivatized controlled-pore-glass

Phosphoramidite

Points of Interest

1. Combinatorial oligonucleotide synthesis is a subject unto itself that has been thoroughly reviewed.[2]

2. A detailed DNA synthesis cycle is described and discussed in the *Applied Biosystems Synthesizer Manual*. The cycle includes detritylation, phosphoramidite coupling, capping, oxidation, and cleavage steps.

3. DNA synthesis is usually performed on controlled-pore glass (CPG) in contrast to peptide and combinatorial synthesis, which is most often performed on polystyrene–divinylbenzene resins.

4. The dimethoxytrityl (DMT) cation formed under acidic conditions is easily detected and quantitated spectrophotometrically.

5. Oligonucleotide cyanoethyl phosphoramidites are typically used as activated monomers in solid-phase DNA synthesis.

6. A summary of the pioneering work on the phosphoramidite approach used in gene synthesis machines has been discussed by Caruthers.[3]

Representative Examples

DNA sequences of up to 70 oligonucleotides are routinely prepared by the phosphoramidite approach. Coupling efficiencies of 98–100%, as measured by the trityl cation assay, enable the synthesis of oligomers of up to 175 bases in length.

Procedures

The loading of nucleoside on controlled-pore glass is typically 27–30 μmol/g of 500-Å support as measured by DMT release. Prefilled columns can contain 0.2, 1, or 10 μmol of initial nucleoside. Detailed procedures for oligonucleotide synthesis accompany specific gene synthesis machines.

4.1.5.2   A Combinatorial Method for the Solid-Phase Synthesis of α-Amino Phosphonates and Phosphonic Acids[4]

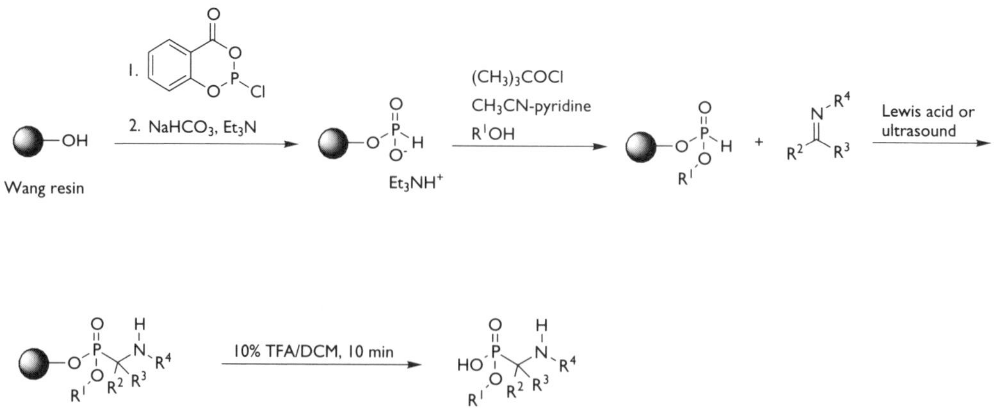

Points of Interest

1. α-Amino phosphonates were synthesized from the condensation of imines with support-bound H-phosphonates.

2. La(OTf)$_3$ and Yb(OTf)$_3$ provided condensation products in higher yield than the other Lewis acids investigated.

3. Sonication of the support-bound H-phosphonates in the presence of imines also afforded the condensation products. To compensate for partial cleavage from resin that occurred during sonication, the esterification procedure was repeated after the sonication step.

Representative Examples

Imines derived from amines, anilines, aldehydes, and a ketone were incorporated into α-amino phosphonates. An α-amino phosphonic acid was also prepared.

Literature Summary

Wang resin was treated with 2-chloro-4H-1,3,2-benzodioxaphosphorin-4-one followed by hydrolysis with NaHCO$_3$–NEt$_3$ to afford the resin-bound H-phosphonate salt. Next, esterification with benzyl alcohol or (p-nitrophenyl)-ethanol (NPE) (5 equiv, 1.0 M in CH$_3$CN) and pivaloyl chloride (5 equiv, 1.0 M in 1:1 CH$_3$CN/pyridine) afforded the corresponding H-phosphonate esters (60% yield). Condensations were performed either with a 2.0 M solution of the imine with either Yb(OTf)$_3$ (0.05 M) at 23°C for 5 h or with sonication at 50–55°C for 3 h followed by a reesterification step. The condensation conditions were compatible with a range of solvents (DCE, toluene, THF, DMF).

4.1.5.3   A Combinatorial Method for the Solid-Phase Synthesis of α-Hydroxy Phosphonates[5]

Points of Interest

1. α-Hydroxy phosphonates were prepared by the reaction of support-bound H-phosphonate ester–DBU salts with aldehydes.

Representative Examples

Aliphatic, aromatic, and heteroaromatic aldehydes and four alcohols were incorporated into α-hydroxy phosphonates in >90% purity without purification.

Literature Summary

To a suspension of Wang resin (89 mmol, 100 mg, 1.0 equiv) in dry DCM (1.0 mL) and pyridine (0.5 mL) at 0° was added a 1.0 M solution of 2-chloro-4H-1,3,2-benzodioxaphosphorin-4-one (267 mg, 267 μL, 3.0 equiv) in dry DCM. After warming to room temperature, the reaction mixture was gently stirred for an additional 30 min. Treatment with a cold solution of 1:1 Et$_3$N–NaHCO$_3$ in H$_2$O provided the triethylammonium phosphonate polymer, which was filtered, washed with H$_2$O, CH$_3$CN, and DCM, and then dried *in vacuo* over P$_2$O$_5$. The dry triethylammonium phosphonate polymer was treated with a 1.0 M solution of the alcohol (5 equiv) in CH$_3$CN followed by the addition of a 1.0 M solution of pivaloyl chloride (5 equiv) in CH$_3$CN–pyridine for 15 min at room temperature. The resulting polymer was filtered, washed with dry CH$_3$CN, and treated with a 1.0 M solution of DBU in CH$_3$CN. The mixture was stirred at room temperature for 10 min, and then a 1.0 M solution of the aldehyde (5 equiv) in CH$_3$CN was added. After 30 min, the polymer was filtered and then washed with CH$_3$CN and DCM. Cleavage with 10% TFA in DCM (2 × 20 min) provided the desired α-hydroxy phosphonates.

4.1.5.4 Other Examples of Condensations Involving Phosphorus Compounds

1. The solid-phase synthesis of phosphonic acid esters (peptidyl phosphonates) under modified Mitsunobu conditions has been reported (see p 109).[6]

REFERENCES FOR SECTION 4.1.5

1. *Applied Biosystems Model 391 DNA Synthesizer Manual;* Section 6: Chemistry for Automated DNA Synthesis.
2. (a) Ecker, D. J.; Vickers, T. A.; Hanecak, R.; Driver, V.; Anderson, K. *Nucleic Acids Res.* **1993,** *21,* 1853–1856. (b) Gold, L.; Polisky, B.; Ulenbeck, O.; Yarus, M. *Annu. Rev. Biochem.* **1995,** *64,* 763–797.
3. Caruthers, M. H. *Science* **1985,** *230,* 281–285.
4. Zhang, C.; Mjalli, A. M. M. *Tetrahedron Lett.* **1996,** *37,* 5457–5460.
5. Cao, X.; Mjalli, A. M. M. *Tetrahedron Lett.* **1996,** *37,* 6073–6076.
6. Campbell, D. A.; Bermak, J. C. *J. Am. Chem. Soc.* **1994,** *116,* 6039–6040.

4.2 CARBON–CARBON BOND FORMATION ON SOLID SUPPORT

Carbon–carbon bond formation has been considered the Holy Grail of organic synthesis (although some would argue the real Holy Grail is carbon–hydrogen insertion reactions). Many fundamental carbon–carbon bond formation reactions have been adapted to solid support. For example, support-bound silyl enol ethers have been employed for solid-phase Mukaiyama aldol and imine reactions. The Suzuki reaction has proven to be a particularly general reaction for carbon–carbon bond formation in the hands of a number of researchers. Kenner's safety-catch linker is well suited for carbon–carbon bond formation on solid support due to its stability under a range of reaction conditions. The acylsulfonamide lends itself to enolate alkylation reactions. In contrast to ester or carboximide enolate alkylations, ketene formation is not observed even when the unreactive alkylating agent isopropoyl iodide is employed, since ketene formation would require that the sulfonamide dianion be the leaving group. The following section surveys the methods in the literature for carbon–carbon bond formation on solid support.

Examples of carbon-carbon bond formation on solid support.

4.2.1 Suzuki Reactions on Solid Support

4.2.1.1 Biaryl Synthesis via Suzuki Coupling on Solid Support[1]

Points of Interest

1. Polymer-bound aryl halides were coupled with arylboronic acids via the Suzuki coupling reaction to form biaryls.

2. Benzoic acid scaffolds were attached to resin by alkylation (see p 129) and cleaved by transesterification.

Representative Examples

Aryl bromides with ortho, meta, para, and poly substituents were coupled with both electron-rich and electron-poor boronic acids in high yield (usually greater than 90% crude yield on the aryl halide loading).

Literature Summary

Catalytic $Pd(Ph_3P)_4$ (29 mg, 0.05 equiv) was added to a degassed suspension of the support-bound aryl halide (500 mg, 1.0 mequiv/g) in 4 mL of DME. The slurry was stirred for 5 min, and then phenylboronic acid (122 mg, 2.0 equiv) and 2 M Na_2CO_3 (0.63 mL, 2.5 equiv) were added. The mixture was refluxed overnight under an argon atmosphere. The mixture was then cooled to room temperature, diluted with 10 mL of 25% NH_4OAc and stirred for an additional 5 min before being filtered. The resin was washed with DME/H_2O, H_2O, 0.2 N HCl, H_2O, DME, EtOAc, EtOAc/MeOH, and MeOH and then dried under high vacuum. The product was obtained by transesterification of the ester linkage. A mixture of polymer (490 mg) in MeOH/THF (1:4, 5 mL) in the presence of NaOMe (0.05 mL of a 1.0 M solution of MeOH, 0.1 equiv) was refluxed overnight. After filtration, the resin was washed with MeOH/THF, THF, and MeOH. Concentration of the filtrate and washings provided the biaryl product (in >90% purity by ^{1}H NMR and HPLC) as a white solid (100mg) in 95% yield (based on loading of the aryl halide to the resin).

4.2.1.2   Suzuki Coupling for Carbon–Carbon Bond Formation on Solid Support[2]

Points of Interest

1. The solid-phase synthesis of substituted arylacetic acid derivatives was developed to evaluate two important carbon–carbon bond-forming reactions on solid support: enolate alkylation (see p 121) and palladium-mediated Suzuki cross-coupling.

2. A variant of Kenner's safety-catch linker[3] was utilized because it is stable to the basic reaction conditions, yet can be cleaved with nucleophiles upon activation.

Representative Examples

Electron-rich, electron-poor, and ortho-substituted arylboronic acids, as well as *B*-alkyl-9-borabicyclo[3.3.1]nonane derivatives that were prepared by *in situ* hydroboration of isobutylenes, were used in the preparation of aryl-acetic acid derivatives. High yields were observed (average 93% purified yield, 12 compounds) after cleavage and chromatography of arylacetic acid derivatives.

Literature Procedure

Derivatized macroreticular resin (Amberlite XE-305 resin, 750 mg, 0.185 mmol) and a phenylboronic acid derivative (1.5 mmol, 8.1 equiv) or *B*-isobutyl-9-borabicyclo[3.3.1]nonane (15 mL, 7.5 mmol) were placed in a flask, and THF was added to a total volume of 20 mL. Aqueous 2 N Na_2CO_3 (1.5 mL) was added and the resulting slurry was stirred for 5 min while the flask was flushed with N_2. The addition of $Pd(PPh_3)_4$ (80 mg, 37 mol %) followed, and the flask was flushed with N_2 for another 5 min. The flask was maintained under positive N_2 pressure, and the reaction mixture was heated at reflux for 24–40 h. The slurry was filtered and the resin was washed with THF, H_2O, THF, 5% TFA in THF (to reprotonate the support-bound acylsulfonamide), and THF again.

4.2.1.3 Additional Examples of Solid-Phase Suzuki Coupling Reactions

1. Two groups have reported solid-phase Suzuki reaction conditions on aryl halides tethered to solid support via silyl linkers.[4,5]
2. Resin capture strategies (see p 260) have been employed that selectively react with vinylboranes generated *in situ* from a mixture of reagents in solution.[6]
3. In a detailed study of the solid-phase Suzuki coupling reaction, the optimal Pd catalyst varied depending on the boronic coupling partner.[7] The aryl- or heterobiarylboronic acids exhibited excellent conversion with $Pd_2(dba)_3$. The $PdCl_2(dppf)$ catalyst worked best for reactions involving the alkenylborane, in sharp contrast to its complete lack of activity in the biphenyl coupling. Suzuki coupling reaction conditions were described for both support-bound aryl halides and a support-bound arylboronic acid.
4. The Suzuki reaction was used to increase the diversity accessible in the solid-phase synthesis of 1,4-benzodiazepine-2,5-diones from support-bound peptoids (see p 152).[8]
5. A palladium-mediated three-component coupling reaction involving both Suzuki and Heck reactions has been developed for the solid-phase synthesis of tropane derivatives (see p 197).[9] The first step in the three-component coupling produces a stable support-bound palladium–arene complex. Because the palladium intermediate is isolated, split synthesis strategies may be employed for the generation of tropane libraries.
6. Two detailed studies of palladium-catalyzed amination of resin-bound aromatic bromides have been reported (see p 126).[10,11]

REFERENCES FOR SECTION 4.2.1

1. Frenette, R.; Friesen, R. W. *Tetrahedron Lett.* **1994**, *35*, 9177–9180.
2. Backes, B. J.; Ellman, J. A. *J. Am. Chem. Soc.* **1994**, *116*, 11171–11172.
3. Kenner, G. W.; McDermott, J. R.; Sheppard, R. C. *J. Chem. Soc., Chem. Commun.* **1971**, 636–637.

4. Chenera, B.; Finkelstein, J. A.; Veber, D. F. *J. Am. Chem. Soc.* **1995**, *117*, 11999–12000.
5. Han, Y.; Walker, S. D.; Young, R. N. *Tetrahedron Lett.* **1996**, *37*, 2703–2706.
6. Brown, S. D.; Armstrong, R. W. *J. Am. Chem. Soc.* **1996**, *118*, 6331–6332.
7. Guiles, J. W.; Johnson, S. G.; Murray, W. V. *J. Org. Chem.* **1996**, *61*, 5169–5171.
8. Golf, D. A.; Zuckermann, R. N. *J. Org. Chem.* **1995**, *60*, 5744–5745.
9. Koh, J. S.; Ellman, J. A. *J. Org. Chem.* **1996**, *61*, 4494–4495.
10. Ward, Y. D.; Farina, V. *Tetrahedron Lett.* **1996**, *37*, 6993–6996.
11. Willoughby, C. A.; Chapman, K. T. *Tetrahedron Lett.* **1996**, *37*, 7181–7184.

4.2.2 Organozinc Reactions on Solid Support: Zinc Organic Reagents in Aryl–Aryl Cross-Coupling on Solid Support[1]

Points of Interest

1. Zinc organic reagents are well suited for the preparation of biaryls. Transmetalation of lithium or magnesium organic reagents allows access to zinc reagents from numerous, commercially available aryl bromides. Thus an alternative is available for aryl–aryl coupling on solid support, especially when the desired boronic acids are not easily accessible.

2. Zinc aryl halogenides were used to form biaryls with support-bound aryl bromides.

Representative Examples

Ortho-, meta-, and para-substituted support-bound bromobenzoic acid derivatives were used to prepare biaryls. Organozinc reagents derived from electron-rich, electron-poor, and ortho-substituted aromatic compounds as well as heteroaromatic compounds were prepared in solution and employed in palladium-catalyzed cross-coupling reactions.

Literature Summary

(3-Fluorophenyl)magnesium bromide (5 mmol) was prepared from 875 mg of 3-fluorobromobenzene and 122 mg of magnesium in 5 mL of THF. The Grignard reagent was transmetalated by adding a solution of 1.13 g (5 mmol) of dried $ZnBr_2$ in 5 mL of THF at $0°C$. (Note: a number of zinc aryl halogenides are commercially available.) After 15 min of stirring at $0°C$ to form the zinc aryl halogenide, 500 mg of the appropriately derivatized resin (1.4 mmol/g) and 15 mg of $PdCl_2(dppf)$ were added. The resulting mixture was stirred at room temperature for 18 h. The polymer was then filtered and washed with THF/H_2O, THF, MeOH, and ether.

REFERENCES FOR SECTION 4.2.2

1. Marquais, S.; Arlt, M. *Tetrahedron Lett.* **1996**, *37*, 5491–5494.

4.2.3   Stille Coupling Reactions on Solid Support

4.2.3.1   Application of the Stille Coupling Reaction with a Support-Bound Aryl Iodide and a Range of Stannanes in Solution [1]

Rink resin

R = aryl, vinyl

Points of Interest

1. Vinyl- and arylstannanes were prepared in solution and coupled to a support-bound aryl iodide.

2. Under optimal conditions, difficult couplings with hindered stannanes proceeded in good yield.

Representative Examples

Vinyl- and arylstannanes were coupled with a support-bound aryl iodide. Cleavage from resin afforded products with >90% purity by HPLC.

Literature Summary

Support-bound 4-iodobenzoic acid was subjected to Stille cross-coupling reactions using a variety of stannanes. To a degassed suspension of support-bound aryl iodide in anhydrous NMP were added $Pd_2(dba)_3$ (5 mol%) and $AsPh_3$ (20 mol%). After 5 min, the appropriate stannane was added and the reaction was heated overnight at 45°C.

4.2.3.2   Stille Coupling Reaction between Support-Bound Aminoarylstannane and Acid Chlorides in Solution in the Solid-Phase Synthesis of 1,4-Benzodiazepines [2]

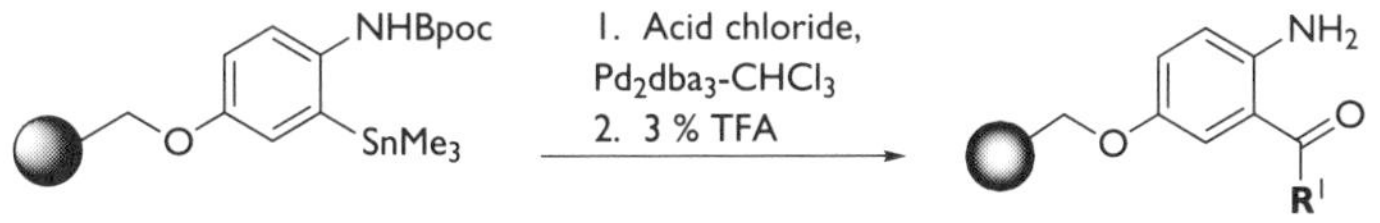

Points of Interest

1. A large number of acid chlorides are commercially available and chemically compatible with the mild reaction conditions.

2. The Stille coupling reaction on solid support was complete after 1 h at room temperature.

3. After the Stille coupling reaction, the resin was rinsed with KCN (*Caution:* toxic!) in DMSO to effectively remove palladium black from the resin.

Representative Examples

A range of aliphatic, heteroaromatic, and aromatic acid chlorides were incorporated into 1,4-benzodiazepines (see p 147). Both electron-rich and electron-poor as well as ortho-substituted acid chlorides were compatible with the reaction conditions developed.

Literature Procedure

To an oven-dried Schlenk flask under N_2 were added the support-bound aminoarylstannane (0.40 g, 0.17 mmol), $Pd_2(dba)_3 \cdot CHCl_3$ (60 mg, 0.060 mmol), K_2CO_3 (10 mg), THF (4.0 mL), and DIEA (40 μL, 0.20 mmol). The resin was stirred for 3 min to ensure complete solvation, at which point the acid chloride (1.00 mmol) was added slowly. The reaction was allowed to proceed for 1 h, after which the mixture was transferred to a peptide flask and rinsed with DCM (3×), DMSO saturated with KCN (1–3×), MeOH, H_2O, MeOH (3×), and DCM (3×). The resin was treated with 97:3 DCM/TFA for 5 min, rinsed with DCM (2×), and treated again for 5 min with the acidic cocktail to provide support-bound 2-aminobenzophenone or 2-aminoacetophenone. The resin was rinsed with DCM (5×) and MeOH (5×) and dried under N_2.

4.2.3.3 Additional Examples of Solid-Phase Stille Coupling Reactions

1. The Stille reaction has been investigated with polymer-supported organotin reagents (see p 273).[3] The polymer-supported organotin byproduct can be separated by simple filtration and regenerated. The organic target molecule is virtually free of toxic organotin materials.

2. Stille couplings with fluorous tin reactants have been reported for liquid-phase combinatorial synthesis (see p 249).[4]

REFERENCES FOR SECTION 4.2.3

1. Deshpande, M. S. *Tetrahedron Lett.* **1994,** *35,* 5613–5614.
2. Plunkett, M. J.; Ellman, J. A. *J. Am. Chem. Soc.* **1995,** *117,* 3306–3307.
3. Kuhn, H.; Neumann, W. P. *Synlett* **1994,** 123–124.
4. Curran, D. P.; Hoshino, M. *J. Org. Chem.* **1996,** *61,* 6480–6481.

4.2.4   Heck Reactions on Solid Support

4.2.4.1   Heck Reactions in Solid-Phase Synthesis [1]

Points of Interest

1. Heck reactions were performed with either the aryl halide or the alkene component on solid support. The flexible reaction conditions allow the incorporation of a larger number of commercially available building blocks into combinatorial libraries.

2. The methodology developed can be applied to generate combinatorial libraries of 1,2-disubstituted olefins.

Representative Examples

Aryl halides were coupled to a support-bound styrene derivative; however, aryl triflates were ineffective. Activated alkenes and alkynes were coupled to a support-bound aryl iodide.

Literature Summary

Reactions between aryl halide and activated alkenes or alkynes proceeded in the highest yield (64–90% crude yields, 90% purity). The resin-bound aryl halide (200 mg) with the activated alkene or alkyne in solution (100 mg) and Et$_3$N (50 μL) in 3 mL of DMF was treated with Pd$_2$(dba)$_3$ (30 mg) and P(2-Tol)$_3$ (20 mg) at 100°C for 20 h.

4.2.4.2 Solid-Phase Synthesis Using the Heck Reaction under Phase Transfer Conditions at 37°C [2]

Points of Interest

1. The Heck reaction proceeded on TentaGel resin at 37°C under phase transfer conditions in DMF/H$_2$O (9:1). The conversion was significantly lower in 100% H$_2$O. Changing the phase transfer agent (PTA) did not significantly improve yields.

2. The Heck reaction was performed at a lower temperature (37 vs 80–100°C) than previously described (see p 96).

Representative Examples

Vinyl amides, esters, aldehydes, or nitriles reacted with support-bound aryl iodides with good conversion (54–95% conversion).

Literature Summary

To the resin-bound aryl iodide was added a 0.3 M solution of the appropriate acrylate in DMF/H$_2$O/Et$_3$N (9:1:1, v/v/v). The resulting mixture was stirred in the presence of 0.05 M Pd(OAc)$_2$, 0.1 M PPh$_3$, 0.1 M Bu$_4$NCl at 37°C for 4–18 h.

4.2.4.3 Additional Examples of Solid-Phase Heck Reactions

1. Palladium-mediated macrocyclization was reported employing an intramolecular Heck reaction under phase transfer conditions.[3]

2. An intramolecular Heck reaction has been employed in the solid-phase synthesis of highly substituted peptoid 1(2H)-isoquinolines (see p 178).[4]

3. A palladium-mediated three-component coupling reaction involving both Suzuki and Heck reactions has been developed for the solid-phase synthesis of tropane derivatives (see p 197).[5]

4. The Heck reaction was reported in the solid-phase synthesis of indole derivatives (see p 206).[6]

REFERENCES FOR SECTION 4.2.4

1. Yu, K.-L.; Deshpande, M. S.; Vyas, D. M. *Tetrahedron Lett.* **1994,** *35,* 8919–8922.
2. Hiroshige, M.; Hauske, J. R.; Zhou, P. *Tetrahedron Lett.* **1995,** *36,* 4567–4570.
3. Hiroshige, M.; Hauske, J. R.; Zhou, P. *J. Am. Chem. Soc.* **1995,** *117,* 11590–11591.
4. Goff, D. A.; Zuckermann, R. N. *J. Org. Chem.* **1995,** *60,* 5748–5749.
5. Koh, J. S.; Ellman, J. A. *J. Org. Chem.* **1996,** *61,* 4494–4495.
6. Yun, W.; Mohan, R. *Tetrahedron Lett.* **1996,** *37,* 7189–7192.

4.2.5   Enolate Alkylation for Carbon–Carbon Bond Formation on Solid Support

4.2.5.1   Enolate Alkylation on Kenner's Acylsulfonamide Linker [1,2]

Points of Interest

1. In contrast to ester or carboximide enolate alkylations, ketene formation is not observed even when an unreactive alkylating agent is employed, since ketene formation would require that the sulfonamide dianion be the leaving group.

2. Minimal overalkylation was observed (<4%).

3. The support-bound acylsulfonamide can be activated with diazomethane or iodoacetonitrile and then cleaved with a range of nucleophiles (see pp 48 and 267). This is particularly useful for combinatorial synthesis because additional diversity is incorporated during the cleavage step. Furthermore, because the linker is highly activated, equal quantities of amides can be obtained in solution from limiting quantities of pools of amines in the cleavage cocktail.

Representative Examples

Treatment of the acylsulfonamide with excess LDA in THF at 0°C results in rapid deprotonation to give the trianion (only the dianion is shown in the reaction scheme). Subsequent addition of activated or unactivated alkyl halides results in rapid alkylation of the enolate trianion. The solid-phase synthesis of arylacetic acid derivatives on the sulfonamide "safety-catch" linker was described.[2]

Literature Procedures

For Loading onto the Linker. Sulfonamide-derivatized macroreticular resin was acylated with the symmetric anhydride of a carboxylic acid and catalytic DMAP as follows (this method has proven successful with a number of carboxylic acids and represents an improvement over the PFP ester method initially reported). To a 100-mL round-bottom flask were added 3-(3,4,5-trimethoxyphenyl)propanoic acid (8.6 g, 36 mmol), DCM (60 mL), and DIC (2.8 mL,

18 mmol). After 8 h of stirring, the solution was cooled with an ice bath and the urea precipitate was filtered. The filtrate was added to a 250-mL round-bottom flask fitted with an overhead stirrer along with sulfonamide-derivatized macro-reticular resin (10 g, 3.9 mmol), DMAP (44 mg, 0.36 mmol), DIEA (2.1 mL, 12 mmol), and DCM (10 mL). The slurry was stirred for 39 h and filtered, and the resin was washed with THF (200 mL), 5% TFA in THF (to protonate the support-bound acylsulfonamide), THF (200 mL), and MeOH (200 mL). The resin was dried with rotary evaporation and then under high vacuum for 24 h. The acylation of the sulfonamide could be monitored by the ninhydrin test.[3]

For Alkylation of the Support-Bound Acylsulfonamide.   To a N_2-flushed 100-mL round-bottom flask placed in a 0°C ice bath and fitted with a stir bar and a fresh rubber septum was added the derivatized resin (750 mg, 0.185 mmol). To a second 100-mL round-bottom flask with a stir bar and a fresh rubber septum were added THF (80 mL) and diisopropylamine (0.49 mL, 3.5 mmol). The second flask was placed in a −78°C acetone/dry ice bath and flushed with N_2. A solution of 2.5 M n-butyllithium in hexanes (1.1 mL, 2.8 mmol) was then added dropwise by syringe. The solution was stirred for 30 min and cannulated to the resin-containing flask. The resin rapidly turned a deep purple color and the slurry was stirred for 1 h. At that time, the appropriate alkylating agent (9.25 mmol, 50 equiv) was added rapidly by syringe. The slurry was then stirred and allowed to come to room temperature over 10 h, after which the solvent was removed and the resin was washed with THF, H_2O, DMF/H_2O, THF, 5% TFA in THF, and THF.

4.2.5.2   Aldol and Imine Reactions with Support-Bound Silyl Enol Ethers [4,5]

Points of Interest

1. In the presence of a catalytic amount of scandium triflate ($Sc(OTf)_3$), support-bound silyl enol ethers reacted with aldehydes to afford the corresponding β-hydroxy thioester derivatives, which were reduced to 1,3-diol and β-hydroxy aldehyde derivatives, or hydrolyzed to β-hydroxy carboxylic acid derivatives.

2. The reaction of support-bound silyl enol ethers (thioketene silyl acetals) with imines was used for the construction of an amino alcohol library. The reaction gave best results employing $Sc(OTf)_3$ as a Lewis acid catalyst.[6]

3. The Mukaiyama aldol reaction proceeded in higher yield with 20% (vs 10%) of $Sc(OTf)_3$ catalyst at low temperature ($-78°C$).

4. A 48 amino alcohol library was prepared from the imine reaction with support-bound silyl enol ethers.[7]

5. Three-component reactions leading to diverse amino ketone, amino ester, and amino nitrile derivatives were performed with a polymer-supported scandium catalyst (see p 278).

Representative Examples

Both substituted and unsubstituted silyl enol ethers underwent aldol reactions with a range of aromatic, heteroaromatic, and aliphatic aldehydes in 55–82% yield. Support-bound thioesters were cleaved under different conditions to provide the corresponding alcohols, aldehydes, or carboxylic acids. Imines prepared from aromatic, heteroaromatic, and aliphatic aldehydes reacted cleanly with support-bound silyl enol ethers at room temperature to afford the corresponding support-bound amino alcohols. Imines prepared from aromatic amines (anilines) reacted with enolates in higher yield than imines prepared from primary amines.

Literature Summary

For the Preparation of Support-Bound Substituted Thiol Esters. (Chloromethyl)copoly(styrene–1% divinylbenzene) resin (Merrifield resin, 1.15 mmol/g) was treated with potassium thioacetate in DMF. Formation of the support-bound thioester was monitored by IR spectroscopy for a strong carbonyl stretching vibration at 1693 cm^{-1}. A chlorine titration indicated that the thioester was obtained in 95% yield. Substituted silyl enol ethers can be directly prepared with support-bound thioacetates. Alternatively, reducing the support-bound thioester (prepared from potassium thioacetate) using $LiBH_4$ in Et_2O at room temperature afforded the support-bound thiol that could then be acylated with substituted acetyl chlorides.

For the Preparation of Support-Bound Silyl Enol Ethers (Thioketene Silyl Acetals). The support-bound thioester was combined with TMSOTf and triethylamine in DCM to afford the corresponding support-bound silyl enol ethers.

For the Reaction of Support-Bound Silyl Enol Ethers with Imines. To a mixture of the support-bound silyl enol ether (234 mg, 0.20 mmol) and

Sc(OTf)$_3$ (9.8 mg, 10 mol %) in DCM (2 mL) was added the imine prepared from benzaldehyde and aniline (36.2 mg, 0.24 mmol). The mixture was stirred for 20 h at room temperature, and the reaction was then quenched with saturated sodium hydrogen carbonate.

For the Support-Bound Mukaiyama Aldol. A mixture of benzaldehyde (25.4 mg, 0.24 mmol), the support-bound silyl enol ether (224 mg, 0.20 mmol), and Sc(OTf)$_3$ (19.6 mg, 20 mol %) in 2 mL DCM was stirred for 20 h at $-78°$C. The reaction was quenched with dioxane/water/HCl. The resin was washed with water, dioxane, and ether and dried *in vacuo*. The resulting product can be cleaved from resin by either reduction or hydrolysis.

4.2.5.3   Other Examples of Carbon–Carbon Bond Formation with Support-Bound Enolates

1. Benzophenone imine α-carbon alkylation in solid-phase unnatural peptide synthesis has been reported (see p 118).[8]

2. 1,3-Diones were alkylated in the solid-phase synthesis of pyrazoles and isoxazoles (see p 122).[9] For the alkylation reaction best results were obtained with TBAF as base to avoid *O*-alkylation and increase nucleophilicity.

3. Alkylation of the dianion of a support-bound β-keto ester was reported in the solid-phase synthesis of 1-phenylpyrazolone derivatives (see p 163).[10]

4. An aldol reaction using a support-bound zinc enolate affords diols in modest yield.[11]

5. Iterative, asymmetric aldol reactions have been adapted to solid-phase synthesis.[12]

6. Aldol condensations between support-bound aryl methyl ketones and aromatic aldehydes have been performed under basic reaction conditions.[13]

REFERENCES FOR SECTION 4.2.5

1. Backes, B. J.; Virgilio, A. A.; Ellman, J. A. *J. Am. Chem. Soc.* **1996**, *118*, 3055–3056.
2. Backes, B. J.; Ellman, J. A. *J. Am. Chem. Soc.* **1994**, *116*, 11171–11172.
3. A positive test for unreacted amine gives an intense blue color and a positive test for unreacted sulfonamide gives a pale red color. Kaiser, E.; Colescot, R. L.; Bossinger, C. D.; Cook, P. I. *Anal. Biochem.* **1970**, *34*, 595–598.
4. Kobayashi, S.; Hachiya, I.; Suzuki, S.; Moriwaki, M. *Tetrahedron Lett.* **1996**, *37*, 2809–2812.
5. Kobayashi, S.; Hachiya, I.; Yasuda, M. *Tetrahedron Lett.* **1996**, *37*, 5569–5572.
6. Kobayashi, S.; Hachiya, I.; Suzuki, S.; Moriwaki, M. *Tetrahedron Lett.* **1996**, *37*, 2809–2812.
7. Kobayashi, S.; Moriwaki, M.; Akiyama, R.; Suzuki, S.; Hachiya, I. *Tetrahedron Lett.* **1996**, *37*, 7783–7786.
8. O'Donnell, M. J.; Zhou, C.; Scott, W. L. *J. Am. Chem. Soc.* **1996**, *118*, 6070–6071.
9. Marzinzik, A. L.; Felder, E. R. *Tetrahedron Lett.* **1996**, *37*, 1003–1006.
10. Tietze, L. F.; Steinmetz, A. *Synlett* **1996**, 667–668.
11. Kurth, M. J.; Ahlberg Randall, L. A.; Chen, C.; Melander, C.; Miller, R. B.; McAlister, K.; Reitz, G.; Kang, R.; Nakatsu, T.; Green, C. *J. Org. Chem.* **1994**, *59*, 5862–5864.
12. Reggelin, M.; Brenig, V. *Tetrahedron Lett.* **1996**, *37*, 6851–6852.
13. Hollinshead, S. P. *Tetrahedron Lett.* **1996**, *37*, 9157–9160.

4.2.6   Solid-Phase Wittig Reactions

4.2.6.1   Solid-Phase Synthesis of Alkenes Using the Horner–Wadsworth–Emmons Reaction [1]

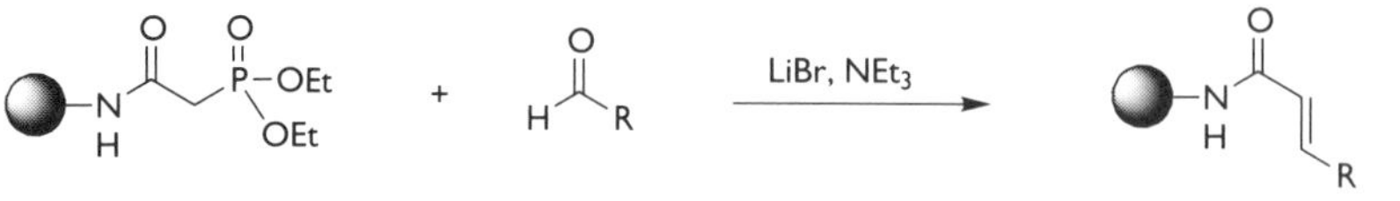

PEG-PAL resin

or

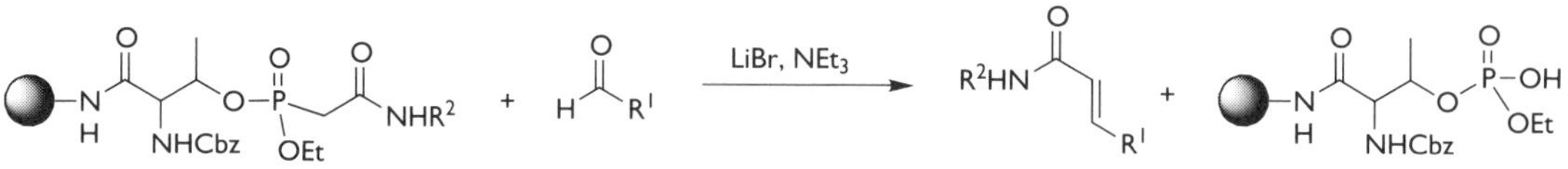

PEG-PAL resin

Points of Interest

1. The Horner–Wadsworth–Emmons variant of the Wittig reaction was adapted to the solid phase. High conversions to alkenes were accomplished under very mild reaction conditions, using amine bases and lithium salts, in the presence of peptide functionality.

2. Attachment to the solid phase allows easy isolation of the olefinic product from the phosphorus-containing byproducts.

3. Alternatively, a phosphono ester was attached to solid support in a one-pot react and release approach.

4. The reaction progress was monitored by gel-phase ^{31}P NMR (see p 221) of the solid support suspended in acetonitrile. ^{31}P NMR showed a resonance of the starting resin-bound diethylphosphonoacetamide as a narrow multiplet at δ 22 ppm. After the reaction was complete the δ 22 ppm peaks were replaced with a new broad resonance for diethyl phosphonate near δ 0 ppm.

Representative Examples

Diethylphosphonoacetamide linked to PEG–PAL resin was reacted with an excess of aliphatic or aromatic aldehyde in the presence of LiBr and triethylamine.

Literature Summary

Fmoc-PEG–PAL (Millipore; 0.1 g, 0.20 mequiv/g) was treated with 20% piperidine in DMF for 20 min and the resin was then washed with DMF (6×), DCM (6×), and DMF (2×). A 0.3 M solution (0.8 mL) of diethylphosphonoacetic acid, PyBOP, HOBt, and NMM was added to the resin and the resulting mixture was shaken for 2 h. The resin was then washed and dried under vacuum. The acylation reaction was monitored by the Kaiser ninhydrin test. The Wittig reaction was monitored in an NMR tube with the resin suspended in 0.6 mL of acetonitrile-d_3. ^{31}P NMR initially showed a resonance at δ 22 ppm. After the solution was treated with 20 μL of isobutyraldehyde, 20 mg LiBr, and 30 μL of triethylamine for 24 h, ^{31}P NMR only showed a resonance at δ 0 ppm.

4.2.6.2 Preparation of a Hydroxystilbene Library with a Solid-Phase Wittig Reaction[2]

Points of Interest

1. A small 23-component hydroxystilbene library was prepared using a solid-phase olefination reaction. The library was screened for estrogenic and antiestrogenic activity. Three of the analogs displayed dose-dependent estrogenic activity with EC_{50} values between 5 and 15 μL.

2. Although yields varied for the individual analogs, a sufficient quantity of pure material was obtained directly from the resin for structural characterization and biological evaluation.

Representative Examples

Four different hydroxybenzaldehydes were derivatized with the acid-cleavable [4-(hydroxymethyl)phenoxy]acetic acid linker and attached to (aminomethyl)polystyrene (PAM) resin as described for loading 2-aminobenzophenones onto resin (see p 145).[3]

Literature Procedure

The hydroxybenzaldehyde-derivatized resins were prepared on a 0.3–1.0-mmol scale and subdivided into six equal portions by weight for parallel treatment with six different benzylphosphonate anions. The anions were generated in separate flasks by treating 1.3 mmol of the benzylphosphonate (10-fold molar excess) with 2 mmol of sodium methoxide in 4 mL of DMF. After the anion solution was stirred for 0.5 h at room temperature, a 0.13-mmol portion of resin aldehyde was added and the reaction mixture was stirred at room temperature overnight. The reaction mixtures were washed with H_2O (5×), DMF (5×), DMF/H_2O, and DCM. The resin was then dried *in vacuo*.

4.2.6.3   Other Examples of Solid-Phase Wittig Reactions

1. Support-bound Horner–Emmons reagents within the pK_a range ca. 6–9 have been supported on a basic anion-exchange resin (Amberlyst A-26) by a simple neutralization process (see p 271). The reagents were prepared immediately before use due to slow decomposition.[4]

2. Horner–Emmons condensation of a support-bound aldehyde was used to prepare support-bound enones.[5] The transformation was conveniently monitored by KBr pellet FT-IR analysis of the polymer.

3. Olefin and hydroxyethylene peptidomimetics were prepared by Wittig olefination of support-bound amino aldehydes.[6]

4. Polymer-bound phosphonium salts have been used as traceless supports for solid-phase synthesis (see p 70).

REFERENCES FOR SECTION 4.2.6

1. Johnson, C. R.; Zhang, B. *Tetrahedron Lett.* **1995**, *36*, 9253–9256.
2. Williard, R.; Jammalamadaka, V.; Zava, D.; Benz, C. C.; Hunt, C. A.; Kushner, P. J.; Scanlan, T. S. *Chem. and Biol.* **1995**, *2*, 45–51.
3. Bunin, B. A.; Ellman, J. A. *J. Am. Chem. Soc.* **1992**, *114*, 10997–10998.
4. Cainelli, G.; Contento, M.; Manescalchi, F.; Regnoli, R. *J. Chem. Soc., Perkin Trans. 1*, **1980**, 2516–2519.
5. Chen, C.; Ahlberg Randall, L. A.; Miller, R. B.; Jones, A. D.; Kurth, M. J. *J. Am. Chem. Soc.* **1994**, *116*, 2661–2662.
6. Rotella, D. P. *J. Am. Chem. Soc.* **1996**, *118*, 12246–12247.

4.2.7   Metathesis Reactions on Solid Support

4.2.7.1   Ruthenium-Catalyzed Metathesis of Polymer-Bound Olefins [1]

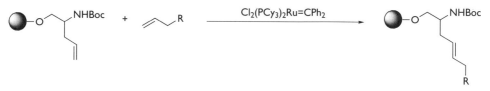

Trityl resin

Points of Interest

1. Ring-closing metathesis was employed to form nitrogen heterocycles and cross-metathesis was performed with a polymer-bound and a dissolved olefin.

2. Both tritylpolystyrene-bound 4-pentenol and allyl alcohol dimerized on the polymer surface by macrocyclization. Therefore, more sterically hindered hydroxy olefins were immobilized on resin and subjected to cross-metathesis conditions.

Representative Examples

A series of five- and six-membered heterocyclic olefins were prepared in 70–91% yield (by HPLC). Cross-metathesis was performed on polymer-bound allylglycinol and *N*-allylglycine derivatives.

Literature Procedures

For Ruthenium-Catalyzed Ring-Closing Metathesis. Under exclusion of moisture and air, the tritylpolystyrene-bound diolefin (200–500 mg) was treated with Grubb's ruthenium metathesis catalyst[2] (10–15 mol %) in benzene or DCM (10 mL/g of resin). After 1–5 days, the resin was washed with $CHCl_3$ (5×) and MeOH (5×).

For Ruthenium-Catalyzed Cross-Metathesis Under exclusion of moisture and air, the tritylpolystyrene-bound diolefin (200–500 mg) was treated with Grubb's ruthenium metathesis catalyst (3–6 mol %) and the soluble olefin (3–6 equiv) in benzene or DCM (10 mL/g of resin). The catalyst was added in two steps: at the beginning of the reaction and after half the quoted reaction time had elapsed. After 12–48 h, the resin was washed with $CHCl_3$ (5×) and MeOH (5×).

REFERENCES FOR SECTION 4.2.7

1. Schuster, M.; Pernerstorfer, J.; Blechert, S. *Angew. Chem., Int. Ed. Engl.* **1996**, *35*, 1979–1980.
2. Nguyen, S. T.; Grubbs, R. H.; Ziller, J. W. *J. Am. Chem. Soc.* **1993**, *115*, 9858. The ruthenium metathesis catalyst is commercially available from Strem (Newburyport, Massachusetts).

4.2.8 Other Examples of Carbon–Carbon Bond Formation on Solid Support

1. Functionalized resins have been prepared by several carbon–carbon bond-forming reactions following bromination and lithiation of polystyrene resin (see p 12).[1]

2. Bicyclo[2.2.2]octane derivatives have been prepared via tandem Michael addition reactions on solid support (see p 195).[2]

3. Domino Knoevenagel–ene reactions were employed in the highly stereoselective solid-phase synthesis of cyclopentane and cyclohexane derivatives.[3]

4. An intramolecular Pauson–Khand cyclization[4] has been reported on solid support (see p 199).

5. Phenylacetylene oligomers consisting entirely of hydrocarbons have been prepared by solid-phase synthesis utilizing a triazene linkage (see p 70).[5]

REFERENCES FOR SECTION 4.2.8

1. Farrall, M. J.; Frechet, J. M. J. *J. Org. Chem.* **1976**, *41*, 3877–3882.
2. Ley, S. V.; Mynett, D. M.; Koot, W. J. *Synlett* **1995**, 1017–1020.

3. Tietze, L. F.; Steinmetz, A. *Angew. Chem., Int. Ed. Engl.* **1996**, *35*, 651–652.

4. Bolton, G. L. *Tetrahedron Lett.* **1996**, *37*, 3433–3436.

5. (a) Nelson, J. C.; Young, J. K.; Moore, J. S. *J. Org. Chem.* **1996**, *61*, 8160–8168. (b) Young, J. K.; Nelson, J. C.; Moore, J. S. *J. Am. Chem. Soc.* **1996**, *116*, 10841–10842.

4.3 MITSUNOBU REACTIONS ON SOLID SUPPORT

A number of researchers have reported on a variety of Mitsunobu reactions on solid support. Mitsunobu reactions are more straightforward to work up on support than the analogous reaction in solution. On solid support the phosphine complexes can simply be removed by filtration, in contrast to the chromatographic purification typically required in solution. Mitsunobu reactions have also been applied to combinatorial synthesis because they are performed under neutral conditions which tolerate other functional groups.

4.3.1 An Efficient Solid-Phase Mitsunobu Reaction in the Presence of a Tertiary Amine [1]

R = Boc, H

Points of Interest

1. Boc-L-tyrosine methyl ester was attached to (hydroxymethyl)polystyrene through the phenolic side chain for head-to-tail cyclizations of peptides on solid support. Mitsunobu conditions were optimized for high loading without racemization.

2. The yield of this Mitsunobu reaction is significantly higher in the presence of an amine, preferably a tertiary amine.

Literature Summary

To a mixture of *N*-Boc-L-tyrosine methyl ester (3.0 equiv), triphenylphosphine (3 equiv), and (hydroxymethyl)polystyrene resin (1.0 equiv, 1.08 mequiv/g) in 10 mL of *N*-methylmorpholine/g of resin was added diethyl azodicarboxylate (3.0 equiv). The resulting mixture was stirred for 16 h at room temperature. The resin was washed with THF and MeOH and dried. The reaction worked equally well without racemization for 1 mmol and 30 mmol of resin (cooling is necessary during DEAD addition!). After removal of the Boc group (45% TFA, 45% DCM, 5% thiocresol, 5% ethanedithiol), the resin was washed with 10% DIEA in DCM, and the substitution level was quantitated by picrate acid titration.[2] A substitution level of 0.90 mequiv/g was obtained with a tertiary amine as solvent, corresponding to quantitative Mitsunobu loading by mass balance after taking the increase in weight into account.

4.3.2 Solid-Phase Synthesis of Aryl Ethers via the Mitsunobu Reaction[3]

or

Points of Interest

1. The Mitsunobu reaction was performed with either the more or less acidic coupling partner on solid support. After some experimentation, which included the examination of several Mitsunobu reagents, the reaction was optimized when a 5-fold excess of TMAD[4] (N,N,N',N'-tetramethylazodicarboxamide) and tributylphosphine were employed along with a 10-fold excess of phenol (or 5-fold excess of alcohol) in THF/DCM (1:1).

2. Very good results were also displayed by DEAD/Ph$_3$P and DIAD/Ph$_3$P as well as 1,1'-(azodicarbonyl)dipiperidine (ADDP)/Bu$_3$P. However, TMAD/Bu$_3$P provided the best purity (89–97% purity) and yield (72–99% crude yield after lyophilization from MeCN/H$_2$O or azeotropic removal of TFA with MeCN).

Representative Examples

A variety of phenols were successfully employed. Ortho substitution on the phenol had no adverse affect. 2,6-Dimethylphenol also coupled well, but 4-nitrophenol provided no Mitsunobu product.

Literature Summary

To approximately 100 mg (0.023 mmol) of phenol-bound dry resin swelled in a 1:1 mixture of THF/DCM (1.5 mL) under a nitrogen atmosphere was added solid TMAD (5 equiv, 0.155 mmol), followed by mixing until dissolution occurred. Excess alcohol (5 equiv, 0.155 mmol) was then added and the reaction vessel was agitated until dissolution occurred. Neat Bu$_3$P (5 equiv, 0.155 mmol) was added last via a gastight syringe and the resulting slurry was mixed for ca. 1 h. The resin was filtered and washed with a 1:1 mixture of THF/DCM (3×), THF (3×), DCM (3×), i-PrOH (3×), and DCM (3×). The product was cleaved from the resin with 90% TFA/H$_2$O (2 × 20 min).

4.3.3 Polymer-Supported Mitsunobu Ether Formation and Its Use in Combinatorial Chemistry [5]

Points of Interest

1. The Mitsunobu reaction was performed between a polymer-bound phenol and a range of commercially available alcohols in solution.

2. In most cases the reaction provided a nearly quantitative yield of alkyl aryl ethers, although minor amounts of "alkylated DEAD" products (the corresponding ethyl ethers) were observed and quantified. Dilute solutions of DEAD in THF were added in portions, because fast addition of undiluted DEAD resulted in significant formation of the ethyl ethers. In cases where the formation of the ethyl ether was substantial, the reaction mixture was cooled to −15°C and diisopropyl azodicarboxylate was substituted for DEAD.

3. To demonstrate the utility for combinatorial synthesis, a number of model compounds were synthesized and a simple, three randomization step library of 4200 different compounds was generated.

Representative Examples

Unprotected phenols (aromatic hydroxy acids) were successfully loaded onto resins with DCC/HOBt. A range of aliphatic, aromatic, and heteroaromatic alcohols as well as diols and N-protected amino alcohols were used as Mitsunobu coupling partners in solution.

Literature Summary

After the support-bound phenols (0.2–0.3 g, loading 0.2 mmol/g) were washed with with dry THF (5×), 0.5 mL of a 1 M solution of PPh$_3$ (262 mg/mL) in THF and 1 mmol of the appropriate alcohol were added, and the resulting mixture was shaken. Then, in portions at 5-min intervals, a total of 0.5 mmol (78 μL) of DEAD in THF (234 μL) was added and the resulting slurry was mixed. After 1 h, the resin was rinsed with DMF (5×) and MeOH (3×).

4.3.4 Mitsunobu Functionalization of Biphenyl Phenols [6]

Points of Interest

1. The biphenyl scaffold displays three or four different functional groups in a variety of spatial arrangements.

2. Functionalization of the free phenolic hydroxyls was accomplished by using Castro conditions.[7] Mitsunobu reactions were conducted by addition of a suspension of the preformed betaine and a solution of the appropriate alcohol in a 96-well format.

Literature Summary

Mitsunobu chemistry was performed in a drilled 96-well microtiter plate fitted with polyethylene frits charged with a series of unique substituted biphenyl phenol resins (ca. 20 mg/well). The charged plate was clamped in an aluminum frame fitted with a Viton seal. Resin-bound biphenyl phenols were treated with the alcohol of choice (10 equiv) in toluene (0.4 mL/well) followed by a suspension of the sulfonamide betaine in DCM (11.9 g in 53 mL, 10 equiv, 0.5 mL/well). The wells were sealed, and the sealed plates were rotated for 3 days. The plates were transferred to a filtration manifold and the resins washed three times each with DCM, THF, MeOH, DCM, THF, DCM, and THF and then dried under vacuum.

4.3.5 The Solid-Phase Synthesis of Phosphonic Acid Esters under Modified Mitsunobu Conditions[8]

Points of Interest

1. The support-bound alcohol was condensed with methyl α-[N-(4-nitrophenethyloxycarbonyl)amino]alkylphosphonic acids using a modified Mitsunobu coupling procedure to provide peptidylphosphonates with inversion of configuration at the carbinol carbon. The modifications included addition of an exogenous base[1] and the use of tris(4-chlorophenyl)phosphine instead of triphenylphosphine.

2. The betaine formed between the phosphine and DIAD was basic enough to cause β-elimination of an Fmoc group, necessitating the use of the more stable 4-nitrophenethyoxycarbonyl (NPEOC) group.[9] Removal of the NPEOC group was accomplished with 5% DBU in NMP. Coupling yields were determined spectrophotometrically (4-nitrostyrene, $\epsilon = 13,200 \text{ M}^{-1} \text{ cm}^{-1}$, 308 nm).

3. Peptidylphosphonates are known transition state analog inhibitors of peptidases and esterases.

Representative Examples

Primary and secondary alcohols and alkylphosphonic acids were incorporated into peptidylphosphonates.

Literature Procedure

After standard Fmoc solid-phase peptide synthesis, the desired support-bound peptide was charged with a solution of NMP that was 0.1 M in α-O-Fmoc-protected hydroxy acid, 0.1 M in HBTU/HOBt or PyBroP, and 0.3 M in DIEA. Deprotection of the Fmoc group (presumably after standard rinses) with 30% piperidine in NMP provided a terminal hydroxyl group. The resin was washed with NMP (5×) and ether (5×) and then dried overnight under vacuum in the presence of P_2O_5. The methyl N-NPEOC-phosphonic acid was suspended in anhydrous THF and then treated with tris(4-chlorophenyl)phosphine and DIAD for 5 min. The mixture was added to a suspension of the resin in anhydrous THF containing DIEA. Concentrations were chosen such that the reaction mixture contained 0.3 M DIEA and 0.1 M phosphonic acid, DIAD, and tris(4-chlorophenyl)phosphine. The vessel was sealed and placed in a temperature-controlled (25°C) sonication bath. The time required for >90% coupling yields varied from 0.5–20 h depending on the sterics of the resin-bound alcohols and the phosphonic acids. Coupling efficiency can be determined at this point by spectrophotometric analysis of the released 4-nitrostyrene (λ = 308 nm, ϵ = 13,200 M^{-1} cm^{-1}). After allowing appropriate coupling time, the resin was washed with NMP (5×) and ether (5×) and then treated with 5% DBU in NMP for 2 h to remove the NPEOC group. Standard peptide chemistry then proceeds without modification. Upon completion of the synthesis of the peptidylphosphonate, demethylation of the phosphonate diester is achieved with 1:2:2 thiophenol:triethylamine:dioxane for 1 h.

4.3.6 Mitsunobu Coupling with a Support-Bound Trifluoroanilide [10]

1. TFAA, 4-methyl-2,6-di-*tert*-butylpyridine
   DCM 37°C, 4 h
2. R^1OH, PPh$_3$, DEAD, THF, -10°C to rt, 6 h

1. R^2NH_2, DMSO,
   70°C, 12 h
2. TFA/H_2O, rt, 1 h

Points of Interest

1. The C-2 amino group of a resin-bound 2-trifluoroacetamido-6-chloropurine was alkylated under Mitsunobu conditions between the trifluoroanilide and alcohols. Subsequent amination of the purine was accompanied by aminolysis of the trifluoroacetamide to provide the fully derivatized 2,6-diaminopurines.

Representative Examples

Six purine analogs were prepared on aminoalkyl-derivatized resin (0.87 mmol/g) with overall yields of 75–85% for a five-step sequence including the Mitsunobu alkylation and amination. A small library of 16 alkylated aminopurines was prepared on polyethylene pins using primary and benzylic alcohols.

Literature Procedure

2,6-Di-*tert*-butyl-4-methylpyridine (1.54 g, 7.50 mmol, 0.3 M) was added to freshly distilled DCM (25 mL) at 0°C under N_2. To this solution was added trifluoroacetic anhydride (706 μL, 5.0 mmol, 0.2 M) and the mixture was stirred for 5–10 min. The solution was transferred to support-bound 2-amino-6-chloro-9-(2-hydroxyethyl)purine (0.67 mmol, substitution 0.87 g/mmol; see p 207) in a peptide reaction flask, and the flask was vortexed and vented several times to relieve pressure. The flask was put on a wrist action shaker for 6–12 h, after which the solvent was removed and the resin was washed with dry DCM (8×), anhydrous DMF (3×), and DCM (5×), with vortexing between each rinse. The rinsed resin was dried under a stream of N_2 or, for rigorous drying, under high vacuum (ca. 1 mmHg). Diethyl azodicarboxylate (394 μL, 2.5 mmol) was added dropwise to triphenylphosphine (1.31 g, 5.0 mmol) dissolved in dry 1:1 THF/DCM (5 mL) at 0°C. After 1 h of stirring at 0°C, the solution was transferred by Teflon cannula to trifluoroacetylated resin (180 mg) simultaneous with the dropwise addition of the appropriate alcohol (2.0 mmol) in THF (100 μL). The flask was vortexed and vented several times and then placed on a wrist action shaker for 12 h. The solution was filtered and the resin washed with DMF (6×) followed by DCM (6×). The rinsed resin was dried under a stream of N_2 or, for rigorous drying, under high vacuum (ca. 1 mmHg). The alkylated resin (88 mg) was transferred to a silanized round-bottom flask and suspended in a solution of the appropriate amine (3.0 mmol, 0.5 M) in dry DMF (5 mL). The mixture was stirred for 1 h at room temperature and then for 12 h at 75°C. The solvent was removed by filtration cannula and the resin was washed with DMF (8×), DCM (5×), MeOH (5×), and DCM (5×). The resin was dried to a constant weight under vacuum.

4.3.7   Additional Studies of Mitsunobu Reactions on Solid Support

1. Valerio and coworkers have reported solid-phase Mitsunobu chemistry on functionalized polyethylene pins.[11] Reaction optimization was performed on polyethylene pins simultaneously by using the multipin approach: the optimal conditions were found to include 0.15 M PPh$_3$/DEAD/alcohol in THF at 37°C for 4 days in the presence of 0.45 M triethylamine.

2. Sarshar and coworkers have reported solid-phase Mitsunobu conditions

that included *N*-ethylmorpholine, DIAD, PPh$_3$, sonication for 1 h, and then stirring at room temperature for 16 h.[12]

REFERENCES FOR SECTION 4.3

1. Richter, L. S.; Gadek, T. R. *Tetrahedron Lett.* **1994**, *35*, 4705–4506.
2. Gisin, B. *Anal. Chim. Acta* **1972**, *58*, 248.
3. Rano, T. A.; Chapman, K. T. *Tetrahedron Lett.* **1995**, *36*, 3789–3792.
4. Tsunoda, T.; Otsukka, J.; Yamamiya, Y.; Ito, S. *Chem. Lett.* **1994**, 539.
5. Krchñàk, V.; Flegelovà, Z.; Weichsel, A. S.; Lebl, M. *Tetrahedron Lett.* **1995**, *36*, 6193–6196.
6. Pavia, M. R.; Cohen, M. P.; Dilley, G. J.; Dubuc, G. R.; Durgin, T. L.; Forman, F. W.; Hediger, M. E.; Milot, G.; Powers, T. S.; Sucholeiki, I.; Zhou, S.; Hangauer, D. G. *Bio. Med. Chem.* **1996**, *4*, 659–666.
7. Castro, J. L.; Matassa, V. G.; Ball, R. G. *J. Org. Chem.* **1994**, *59*, 2289–2291.
8. Campbell, D. A.; Bermak, J. C. *J. Am. Chem. Soc.* **1994**, *116*, 6039–6040.
9. Pfleiderer, W.; Schwarz, M.; Schirmeister, H. *Chem. Scr.* **1986**, *26*, 147.
10. Norman, T. C.; Gray, N. S.; Koh, J. T.; Schultz, P. G. *J. Am. Chem. Soc.* **1996**, *118*, 7430–7431.
11. Valerio, R. M.; Bray, A. M.; Patsiouras, H. *Tetrahedron Lett.* **1996**, *37*, 3019–3022.
12. Sarshar, S.; Siev, D.; Mjalli, A. M. M. *Tetrahedron Lett.* **1996**, *37*, 835–838.

4.4   SUBSTITUTION AND ADDITION REACTIONS ON SOLID SUPPORT

Many substitution and addition reactions have been performed on solid support. These reactions are well suited for the addition of commercially available building blocks, such as amines, to different scaffolds. In a remarkable display of the recent, rapid development of solid-phase synthesis, the palladium-catalyzed amination of aryl bromides was adapted to solid support by two groups (within weeks of each other). The truly remarkable fact is that this palladium-catalyzed reaction was applied to solid-phase combinatorial synthesis only a few years after it was developed by Buchwald and Hartwig.

4.4.1   Nucleophilic Displacement of Support-Bound α-Bromo Amides in the Submonomer Preparation of Peptoids[1]

Points of Interest

1. The submonomer method allowed peptoids to be constructed from a large number of commercially available amines. The original method for the synthesis of peptoids required the individual preparation of Fmoc-protected N-substituted glycines in solution.

2. The preparation of peptoids by the submonomer method in a solid-phase mode has been adapted to a robotic synthesizer.[2]

3. Haloacetic acids facilitate acylation of a secondary amine, which can be difficult, especially when coupling a bulky amino acid.

4. The efficiency of the nucleophilic displacement with an excess of primary amines is modulated by the choice of halide (e.g., I > Cl): bromides are typically employed.

Representative Examples

Support-bound α-bromoacetamides were displaced with a range of primary and secondary amines. Protection of carboxyl, thiol, amino, and other reactive side-chain functionalities is required to minimize undesired side reactions. However, the authors mention that the mild reactivity of some side-chain moieties toward displacement or acylation may allow their use without protection (e.g., indole, imidazole, and phenol).

Literature Procedure

Peptoid synthesis by the submonomer method was performed on Rink amide resin[3] (50 μmol, 0.45 mmol/g) to avoid diketopiperazine formation. Acylation reactions were performed by addition of bromoacetic acid (600 μmol, 83 mg) in DMF (0.83 mL), followed by the addition of DIC (600 μmol, 103 μL) in DMF (170 μL). Reaction mixtures were agitated at room temperature for 30 min. Each acylation was repeated once. Displacement reactions were performed by addition of primary amine (2.5 mmol) as 2.5 M solutions in DMSO (1.0 mL), followed by agitation for 2 h at room temperature. Optimization of displacement reactions was performed by varying amine concentrations from 0.25 to 2.5 M. Side-chain protecting groups were removed, and the oligomer was released from support by treatment of the oligomer-resin with 95% TFA in H$_2$O (10 mL) for 20 min at room temperature, followed by filtration, dilution, and lyophilization.

4.4.2   Thiol Alkylation in the Solid-Phase Synthesis of β-Turn Mimetics [4,5]

Derivatized Rink resin

Points of Interest

1. Both $i + 1$ and $i + 2$ side chains were incorporated into nine- and ten-membered cyclic β-turn mimetics on solid support. The flexibility of the turn mimetics and the relative orientations of the side chains can be varied by either the ring size or the stereocenters introduced by the α-halo acid or amino acid.

2. β-Turn mimetics were synthesized on Rink derivatized PEG–PS resin and Chiron Mimotope pins in 59–93% purity by HPLC. *p*-Nitrophenylalanine was introduced in the first step as a UV tag to ensure that all products (and impurities) would be detected by UV-HPLC.

3. Barker and coworkers have reported the synthesis of 15-membered ring RGD mimetics by thioalkylation.[6]

4. For all of the turn mimetics synthesized, cyclization provided the desired cyclic monomer with no cyclic dimer detected (<5%). Also less than 5% of the epimer resulting from racemization at the α-bromo site was observed.

Representative Examples

A range of functionalized and sterically hindered side chains, including phenols, amines, carboxylic acids, and isopropyl groups, were incorporated into β-turn mimetics. α-Bromo acids derived from Ala, Val, Asp, Gly, and Tyr were incorporated at the site of intramolecular alkylation.

Literature Procedures

General Procedures. The β-turn mimetics were synthesized on a poly-(ethylene glycol)–polystyrene (PEG–PS) graft copolymer (Rapp TentaGel) with a loading level of 0.28 mequiv/g. All reaction sequences were performed in peptide flasks except for the final cleavage, which was performed in a round-bottom flask. The synthesis of the β-turn mimetics proceeded from a single batch of resin that was separated into portions of approximately 500 mg of resin per mimetic as the syntheses diverged. All reactions employed at least 2 and typically 3–5 equiv of reagent relative to the resin loading level. To ensure efficient resin mixing during rinses or during reaction steps, nitrogen was gently bubbled through the mixture in a peptide flask. For reaction times greater than 1 h, the peptide flasks were stoppered and shaken at low speed on a Fisher wrist action shaker. A volume of 10 mL of solvent per gram of resin was used for each rinse, which lasted approximately 1 min.

Acylation and Intramolecular Alkylation. A solution of the symmetric preformed anhydride (corresponding to the side chain at the $i + 1$ position) was prepared in DCM by reacting the α-bromo acid (0.4 M) with 0.5 equiv of DIC. After 15 min, the solution of anhydride in DCM was filtered to remove the precipitated urea followed by dilution with an equal volume of DMF to afford a 0.1 M solution of the anhydride in a 1 : 1 DCM/DMF mixture. The anhydride solution was then added to the support and the mixture was gently shaken for 8 h. The resin was then rinsed with DMF (4×), DCM (5×), and MeOH (2×). The *tert*-butyl disulfide resin was solvated in a 5 : 3 : 2 *i*-PrOH/DMF/H$_2$O solvent comixture (10 mL/g of resin) that had been purged with Ar for 20 min. Bu$_3$P (250 μL/10 mL of solvent) was then added. After 1–3 h, the reaction was drained and the resin was washed with DCM (4×) and MeOH (2×). The resin-bound free thiol was solvated in THF (10 mL/g of resin) that had been purged with Ar for 20 min. TMG (250 μL/10 mL of solvent) was then added. After gentle shaking for 8 h, the resin was rinsed with 1% AcOH in DMF, DMF (3×), and MeOH (3×) and then dried *in vacuo*. The β-turn mimetics were liberated from support as primary amides by treatment of the resin with 90:5:5 TFA/H$_2$O/Me$_2$S (10 mL/g of resin). The combined cleavage and rinse solutions were concentrated and the crude product was dissolved in 500 μL of DMSO. The purity of the β-turn mimetics was then ascertained by reverse-phase HPLC analysis (270 nm). The β-turn mimetics were purified by preparative HPLC prior to analytical characterization.

4.4.3 Anilide Alkylation with Primary Alkyl Halides in the Solid-Phase Synthesis of 1,4-Benzodiazepines[7,8]

Points of Interest

1. Employing lithiated 5-benzyl-2-oxazolidinone as a base allows selective alkylation of the anilide in the presence of protected amino acid functionality such as *tert*-butyl esters and carbamates. No racemization was detected under these alkylation conditions.

2. The synthetic methodology provides high yields of fully purified and characterized 1,4-benzodiazepines (see p 144).

Representative Examples

A range of amino acids with protected functionality including amines, carboxylic acids, amides, phenols, indoles, and alcohols were incorporated into 1,4-benzodiazepines. Alkylation is compatible with aromatic and primary aliphatic alkylating agents.

Literature Procedure

After cyclization the support-bound 1,4-benzodiazepine was alkylated as follows. The solvent was removed and the support was rinsed with DMF (3×), MeOH (2×), DCM (3×), and THF (3×). The reaction flask was sealed with a fresh rubber septum (to remove residual acetic acid from the cyclization which can quench the deprotonated 1,4-benzodiazepine and lower the conversion during the alkylation step), flushed with nitrogen, and cooled to 0°C. A solution of freshly prepared lithium benzyloxazolidinone in THF (0.1 M, 10 equiv) was transferred by cannula to the solid support with stirring at 0°C. The resulting slurry was stirred for 1.5 h, at which point 15 molar equiv of the appropriate alkylating agent was added, followed by anhydrous DMF to reach a final solvent ratio of ca. 70:30 THF/DMF. The slurry was allowed to warm to room temperature. After 12 h of stirring, the solvent was removed by filtration. The support was then washed with THF, THF/H$_2$O (2×), MeOH, THF, and DCM (2×). To the fully derivatized 1,4-benzodiazepine on the support was added 15 mL of 85:5:10 TFA/H$_2$O/DMS and the slurry was stirred for 12–24 h. The cleavage solution was removed by filtration and the resin was rinsed with 20 mL of MeOH/DCM (3×). Concentration of the combined filtrates then provided the unpurified product (generally a single compound by TLC), which was purified by silica gel chromatography and fully characterized.

4.4.4 One-Pot Cyclization and Anilide Alkylation in the Solid-Phase Synthesis of 1,4-Benzodiazepine-2,5-diones [9]

Points of Interest

    1. Deprotonation, cyclization, and alkylation of 1,4-benzodiazepine-2,5-diones are performed in a one-pot reaction. An acyclic tertiary amide precursor was critical to achieve efficient lactamization.

    2. 1,4-Benzodiazepine-2,5-diones were prepared from a range of anthranilic acids and amino esters on support (see p 150).

Representative Examples

    Both sterically hindered (Val) and functionalized (Tyr, Lys) amino acids were incorporated into 1,4-benzodiazepine-2,5-diones. Electron-rich and electron-poor anthranilic acids were incorporated; 7-bromoanthranilic acid was incorporated and further modified by a Suzuki cross-coupling. Five different alkylating agents were used.

Literature Procedure

    The support-bound acyclic precursor was prepared by reductive alkylation with amino esters followed by acylation with unprotected anthranilic acids. The derivatized resin was rinsed with DMF (7×), DCM (7×), and finally MeOH (3×). The acylation and rinsing were repeated (2 h is adequate) to ensure reaction completion and the resin was dried to a constant weight.

    Cyclization and alkylation were accomplished in one pot: The acylated resin (0.500 g, 0.175 mmol) was placed in a round-bottom flask and purged with Ar for 5 min. The flask was charged with 5 mL of DMF and 5 mL of THF. In a separate flame-dried flask p-methoxyacetanilide or acetaniline (4.44 mmol) was dissolved in THF, the flask was purged with Ar for 5 min, and the solution was cooled to −78°C. n-Butyllithium (1.48 mL, 2.5 M in hexanes, 3.7 mmol) was added dropwise over 10 min and the suspension was mixed for 45 min. Dimethylformamide (5 mL) was then added for solubility, and the mixture was stirred for an additional 15 min and then transferred via cannula into the flask containing the resin. The suspension was stirred gently at room temperature for 30 h under an Ar atmosphere. Alkylating agent (7.40 mmol) was added via syringe and stirring was continued until the suspension no longer turned pH paper dark green or blue (typically 3–6 h). The alkylating solution was removed and the resin was rinsed with DMF (7×), DCM (7×), and MeOH (3×) and dried *in vacuo* to a constant weight. The benzodiazepines were cleaved from support with 20 mL of TFA/Me$_2$S/H$_2$O (90:5:5) for 36 h, and the resin was filtered and rinsed with DCM (4×) and MeOH (2×). The benzodiazepines were purified by silica gel chromatography.

4.4.5 Successive Amide Alkylations: Libraries from Libraries [10]

Points of Interest

1. Libraries of libraries have been prepared by the chemical transformation of pools of peptides (i.e., permethylation, reduction) to prepare new libraries with different properties.[11]

2. In contrast to permethylation of all amide sites at the end of the synthesis, successive amide alkylations allow selective choice of alkylating reagents (or none at all) at each step in the iterative synthesis.

3. At each stage of combinatorial Fmoc-amino acid synthesis, directly after Fmoc deprotection, the free amine was treated with trityl chloride/DIEA. Trityl-protected terminal amines assured that alkylation occurred at the amide position.

4. In a model dipeptide, <1% racemization was found following repeated base and methylation treatments.

5. Individual model dipeptides were used to study the modification of amino acid side chains during the alkylation conditions. Aspartic acid and glutamic acid were excluded from the 20 proteinogenic amino acids, since multiple products were formed following base treatment and alkylation. Methyl iodide, allyl bromide, and benzyl bromide were initially used as alkylating agents. Up to nine separate treatments with base and alkylating agents were used in some cases. During the alkylation procedure the functional groups of six amino acid side chains were reproducibly modified.

Representative Examples

A library of 57,000 pooled compounds was prepared from a dipeptide scaffold with each amide hydrogen replaced with five different alkyl groups (methyl,

ethyl, allyl, benzyl, or naphthylmethyl). A range of protected natural and un-natural amino acids were incorporated into alkylated dipeptides.

Literature Summary

An alkylation sequence was combined with Fmoc-amino acid synthesis in polypropylene mesh packets (teabags) using a split synthesis protocol.[12] After removal of the first Fmoc group, 270 resin packets (a total of 55 mmol of support-bound free N-α-amino groups) were shaken for 3 h in a solution of trityl chloride (276.8 mmol, 0.077 M) in DCM/DMF (9:1, 3.6 L) containing DIEA (1.6 mol, 280 mL). After washes with DMF, DMF with 5% DIEA, and DCM, the tritylation procedure was repeated twice overnight in a 0.05 M solution of trityl chloride in DCM (5.5 L) containing the same amount of base. Washing with DMF (2×), DMF with 5% DIEA, DCM (3×), and MeOH was repeated. The trityl coupling was verified for each of the 46 different amino acid resins using the bromophenol blue color test.[13] Alkylation proceeded under a N$_2$ atmosphere and strictly anhydrous conditions. The 270 resin packets were dried overnight at 50 mTorr. The resin was split into five separate round-bottom flasks for each alkylating reagent. Each flask contained 11.1 mmol of amide groups. Lithium *tert*-butoxide (1 M) in THF (220 mmol, 220 mL) and THF (220 mL) were added and the vessels were shaken for 15 min at room temperature. Excess base solution was removed, followed by the addition of DMSO (440 mL) and then the appropriate individual alkylating agent (665 mmol). The reaction mixture was shaken for 2 h at room temperature. The alkylation solution was removed and the entire procedure was repeated twice. The resulting resin packets were washed with DMF (3×), IPA (2×), DCM (3×), and MeOH and dried overnight under vacuum. (For allylation, benzylation, and naphthyl-methylation the entire process just described was repeated three times.)

4.4.6   Benzophenone Imine α-Carbon Alkylation in Solid-Phase Unnatural Peptide Synthesis[14]

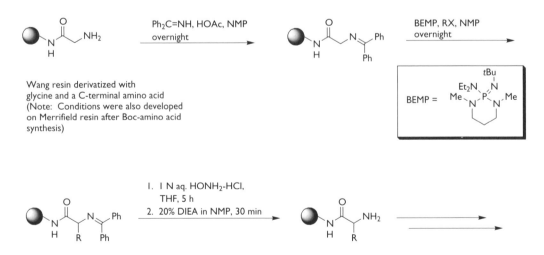

Wang resin derivatized with
glycine and a C-terminal amino acid
(Note: Conditions were also developed
on Merrifield resin after Boc-amino acid
synthesis)

Points of Interest

1. A mild methodology for carbon–carbon bond formation was developed to selectively introduce the side chain at the α-carbon of a particular residue in a growing peptide chain. Three new steps were added to the normal solid-phase peptide synthesis (SPPS) sequence: imine formation/activation, selective alkylation, and imine hydrolysis of the resin-bound Schiff base.

2. Attempts at α-carbon alkylation of the resin-bound Schiff base with melted KOH/K_2CO_3 resulted in incomplete alkylation; however, strong ionic bases led to racemization of the support-bound peptide. Mild and selective alkylation conditions were found when organic-soluble, nonionic "Schwesinger bases" such as BEMP were employed.

Representative Examples

Solid-phase unnatural peptide synthesis allows the preparation of peptides (or other classes of compound) with unnatural amino acids "on the run" as opposed to incorporating individual unnatural amino acids prepared in solution. The mild reagents and conditions are compatible with both Boc- and Fmoc-based SPPS strategies. A variety of hydantoins were prepared by solid-phase unnatural peptide synthesis followed by the Parke-Davis cyclization methodology.[15]

Literature Procedures

For Imine Formation. (A) After standard removal of the support-bound Boc protecting group with TFA, TFA-H_2N-Gly-Merrifield resin in a manual SPPS flask was washed with DCM (3×). The resulting resin was rocked overnight with 1.5 equiv of benzophenone imine in NMP (10 mL/g), filtered, washed with NMP (3×), THF (3×), and DCM (3×), and finally dried *in vacuo* for 5 h. The ^{13}C NMR of the Merrifield resin-bound Schiff base was obtained.

(B) After standard removal of the support-bound Fmoc protecting group with 20% piperidine, H_2N-Gly-Wang resin in a manual SPPS flask was washed with NMP (3×). The resulting resin was rocked overnight with 1.5 equiv of benzophenone imine and 1.3 equiv of acetic acid in NMP (10 mL/g), filtered, washed with NMP (3×), THF (3×), and DCM (3×), and finally dried *in vacuo* for 5 h. The ^{13}C NMR of the Merrifield resin-bound Schiff base was obtained.

For Alkylation of the Resin-Bound Schiff Base. To a round-bottom flask (25 mL) containing the Merrifield or Wang resin-bound Schiff glycinate (1.0 g) were added BEMP (2.0 equiv) and the appropriate alkylating agent (3.0 equiv) followed by NMP (8–10 mL/g). The flask was sealed with a septum and rocked overnight (most of the active alkyl halides took 5–7 h for complete reaction). The resin was filtered and washed with THF (3×) and DCM (3×). In order to determine the extent of alkylation, alkylated resin (10 mg) in 6 N aq HCl (0.2 mL) was heated to 100°C in a pressure tube for 2 h. TLC analysis (silica gel, *n*-butanol/HOAc/H_2O 4/1/1, ninhydrin stain) of the resulting aliquots showed, in most cases, complete disappearance of the starting glycine and formation of the product α-alkylated glycine. When alkylation was incomplete, glycine could be easily detected as a red spot on TLC. This was the case in a single experiment with the relatively unreactive octyl iodide, where both product and glycine were observed by TLC following hydrolysis.

For Imine Hydrolysis and Neutralization.   (A) A round-bottom flask containing the alkylated Schiff base on Merrifield resin (1 g), 1 N aq HCl (6 mL), and THF (15 mL) was rotated on a rotary evaporator for 5 h. The resin was filtered, washed with THF (3×) and NMP (3×), rocked with 10% TEA/NMP (15 mL) for 5 min, and then washed with NMP to provide material ready for further coupling or termination reactions. Completeness of hydrolysis was confirmed by treating the resulting resin with 1 N aq HCl/THF and observing no further formation of benzophenone.

(B) A round-bottom flask containing the alkylated Schiff base on Wang resin (1 g), 1 N aq $NH_2OH–HCl$ (6 mL), and THF (15 mL) was rotated on a rotary evaporator for 5 h. The resin was filtered, washed with MeOH (3×) and NMP (3×), rocked with 10% DIEA/NMP (15 mL) for 5 min, and then washed with NMP to provide material ready for further coupling or termination reactions. Completeness of hydrolysis was confirmed by treating the resulting resin with 1 N aq $NH_2OH–HCl$/THF and observing no further formation of benzophenone oxime.

4.4.7   Alkylation or Sulfonylation of a Support-Bound Phenol[16]

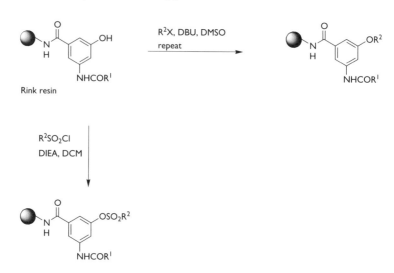

Points of Interest

1. 3-Amino-5-hydroxybenzoic acid was used as a template to prepare a library of 2001 compounds by split synthesis. Aminohydroxybenzoic acids or their derivatives have shown biological activity in a number of different assays.

2. 3-Amino-5-hydroxybenzoic acid could be loaded onto resin through an amino acid to introduce additional diversity.

Representative Examples

The support-bound phenols were alkylated with a variety of alkyl bromides or iodides (28 alkyl halides were used for the library) or sulfonated with a sulfonyl chloride. In a few cases a substantial amount of nonalkylated material was observed by HPLC.

Literature Procedures

For Alkylation of Support-Bound Phenol. Rink resin-bound 3-benzamido-5-hydroxybenzamide (100 mg, 0.045 mmol) was suspended in 1:1 DMSO/NMP (1.1 mL). 1-Iodobutane (31 μL, 0.276 mmol) and DBU (41 μL, 0.276 mmol) were added and the reaction mixture was stirred for 20 h. The alkylation sequence was repeated. The resin was filtered and washed with DCM, MeOH, DCM, MeOH, and DCM. The resin was then suspended in TFA/DCM (1:1) for 1 h, after which it was filtered off and washed well with DCM.

For Sulfonylation of Support-Bound Phenol. Rink resin-bound 3-benzamido-5-hydroxybenzamide (100 mg, 0.045 mmol) was suspended in DCM (1.0 mL). DIEA (16 μL, 0.09 mmol) and benzenesulfonyl chloride (11 μL, 0.09 mmol) were added and the reaction mixture was stirred for 16 h. The resin was filtered and washed with DCM, MeOH, DCM, MeOH, and DCM. The resin was then suspended in TFA/DCM (1:1) for 1 h, after which it was filtered off and washed well with DCM.

4.4.8 Enolate Monoalkylation for Carbon–Carbon Bond Formation on Solid Support [17]

Points of Interest

1. The support-bound acylsulfonamide was treated with excess LDA in THF at 0°C to rapidly give the trianion (only the dianion is shown, since the amide linkage to (aminomethyl)polystyrene is not shown).
2. The acylsulfonamide linker effectively eliminates ketene formation associated with ester or carboximide enolate alkylation. Ketene formation is not observed even when the unreactive alkylating agent isopropyl iodide is employed, since ketene formation would require that the sulfonamide dianion be the leaving group.
3. Carbon–carbon bond formation via the Suzuki reaction on support-bound acylsulfonamide derivatives was also reported (see p 92).

Representative Examples

Addition of activated or unactivated alkyl halides results in rapid alkylation of the enolate trianion with minimal overalkylation (<4%).

Literature Procedure

Acylsulfonamide-derivatized Amberlite XE-305 macroreticular resin (750 mg, 0.185 mmol) was placed in a 100-mL round-bottom flask under N_2 and cooled to 0°C in an ice bath. To a second 100-mL round-bottom flask fitted with a stir bar and a fresh rubber septum were added distilled THF (80 mL) and diisopropylamine (0.49 mL, 3.5 mmol). The second flask was placed in a −78°C acetone–dry ice bath and flushed with N_2. A solution of 2.5 M n-butyllithium in hexanes (1.1 mL, 2.8 mmol) was then added dropwise by syringe. The solution was stirred for 30 min and cannulated to the resin-containing flask. The resin rapidly turned a deep purple color and the slurry was stirred for 1 h. At that time, the appropriate alkylating agent (9.25 mmol, 50 equiv) was added rapidly by syringe. The slurry was then stirred and allowed to come to room temperature over 10 h, after which the solvent was removed and the resin was washed with THF, H_2O, DMF/H_2O, THF, 5% TFA in THF, and THF.

4.4.9 Alkylation of Support-Bound 1,3-Diketones in the Solid-Phase Synthesis of Pyrazoles and Isoxazoles [18]

Rink amide linker Ar^1 = aromatic spacer

$Y = NR^4$, O

Points of Interest

1. For the alkylation reaction best results were obtained with TBAF as base to avoid O-alkylation and increase nucleophilicity. Water traces were detrimental to the yield, which otherwise lies around 90% of the C-monoalkylated product. In addition to the alkyl iodides previously described in analogous solution chemistry,[19] ethyl bromoacetate and allyl bromide reacted without side reactions. With iodoacetonitrile 35% of starting material was observed and bromoacetophenone did not convert cleanly. The failure with benzyl bromide was rather unexpected. The alkylation step is incompatible in the presence of acid or basic heteroaromatic Ar^1 or R^2 residues. Dispensing with the alkylation altogether enables a broader choice of building blocks for the previous Claisen condensation.

2. The Claisen condensation was optimized with the prototypical aromatic ester ethyl benzoate. As expected, carboxylic ester with α-hydrogens was unsuited for the formation of diketones, and also weakly acidic heteroaromatic compounds (e.g. containing unprotected indole) could not be applied.

Representative Examples

For the labels shown in the diagram, Ar^1 = 4-Ph, R^2 = aliphatic, aromatic, or heteroaromatic alkyl ester, R^3 = aliphatic, allyl, or α-carbonyl alkylating agent, and Y = NMe or NPh. Also see the immediately preceding points of interest regarding the scope and limitations of the alkylation conditions.

Literature Summary

Commercially available Fmoc-protected Rink amide resin[20] (see p ■) was deprotected with 20% piperidine/DMF, rinsed, and acylated with 0.3 M solution of acetyl carboxylic acid (3 equiv, preactivated with 3.3 equiv DIC/HOBt for 40 min) until the Kaiser ninhydrin test (see p ■) was negative.[21] Rinses between reactions were not given (typically ca. 12 rinses between reactions is sufficient). To the acylated resin (50 mg, 22.5 μmol) was added a solution of 675 μmol of carboxylic ester in 670 μL of DMA. Under inert gas, 18 mg (0.45 mmol) of 60% sodium hydride was added, and the reaction mixture was shaken well and heated at 90°C for 1 h. The derivatized resin was filtered, washed with 30% acetic acid/water (v/v), DMA, DMSO, and i-PrOH, and then dried *in vacuo*. The resin (20 mg, 8.6 μmol) was treated with 86 μL of 1 M TBAF in THF for 2 h at room temperature and then with 150 μL of a 2.5 M solution of the appropriate alkylating agent for an additional 2 h. The resin was filtered and washed extensively with DCM and THF.

4.4.10 Tosyl Displacement with Primary or Secondary Amines on Solid Support[23]

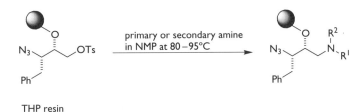

THP resin

Points of Interest

1. The tosyl displacement was performed within the context of the solid-phase synthesis of aspartic acid protease inhibitors.

2. After displacement, the support-bound amines were further derivatized with isocyanates to afford the corresponding ureas.

3. A support-bound alcohol was converted to the corresponding support-bound triflate (see p 159) and then displaced in the solid-phase synthesis of 5-alkoxyhydantoins.[24]

Representative Examples

A range of nonpeptide functionality was incorporated into (hydroxyethyl)amine and (hydroxyethyl)urea isosteres. Many of the functional groups and structures that are present in known inhibitors of HIV-1 protease and renin were incorporated to demonstrate the generality of the synthesis sequence.

Literature Procedures

For Displacement of Support-Bound Tosyl Alcohol with a Primary Amine. The support-bound tosyl ether was added to a 50-mL round-bottom flask fitted with a stir bar. The resin was solvated in NMP and the primary amine was added to provide a 1.0 M solution. The reaction mixture was heated to 75°C for 36 h. Reactions using volatile amine (i.e., isopropylamine) were performed in sealed pressurized tubes. After reaction completion, the solution was removed by filtration cannula, and the resin was rinsed with DMF (3×) and DCM (3×) to afford the support-bound secondary amine.

For Displacement of Support-Bound Tosyl Alcohol with a Secondary Amine. The support-bound tosyl alcohol (0.40 g, 0.12 mmol) was solvated in 3.2 mL of NMP, and the appropriate secondary amine (3.20 mmol) was added. The reaction was heated to 75–95°C for 45–48 h as necessary for complete displacement. The solution was then removed by filtration and the resin was washed with DMF (3×) and DCM (3×) to afford the support-bound tertiary amine.

4.4.11   Grignard Additions to Support-Bound Esters[25]

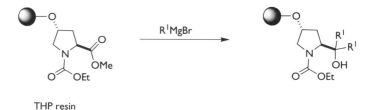

THP resin

Points of Interest

1. Pyrrolidinemethanol ligands (shown) were prepared on support, cleaved from the resin, and then examined as asymmetric catalysts for the addition of diethylzinc to aldehydes.

2. After Grignard addition, the ethyloxycarbonyl (EtOC) protecting group was removed under strongly basic conditions (KOH, BuOH/1,4-dioxane at reflux). Alloc and TMSEtOC protecting groups were also examined, but these more labile protecting groups were not stable to the Grignard reaction conditions.

3. Grignard additions to support-bound acylimidazole have been reported.[26]

Representative Examples

Both aliphatic and aromatic Grignard reagents were added to the support-bound ester to provide the corresponding support-bound tertiary alcohols.

Literature Procedure

To the support-bound *N*-ethyloxycarbonyl methyl ester of 4-hydroxyproline (1.00 g, 0.43 mmol) was added THF (25 mL) followed by the addition of a

1.0 M solution of the appropriate Grignard reagent in THF (21 mL, 50 equiv) at 0°C. The resulting slurry was stirred at room temperature for 12 h. The solvent and excess reagents were filtered away. The resin was rinsed with THF (5×), DMF/H$_2$O (5× to remove salts), DMF (5×), and DCM (5×).

4.4.12 S$_N$Ar Reaction on Solid Support[27]

Rink resin

Points of Interest

1. Piperidines were reacted with a support-bound aryl fluoride to produce the corresponding arylamines in the solid-phase synthesis of arylpiperazines. The alkylation of piperazines with support-bound benzylic halides was also examined.

2. A small library was prepared from the 5 halide core structures and 38 arylpiperazines.

Representative Examples

Support-bound aryl fluorides reacted readily with a large variety of N-substituted piperazines when the fluorides were activated by a nitro group either in the ortho or para positions. However, similarly substituted aryl chlorides or pyridyl chlorides did not react under the same conditions. Aryl fluorides with other electron-withdrawing groups (such as trifluoromethyl) substituted in the ortho or para positions did not react or gave incomplete reaction.

Literature Summary

The support-bound halide cores (100 mg, 0.048 mmol) were treated with 10 equiv of alkylpiperazine (0.48 mmol) in NMP (1 mL) for 48 h at room temperature. The resin was filtered and washed with DCM, MeOH, DCM, MeOH, and DCM.

4.4.13   Palladium-Catalyzed Amination of Resin-Bound Aromatic Bromides [28,29]

Rink resin

Points of Interest

1. The scope and limitations of the palladium-catalyzed amination of support-bound aromatic bromides were investigated with a range of aromatic and aliphatic amines.

2. With the P(o-Tol)$_3$/Pd system both simple aromatic and aliphatic amines could be coupled. However, primary and secondary aliphatic amines gave significant reduction of the bromide. Utilization of Buchwald's improved method [30] with BINAP decreased this side reaction to a level undetectable by NMR. More functionalized amines that were nearly completely inactive in couplings using P(o-Tol)$_3$/Pd were successfully coupled with BINAP (or dppf) and longer reaction times (70 h, 100°C).

3. The use of BINAP as a ligand also allowed for the successful coupling of primary amines.

Representative Examples

A model bromide linked to Rink resin (p-bromobenzamidyl) was coupled with a variety of amines (including anilines and secondary amines). A variety of resin-bound bromides were coupled with two model amines (o- and p-anisidine). A number of amines were found to be unreactive under all conditions tried. All of the bromides studied couple cleanly on Rink resin except for those derived from 2-bromobenzoic acid, 3,5-dibromo-4-hydroxybenzoic acid, and 5-bromofuroic acid, which are recovered unreacted after cleavage from resin. Alternatively, a urea linkage was used for loading to resin.

Literature Summary

To bromobenzyl-derivatized resin (1.0 equiv) in a 25-mL Schlenk tube equipped with a stir bar was added a solid mixture of $Pd_2(dba)_3$ (0.05 equiv), phosphine (0.20 equiv of monophosphine or 0.15 equiv of diphosphine), t-BuONa (10–20 equiv; the use of a large excess of base was found to be necessary to get reproducible results in cases without *rigorously* dried solvents and reagents), and amine (3.0 equiv) followed by 3.0 mL of dry toluene. The tube was sealed either under vacuum or under oil bubbler pressure of N_2 and equipped with reflux condenser. The reaction mixture was heated with an oil bath maintained at 100°C for 20 or 70 h under N_2. The reaction mixture was filtered on a coarse frit and the resin was washed with MeOH (3×) and DCM (3×).

4.4.14   1,4-Michael Addition of Thiols to Support-Bound Enones[31]

Points of Interest

1. The addition to the support-bound enone could be monitored by FT-IR analysis of the polymer (1674–1712 cm^{-1}).
2. A small library of three enones and three thiols was prepared by a split synthesis.

Representative Examples

Three thiol phenols were added to a support-bound enone as part of a four-reaction sequence on solid support that proceeded in 7–27% yield. Clean chromatograms for mixtures suggested a clean Michael addition reaction.

Literature Procedure

A mixture of three different support-bound enones (in equal amounts) was placed into three flasks. Sodium methoxide, thiol, and 3 mL of THF were added and each flask was swirled on a shaker plate. *Flask 1:* 0.382 g of thiophenol, 0.039 g of sodium methoxide. *Flask 2:* 0.260 g of *p*-thiocresol, 0.039 g of sodium methoxide. *Flask 3:* 0.280 g of 4-chlorothiophenol, 0.039 g of sodium methoxide. Each mixture was swirled for 5 days, filtered, and washed three times each with THF, MeOH, THF, and ether.

4.4.15   The Iodoetherification Reaction in the Solid-Phase Synthesis of Miconazole Analogs[32]

Merrifield resin

Points of Interest

1. Miconazole is a member of a class of imidazole agents that elicit broad-spectrum antifungal activity. A novel solid-phase iodoetherification reaction was employed in the solid-phase synthesis of 45 miconazole analogs.

2. A *p*-carboxyl group was installed as a handle for reversible attachment to the solid support. Commercially available Merrifield resin was alkylated with the free carboxylate anion (in DMF at 80°C in the presence of KI).

3. Iodoetherification reaction conditions included a styrene derivative, *N*-iodosuccinimide, and triflic acid. The reaction is incompatible with acid-labile linkers. Displacement of the support-bound iodide was effected with 1-(trimethylsilyl)imidazole in the presence of silver triflate. These conditions minimized a competing elimination pathway that produced considerable amounts of enol ether under a variety of other conditions.

4. The synthesis of primary amines was carried out from the intermediate iodide by displacement of the iodide with tetrabutylammonium azide, followed by reduction to the amine using thiophenol/Et_3N/$SnCl_2$ (see p 140).[33]

Representative Examples

A range of 24 substituted styrene derivatives were incorporated into miconazole analogs via an iodoetherification reaction. The resulting support-bound iodide was then displaced with either 1-(trimethylsilyl)imidazole or tetrabutylammonium azide.

Literature Procedures

For the Iodoetherification.   To a suspension of the support-bound alcohol (500 mg, 1.05 mmol/g) in DME (4.5 mL) were added 2,4,6-trimethylstyrene (595 mg, 4.07 mmol, 7.7 equiv), *N*-iodosuccinimide (916 mg, 4.07 mmol, 7.7 equiv), and trifluoromethanesulfonic acid (7 μL). The reaction mixture was shaken at room temperature for 16 h, washed with DME (5×), 1% Et_3N in DME (2×), DME/H_2O (5×), DMF (3×), and MeOH (5×), and dried under high vacuum. A second treatment was performed to ensure high conversion.

For the Displacement.   To the support-bound iodo ether (621 mg, 0.85 mmol/g) in DMF (5 mL) were added silver triflate (167 mg, 0.69 mmol, 1.3 equiv) and 1-(trimethylsilyl)imidazole (1.47 g, 10.5 mmol, 20 equiv). The reaction mixture was shaken at 85°C for 16 h, washed with DMF (5×), DMF/H$_2$O (5×), DMF (3×), and MeOH (5×), and dried under high vacuum. A second treatment was performed to ensure high conversion.

4.4.16   Reactions with Support-Bound Alkyl Halides in Solid-Phase Peptide and Combinatorial Synthesis

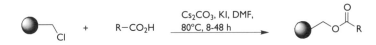

Merrifield resin

Points of Interest

1. Reactions between a range of support-bound alkyl halides (including support-bound benzyl chlorides, alkyl bromides, chlorotrityl chlorides, etc.) and carboxylate anions are quite common in solid-phase peptide and combinatorial synthesis.[34–37]

2. Alkylation reactions are often used for loading C-terminal amino acids to resin because racemization is minimized.

3. A polymer-supported chiral auxiliary based on "Evans oxazolidinones" was prepared by S$_N$2 displacement of (chloromethyl)polystyrene (Merrifield resin) with a chiral potassium alkoxide.[38]

4. Trityl resins react with alcohols and amines via an S$_N$1 mechanism (see p 52).

Representative Examples

A range of nucleophiles, including both Boc- and Fmoc-amino acids, have been alkylated with support-bound alkyl halides.

Literature Procedure

Commercially available Merrifield resin was alkylated with the free carboxylate anion in DMF at 80°C in the presence of KI. TBAI and NaI have also been used to catalyze the alkylation of resin.

REFERENCES FOR SECTION 4.4

1. Zuckermann, R. N.; Kerr, J. M.; Kent, S. B. K.; Moos, W. H. *J. Am. Chem. Soc.* **1992,** *114,* 10646–10647.
2. Zuckermann, R. N.; Kerr, J. M.; Siani, M. A.; Banville, S. C. *Int. J. Pept. Protein Res.* **1992,** *40,* 498–507.
3. Rink, H. *Tetrahedron Lett.* **1987,** *28,* 3787–3790.
4. Virgilio, A. A.; Ellman, J. A. *J. Am. Chem. Soc.* **1994,** *116,* 11580–11581.

5. Virgilio, A. A.; Schürer, S. C.; Ellman, J. A. *Tetrahedron Lett.* **1996**, *37*, 6961–6964.

6. Barker, P. L.; *et al. J. Med. Chem.* **1992**, *35*, 2040–2048.

7. Bunin, B. A.; Ellman, J. A. *J. Am. Chem. Soc.* **1992**, *114*, 10997–10998.

8. Bunin, B. A.; Plunkett, M. J.; Ellman, J. A. *Proc. Natl. Acad. Sci. U.S.A.* **1994**, *91*, 4708–4712.

9. Boojamra, C. G.; Burrow, K. M.; Ellman, J. A. *J. Org. Chem.* **1995**, *60*, 5742–5743.

10. Dörner, B.; Husar, G. M.; Ostresh, J. M.; Houghten, R. A. *Bioorg. Med. Chem.* **1996**, *4*, 709–715.

11. Ostretch, J. M.; Husar, G. M.; Blondelle, S. E.; Dörner, B.; Weber, P. A.; Houghten, R. A. *Proc. Natl. Acad. Sci. U.S.A.* **1994**, *91*, 11138–11142.

12. Houghten, R. A. *Proc. Natl. Acad. Sci. U.S.A.* **1985**, *82*, 5131.

13. Krchñàk, V.; Vàgner, J.; Safar, P.; Lebl, M. *Collect. Czech. Chem. Commun.* **1988**, *53*, 2542–2548.

14. O'Donnell, M. J.; Zhou, C.; Scott, W. L. *J. Am. Chem. Soc.* **1996**, *118*, 6070–6071.

15. DeWitt, S. H.; Kiely, J. S.; Stankovic, C. J.; Schroeder, M. C.; Cody, D. M. R.; Pavia, M. R. *Proc. Natl. Acad. Sci. U.S.A.* **1993**, *90*, 6909–6913.

16. Dankwardt, S. M.; Phan, T. M.; Krstenansky, J. L. *Mol. Diversity* **1995**, *1*, 113–120.

17. Backes, B. J.; Ellman, J. A. *J. Am. Chem. Soc.* **1994**, *116*, 11171–11172.

18. Marzinzik, A. L.; Felder, E. R. *Tetrahedron Lett.* **1996**, *37*, 1003–1006.

19. Clark, J. H.; Miller, J. M. *J. Chem. Soc., Perkin Trans. 1* **1977**, 1743.

20. Rink, H. *Tetrahedron Lett.* **1987**, *28*, 3787–3790.

21. Kaiser, E.; Colescott, R. L.; Bossinger, C. D.; Cook, P. I. *Anal. Biochem.* **1970**, *34*, 595–598.

22. Rink, H. *Tetrahedron Lett.* **1987**, *28*, 3787–3790.

23. Kick, E. K.; Ellman, J. A. *J. Med. Chem.* **1995**, *38*, 1427–1430.

24. Hanessian, S.; Yang, R. Y. *Tetrahedron Lett.* **1996**, *37*, 5835–5838.

25. Liu, G.; Ellman, J. A. *J. Org. Chem.* **1995**, *60*, 7712–7713.

26. Hauske, J. R.; Dorff, P. *Tetrahedron Lett.* **1995**, *36*, 1589–1592.

27. Dankwardt, S. M.; Newmann, S. R.; Krstenansky, J. L. *Tetrahedron Lett.* **1995**, *36*, 4923–4926.

28. Ward, Y. D.; Farina, V. *Tetrahedron Lett.* **1996**, *37*, 6993–6996.

29. Willoughby, C. A.; Chapman, K. T. *Tetrahedron Lett.* **1996**, *37*, 7181–7184.

30. Wolfe, J. P.; Wagaw, S.; Buchwald, S. L. *J. Am. Chem. Soc.* **1996**, *118*, 7215–7216.

31. Chen, C.; Ahlberg Randall, L. A.; Miller, R. B.; Daniel Jones, A. D.; Kurth, M. J. *J. Am. Chem. Soc.* **1994**, *116*, 2661–2662.

32. Tortolani, D. R.; Biller, S. A. *Tetrahedron Lett.* **1996**, *37*, 5687–5690.

33. Kick, E. K.; Ellman, J. A. *J. Med. Chem.* **1995**, *38*, 1427–1430.

34. Baleux, F.; Calas, B.; Mery, J. *Int. J. Pept. Protein Res.* **1986**, *28*, 22–28.

35. Barlos, K.; Chatzi, O.; Gatos, D.; Stavropoulos, G. *Int. J. Pept. Protein Res.* **1991**, *37*, 513–520.

36. Kurth, M. J.; Ahlberg Randall, L. A.; Chen, C.; Melander, C.; Miller, R. B.; McAlister, K.; Reitz, G.; Kang, R.; Nakatsu, T.; Green, C. *J. Org. Chem.* **1994**, *59*, 5862–5864.

37. Hanessian, S.; Yang, R. Y. *Tetrahedron Lett.* **1996**, *37*, 5835–5838.

38. Allin, S. M.; Shuttleworth, S. J. *Tetrahedron Lett.* **1996**, *37*, 8023–8026.

4.5 OXIDATIONS ON SOLID SUPPORT

Relatively few examples of oxidations on solid support have been published considering the fundamental nature of oxidations in synthesis. This is not surprising for two reasons. First, with solid-phase synthesis one can choose the starting materials to use for a combinatorial synthesis; thus the building blocks are often in the correct oxidation state. Second, unlike the reductive amination commonly employed for combinatorial synthesis, it is difficult to add diversity during an oxidation step. Nonetheless, in certain cases oxidations are required. Representative examples of oxidations on solid support are provided in this section.

4.5.1   Oxidation of Alcohols to Aldehydes and Ketones [1,2]

R = H, aryl

Points of Interest

1. Trityl resin was derivatized with excess 1,4-butanediol. The large excess in solution effectively minimizes bis-protection. The support-bound free hydroxyl was then oxidized to the corresponding aldehyde with sulfur trioxide–pyridine.[3]

2. Standard oxalyl chloride/Et$_3$N/DMSO conditions resulted in acid-catalyzed deprotection of the sensitive trityl moiety.

3. The transformation of an alcohol (OH stretch at 3572 cm^{-1}) to an aldehyde (C=O stretch at 1724 cm^{-1}) was monitored by FT-IR (KBr pressed windows of ground polystyrene beads).

4. The catalytic oxidation of support-bound alcohols to aldehydes and ketones with NMO/TPAP was monitored by single-bead FT-IR.

Literature Procedures

For Sulfur Trioxide Oxidation.   Polymer-bound 4-(trityloxy)-1-butanol (11.71 g, 1.57 mmol/g) was purged with argon in a 500-mL round-bottom flask for 1 h and dry DMSO (75 mL) was added. In another 500-mL round-bottom flask, Pyr·SO$_3$ complex (24.70 g) was purged with argon for 30 min and DMSO (100 mL) and Et$_3$N (100 mL) were added via syringe. After 15 min of stirring, the Pyr·SO$_3$ solution was cannulated to the suspension of polymer-bound 4-(trityloxy)-1-butanol in DMSO and the mixture was stirred for 3 h. The reaction mixture was filtered and washed with dry DMSO (3×), THF (6×), and ether (3×). IR (KBr) 1724 cm^{-1}.

For NMO/TPAP Oxidation.   To a support-bound alcohol (0.30 mmol) was added a solution of *N*-methylmorpholine *N*-oxide (NMO) (352 mg, 3.00 mmol) in 10 mL of dry DMF, and then catalytic tetrapropylammonium perruthenate (TPAP) (21.1 mg, 0.06 mmol) was added to the resin suspension. The reaction was monitored by single-bead FT-IR until the reaction was completed after 80 min. The reaction mixture was drained with DMF (4×), THF (4×), and DCM (4×) and dried under vacuum for 2 h.

4.5.2   Oxidation of (Chloromethyl)polystyrene Resin to Formylpolystyrene and Carboxypolystyrene Resins[4]

Points of Interest

1. The formyl and carboxylate resins have also been prepared by quenching lithiated polystyrene with DMF and CO_2, respectively (see p 12, derivatized resins).[5]

2. Formyl resins are commercially available from Novabiochem.

Literature Procedures

For Oxidation to Formylpolystyrene Resin.   Chloromethyl resin (5.0 g) and $NaHCO_3$ (7.0 g) were stirred in DMSO (100 mL) for 6 h at 155°C.

For Oxidation with Chromic Acid to the Carboxylic Acid.   The aldehyde resin (1 g) and concentrated H_2SO_4 were stirred in 50 mL of saturated acetous $Na_2Cr_2O_7$ for 48 h at 60°C.

4.5.3   Other Examples of Oxidation on Solid Support

1. A number of different methods for the oxidation of support-bound alcohols to aldehydes have been reported.[6,7]

2. Oxidation of the support-bound aldehyde to the carboxylic acid with *m*CPBA has also been reported.[8]

3. The oxidation of *N*-acylthiazolidine thioethers (see p 188) to the sulfoxide was accomplished in good yield using *m*CPBA, providing acid-stable products. The analogous desired sulfones were not obtained under a variety of conditions.[9]

4. The oxidation of an aryl methyl sulfide to the corresponding sulfone with *m*CPBA (2.5 equiv, DCM, room temperature, 6 h) has been reported in the solid-phase synthesis of 1,2-diazines (see p 209).[10]

5. Alkenes have been converted to epoxides in the solid-phase synthesis of oligosaccharides.[2] Support-bound oxiranes have been prepared by epoxidation of support-bound alkenes with purified *m*CPBA in buffered $NaHCO_3$/DCM at reflux.[11]

6. Support-bound selenium compounds have been oxidized with $NaIO_4$ in an oxidation/elimination sequence.[12]

REFERENCES FOR SECTION 4.5

1. Chen, C.; Ahlberg Randall, L. A.; Miller, R. B.; Jones, A. D.; Kurth, M. J. *J. Am. Chem. Soc.* **1994,** *116,* 2661–2662.

2. Yan, B.; Sun, Q.; Wareing, J. R.; Jewell, C. F. *J. Org. Chem.* **1996**, *61*, 8765–8770.
3. Parikh, J. R.; Doering, W. V. *J. Am. Chem. Soc.* **1967**, *89*, 5505–5507.
4. Ayres, J. T.; Mann, C. K. *J. Polym. Sci., Polym. Lett. Ed.* **1965**, *3*, 505–508.
5. Farrall, M. J.; Fréchet, J. M. J. *J. Org. Chem.* **1976**, *41*, 3877–3882.
6. Leznoff, C. C. *Acc. Chem. Res.* **1978**, *11*, 327–333 and references cited therein.
7. Yan, B.; Sun, Q.; Wareing, J. R.; Jewell, C. F. *J. Org. Chem.* **1996**, *61*, 8765–8770.
8. Beebe, X.; Schore, N. E.; Kurth, M. J. *J. Am. Chem. Soc.* **1992**, *114*, 10061–10062.
9. Holmes, C. P.; Chinn, J. P.; Look, G. C.; Gordon, E. M.; Gallop, M. A. *J. Org. Chem.* **1995**, *60*, 7328–7333.
10. Panek, J. S.; Zhu, B. *Tetrahedron Lett.* **1996**, *45*, 8151–8154.
11. Rotella, D. P. *J. Am. Chem. Soc.* **1996**, *118*, 12246–12247.
12. Kurth, M. J.; Ahlberg Randall, L. A.; Takenouchi, K. *J. Org. Chem.* **1996**, *61*, 8755–8761.

4.6   REDUCTIONS ON SOLID SUPPORT

The most common reduction performed on solid support is a reductive amination. Reductive amination has been performed with either a support-bound aldehyde or a support-bound amine. Because of the large number of commercially available aldehydes, amines, and support-bound amino acids, reductive aminations have been widely used to generate molecular diversity.

4.6.1   Rapid Optimization of Oxidation and Reduction Reactions on a Solid Phase Using the Multipin Approach: Synthesis of 4-Aminoproline Analogs [1]

polyethylene pins

Points of Interest

1. Support-bound 4-hydroxyproline was oxidized to the ketone and then reductively aminated with a range of amines.

2. The multipin method was used to simultaneously evaluate different reaction parameters. High-throughput characterization was performed by ion spray mass spectrometry and HPLC.

Representative Examples

Reductive amination proceeded with high conversion for primary and secondary amines as well as amino esters and a range of anilines.

Literature Summary

The most efficient conditions for oxidizing pin-bound 4-hydroxyproline into the corresponding ketone employed 0.2 M pyridinium dichromate in DMF at 37°C. The following trends were observed in the course of optimizing the reductive amination for a range of amines. In all instances, the higher amine concentration was preferable (2.0 vs 0.05 M) and MeOH was the preferred solvent. High conversions were favored by low pH (pH 5 or 6 vs pH 7) in the cases of benzylamine and cyclohexylamine. Reductive amination with aniline was largely independent of pH; hence studies with this amine were performed at pH 7 (with acetic acid/N-methylmorpholine in the reaction mixture). Optimum conditions employed an amine (2.0 M) and $NaBH_3CN$ (0.05 M) in MeOH at pH 5 for nonaromatic amines or pH 7 for aromatic amines.

4.6.2 Reductive Alkylation on a Solid Phase: Synthesis of a Piperazinedione Combinatorial Library[2]

Points of Interest

1. Sodium triacetoxyborohydride was employed for reductive alkylation to avoid the pH sensitivity and potential reagent toxicity associated with the well-established procedure for the reductive alkylation of peptide resins using sodium cyanoborohydride.[3]

2. The reductive amination procedure included brief sonication and was performed twice to ensure high conversion.

3. Partial (5%) racemization was observed after reductive amination. Aliphatic aldehydes also gave minor, but significant amounts of bisalkylation (1–10%), generating tertiary amines.

4. Two purified diketopiperazines were prepared in 24% yield (DKP from Val, 2,4,6-trimethoxybenzaldehyde, and Leu) and 42% yield (DKP from Phe, p-methoxybenzaldehyde, and Ala). A library of 1000 diketopiperazines was prepared with the split and mix method (see p 167).

Representative Examples

A range of functionally and sterically diverse support-bound amino acids were reductively alkylated with aliphatic, aromatic, and heteroaromatic aldehydes.

Literature Summary

Amino acid-derivatized Wang resin was suspended in DCM (0.5 mL). One of the 10 aldehydes (0.24 mmol) in DCM (0.5 mL) was added. The reaction mixture was sonicated in an ultrasound bath for 20 min followed by addition of a presonicated solution of sodium triacetoxyborohydride (0.28 mmol) in DCM (0.5 mL). The resulting mixture was sonicated for 5 min and then stirred vigorously for 16 h. Resin was filtered, washed (water, aqueous bisulfate, water/THF, 3 × 2 mL each), and dried. The reductive alkylation procedure was repeated.

4.6.3 Reductive Alkylation of Sieber's Xan Linker[4]

Points of Interest

1. Reductive *N*-alkylation of the amide anchor resin [((9-aminoxanthen-3-yl)oxy)methyl]polystyrene was accomplished by treatment with alkyl aldehydes under mild conditions, followed by reduction with NaBH₄ in MeOH/DMF. The modified resin was used for the preparation of peptidyl N-alkylamides.

2. A two-step procedure was necessary for efficient monoalkylation.

3. It was crucial that the solvent mixture for the reduction contained MeOH. Negligible reduction was observed in straight DMF and the substitution of *i*-PrOH for MeOH also resulted in a poor reduction.

4. Reductive alkylation of Rink amide linker[5] (see p 38) was also accomplished, but stable imines were not obtained with the PAL linker[6] (see p 39), prohibiting reductive alkylation of the PAL linker.

5. HATU/HOAt/DIPEA[7,8] were employed as reagents for the acylation of the sterically hindered support-bound secondary amines.

Representative Examples

Rink or Xan linkers were reductively alkylated with aliphatic aldehydes and acylated with a range of Fmoc-amino acids.

Literature Procedure

Sieber's Xan resin (0.05 mmol) suspended in DMF (ca. 1 mL) was treated with the aldehyde (0.25 mmol) and acetic acid (10 μL) with agitation over 45 min. The resin was washed with DMF, and the chemical reaction was repeated to ensure complete formation of the imine. The imine-derivatized resin is stable to DMF wash as confirmed by the negative amine test after washing with DMF at 2.8 mL/min for 5 min. To a stirred suspension of the DMF-washed resin in DMF (1 mL) was then added MeOH (0.4 mL) followed immediately by portionwise addition of NaBH₄ (0.40 mmol) over 1 h. The reaction mixture was stirred for a further 1 h, and the resin was then extensively washed with DMF to afford the support-bound secondary amine. Acylation was carried out using HATU-activated Fmoc-amino acids (4–8 equiv) in DMF (ca. 0.3 M) overnight.

4.6.4 Loading Amino Esters onto Resin via Reductive Amination [9]

Points of Interest

1. Amino esters are loaded onto a support-bound aldehyde via a reductive amination. The imines must be reduced immediately upon formation to avoid racemization. Conversely, racemic products can easily be prepared if desired without purchasing L- and D-amino acids.

2. Support-bound secondary amines were incorporated into 1,4-benzodiazepine-2,5-diones (see p 150).

Representative Examples

Both sterically hindered (Val) and functionalized (Tyr, Lys) amino acids were incorporated into 1,4-benzodiazepine-2,5-diones.

Literature Procedure

Commercially available Merrifield resin [10] was derivatized with the aldehyde linker under Ar as follows. A solution of 4-hydroxy-2,6-dimethoxybenzaldehyde [11] (15.3 g, 84.2 mmol) in DMF (800 mL) was degassed with Ar for 30 min. Under positive Ar flow, 95% NaH (1.92 g, 79.9 mmol) was added slowly in 0.5-g aliquots over 15 min. After 30 min, Merrifield resin (31.7 g, 24.1 mmol, loading 0.76 mmol/g) was added and the contents were stirred mechanically for 36 h at 50°C. The DMF suspension was then diluted with MeOH and filtered. The resin was then rinsed with DMF/MeOH (2×), DMF (5×), DCM (7×), and MeOH (4×). The resin was dried *in vacuo* to a constant weight of 33.4 g. A strong aldehyde carbonyl stretch was observed in the IR (KBr) spectrum of this resin: 1690 cm^{-1}.

Amino esters can be loaded onto the resin by reductive amination either with or without racemization under different experimental conditions. For racemization-free loading, an amino methyl ester hydrochloride (2.59 mmol) was dissolved in 1% AcOH in DMF (20 mL) and then NaBH(OAc)$_3$ was added. To this mixture was immediately added 0.700 g (0.518 mmol) of aldehyde-derivatized resin. The reaction was complete at 60 min as determined by the IR spectrum of an aliquot of resin. The bulk of the resin was then rinsed with MeOH (1×), 10% DIEA in DMF (2×), DMF (7×), DCM (7×), and finally MeOH (3×). The resin was dried *in vacuo* to constant weight and characterized by IR spectroscopy. IR (KBr): 1736 cm⁻¹. For racemized benzodiazepine products the aldehyde-derived resin was added before the other reagents and 1 equiv of DIEA was added with the amino ester hydrochloride.

The loading of resin-bound amino esters was quantitated by acetylation of the free amine followed by cleavage with subsequent determination of the mass balance of the silica gel-purified product. The loading with different amino esters was consistently 0.36–0.37 mmol/g of resin.

4.6.5   Reductive Amination of a Support-Bound Aniline [12]

Rink resin:
X = NH, O

Points of Interest

1. Reductive amination of a support-bound aniline was developed to generate analogs of lavendustin A as potential tyrosine kinase inhibitors.
2. A large excess of NaCNBH₃ (ca. 20 equiv) and overnight reaction at room temperature were necessary for complete reduction.
3. HPLC analysis gave a clear indication of the progress of the reductive amination. Incomplete reduction of the imine intermediate was apparent by the presence of aldehyde in the chromatogram, since the TFA treatment also cleaves the imine intermediate.

Representative Examples

Tetramethyl-lavendustin A and lavendustin A were prepared. Sixty analogs were prepared incorporating three support-bound anilines and five electron-rich aldehydes.

Literature Procedure

2,5-Dimethoxybenzaldehyde (102 mg, 0.61 mmol) was added to the aniline-derivatized resin (0.193 mmol) followed by 1% AcOH in DMA (2 mL) and the mixture was shaken for 5 min. NaCNBH₃ (215 mg, 3.42 mmol) was added in portions over a 2-h period and the mixture shaken overnight. The resin was filtered and washed with 1% AcOH in DMA (5×) and MeOH (4×).

4.6.6   Solid-Phase Reductive Alkylation of Secondary Amines Using Borane–Pyridine Complex [13]

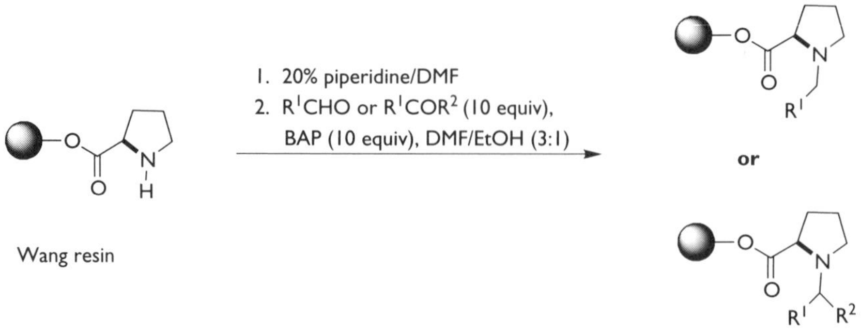

Wang resin

1. 20% piperidine/DMF
2. R^1CHO or R^1COR^2 (10 equiv),
   BAP (10 equiv), DMF/EtOH (3:1)

or

Points of Interest

1. Although the reductive alkylation of primary amines on solid support is commonly used to introduce diversity to small-molecule libraries, the reductive alkylation of secondary amines involves a less stable iminium ion intermediate.

2. Other reagents for the reductive alkylation of secondary amines on solid support, including $NaCNBH_3$, $NaBH(OAc)_3$, or $Ti(O\text{-}iPr)_4$, were less effective than the borane–pyridine complex (BAP).[14]

Representative Examples

Support-bound proline was reductively alkylated with aromatic and aliphatic aldehydes and ketones in high crude yield and purity. Reductive alkylation of support-bound primary amino acids using BAP with a twofold excess of aldehyde gave predominantly the dialkylated product in quantitative yield.

Literature Summary

Anhydrous solvents were used throughout the procedure. Fmoc-Pro-Wang resin was treated with 20% piperidine/DMF. The resin was then washed thoroughly with DMF and DMF/EtOH (3:1). The resin was preswollen in DMF/EtOH (3:1) and treated with 10 equiv of aldehyde/ketone and 10 equiv of BAP. The reaction mixture was shaken under Ar at room temperature for 4 days. Then the resin was washed with DMF/EtOH, DCM, and MeOH, dried under vacuum, and cleaved with 95% TFA.

4.6.7   Reduction of Support-Bound Amides with Red-Al [15]

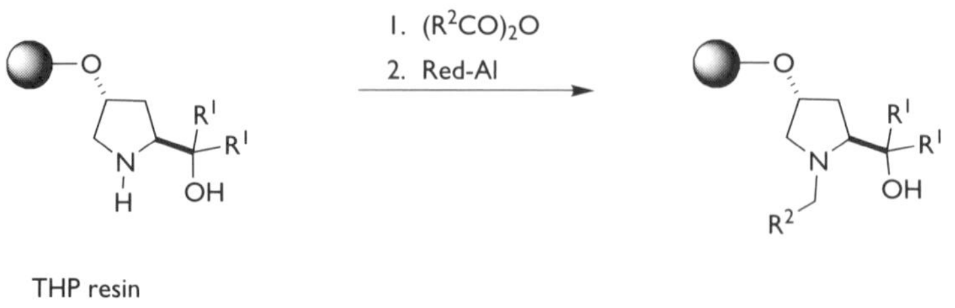

1. $(R^2CO)_2O$
2. Red-Al

THP resin

Points of Interest

1. Pyrrolidinemethanol ligands (shown in the reaction scheme) were prepared on support, cleaved from the resin, and then examined as asymmetric catalysts for the addition of diethylzinc to aldehydes.

2. The support-bound secondary amine was acylated to afford a secondary amide and then reduced with Red-Al to the corresponding tertiary amine.

Literature Procedure

To the support-bound secondary amine (1.0 g, 0.40 mmol) was added pyridine (40 mL) followed by the appropriate anhydride (4.0 mmol, 10 equiv). The mixture was stirred at room temperature for 3 h. The solvent and excess reagent were filtered away. The resin was rinsed with DCM (8×) and toluene (4×) and dried *in vacuo*. To the resulting dry resin was added toluene (40 mL) followed by the addition of 65 wt % Red-Al in toluene (8.4 mL, 28 mmol). The slurry was stirred at room temperature overnight. The solution was filtered and the resin was rinsed with toluene (5×) and DCM (5×).

4.6.8 Reduction of Support-Bound Ketones to Alcohols on PEG–PS [16]

Points of Interest

1. Support-bound ketones were converted to alcohols or dithioacetals.

2. Dihydrobenzopyrans were prepared from the condensation of ketones with hydroxyacetophenones attached to PEG–PS via a photolabile linker (see p 194).

3. The authors report that in general the dihydrobenzopyrans were isolated with purities of >80%; however, the exact number of compounds and the structures of the compounds were not provided.

Literature Procedure

The ketones were reduced with $NaBH_4$, converted into the corresponding dithiolanes with ethane-1,2-dithiol, or left unaltered to produce a library of 1143 distinct compounds.

4.6.9 Reduction of a Support-Bound Nitro Group to an Aniline [17]

Points of Interest

1. A library of phenol derivatives was generated based upon the structures of the known kinase inhibitors lavendustin A and balanol.

2. Nitro reduction was accomplished with $SnCl_2$ in DMF/H_2O, and the resulting aniline was coupled with an acid chloride or isocyanate to introduce functional diversity.

3. $SnCl_2$ (in MeOH, reflux, 3 h) has also been employed to reduce a nitro group to the corresponding aniline in the solid-phase synthesis of 1,4-benzodi-azepine-2,5-diones (see p 152).[18,19]

Literature Procedure

Nitro groups were reduced with 10 equiv of 2 M $SnCl_2 \cdot 2H_2O$ in DMF at 25°C for 15 h.

4.6.10   Reduction of a Support-Bound Azide to an Amine[20]

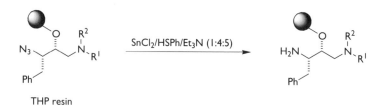

THP resin

Points of Interest

1. The azide reduction was performed in the context of generating libraries of aspartic acid protease inhibitors.

2. The support-bound azide was reduced with thiophenol/Et_3N/$SnCl_2$ (4:5:1) as described by Bartra and coworkers.[21]

3. The reduction was relatively rapid (<4 h at room temperature), and the reaction progress was easily monitored by IR by following the disappearance of the azide stretch.

4. Aryl azides were reduced with Bu_3P to give the corresponding imino-phosphoranes in the solid-phase synthesis of 1,4-benzodiazepine-2,5-diones (see p 152).[22]

Representative Examples

Good yields (47–86%) were obtained of molecules incorporating the (hydroxyethyl)amine and (hydroxyethyl)urea isosteres after four to six chemical transformations (including the azide reduction) on solid support. Many of the functional groups and structures that are present in known inhibitors of HIV-1 protease and renin were incorporated to demonstrate the generality of the synthesis sequence.

Literature Procedure

To the support-bound azide solvated in THF were added $SnCl_2$, thiophenol, and triethylamine to provide a solution that is 0.2, 0.8, and 1.0 M in the reagents, respectively. After 5 h the solution was removed by filtration cannula, and the resin was washed with DMF (3×) and DCM (3×).

4.6.11 Solid-Phase Synthesis of (RS)-1-Aminophosphinic Acids[23]

Points of Interest

1. The resin-bound (RS)-1-aminophosphinic acids are suitable for further elaboration and have the potential for incorporation into other combinatorial libraries.

2. The solid-phase synthesis of 1-aminophosphinic acids exploits the facile procedure for the synthesis of resin-bound imines (see p 135).[24] In certain cases, improved yields were obtained if trimethyl orthoformate was used as a solvent instead of DMF for imine formation.[25]

Representative Examples

Aliphatic, aromatic, and heteroaromatic aldehydes were incorporated into 1-aminophosphinic acids in 71–100% yield (after cleavage and diethyl ether trituration).

Literature Summary

Fmoc-9-amino-xanthen-3-yloxymethylpolystyrene (150 mg, 0.05 mmol) was treated with 20% piperidine in DMF. The resin was washed extensively with DMF, and an appropriate aldehyde (0.2 mmol) and glacial AcOH (1% v/v) were added to the resin in DMF. The resin suspension was then gently agitated for 30 min and washed with DMF. The aldehyde/glacial AcOH treatment was performed twice to ensure complete imine formation. A preformed solution of bis(trimethylsilyl)phosphonite (BTSP)[26] in dry DCM (1.0 M, 1 mL) was then added to a suspension of the derivatized resin in minimal DMF. After the resulting mixture was stirred gently overnight, the solution was removed and the resin was treated with MeOH (10 mL) and washed with DCM (25 mL) to yield the support-bound aminophosphinic acids. Cleavage with TFA:DCM:i-Pr$_3$SiH (50:50:2 v/v, 10 mL) for 30 min afforded the expected product after filtration of the resin. Trituration of the residual material with diethyl ether afforded the desired 1-aminophosphinic acids as crystalline solids.

4.6.12   Additional Studies of Reductions on Solid Support

1. Dihydroisoquinolines have been reduced to tetrahydroisoquinolines on solid support with NaCNBH$_3$.[27]

2. DIBAL has been used in the reductive cleavage of support-bound esters to alcohols.[28]

3. NaBH$_4$ in THF has been used for the reduction of support-bound acid chlorides to alcohols.[29]

4. "Libraries from libraries" have been produced by the reduction of support-bound peptides to polyamines.[30]

5. Rhodium complexes with chiral diphosphine ligands were used as catalysts for the asymmetric hydrogenation of dehydrotripeptides linked to Wang resin.[31]

6. Reductive amination of support-bound ketones has been reported in an application of combinatorial chemistry to the discovery of matrix metalloproteinase inhibitors.[32]

REFERENCES FOR SECTION 4.6

1. Bray, A. M.; Chiefari, D. S.; Valerio, R. M.; Maeji, N. J. *Tetrahedron Lett.* **1995**, *36*, 5081–5084.
2. Gordon, D. W.; Steele, J. *Bioorg. Med. Chem. Lett.* **1995**, *5*, 47–50.
3. (a) Coy, D. H.; Hocart, S. J.; Sasaki, Y. *Tetrahedron* **1988**, *44*, 835. (b) Hocart, S. J.; Nekola, M. V.; Coy, D. H. *J. Med. Chem.* **1987**, *30*, 739. (c) Tourwé, D.; Piron, J.; Defreyn, P.; Van Binst, G. *Tetrahedron Lett.* **1993**, *34*, 5499.
4. Chan, W. C.; Mellor, S. L. *J. Chem. Soc., Chem. Commun.* **1995**, 1475–1477.
5. Rink, H. *Tetrahedron Lett.* **1987**, *28*, 3787–3790.
6. Albericio, F.; Kneib-Cordonier, N.; Biancalana, S.; Gera, L.; Masada, R. I.; Hudson, D.; Barany, G. *J. Org. Chem.* **1990**, *55*, 3730–3743.
7. Carpino, L. A. *J. Am. Chem. Soc.* **1993**, *115*, 4397–4398.
8. Carpino, L. A.; El-Faham, A.; Minor, C. A.; Albericio, F. *J. Chem. Soc., Chem. Commun.* **1994**, 201–203.
9. Boojamra, C. G.; Burrow, K. M.; Ellman, J. A. *J. Org. Chem.* **1995**, *60*, 5742–5743.
10. Feinberg, R. S.; Merrifield, R. B. *Tetrahedron* **1974**, *30*, 3209–3212.
11. Landi, J.; Ramig, K. *Synth. Commun.* **1991**, *21*, 167–171.
12. Green, J. *J. Org. Chem.* **1995**, *60*, 4287–4290.
13. Khan, N. M.; Arumugam, V.; Balasubramanian, S. *Tetrahedron Lett.* **1996**, *37*, 4819–4822.
14. (a) Pelter, A.; Rosser, R. M. *J. Chem. Soc., Perkin Trans. 1* **1984**, 717–720. (b) Bomann, M. D.; Guch, I. C.; Dimare, M. *J. Org. Chem.* **1995**, *60*, 5995–5996. (c) Moormann, A. E. *Synth. Commun.* **1993**, *23*, 789–795.
15. Liu, G.; Ellman, J. A. *J. Org. Chem.* **1995**, *60*, 7712–7713.

16. Burbaum, J. J.; Ohlmeyer, M. H. J.; Reader, J. C.; Henderson, I.; Dillard, L. W.; Li, G.; Randle, T. L.; Sigal, N. H.; Chelsky, D.; Baldwin, J. J. *Proc. Natl. Acad. Sci. U.S.A.* **1995**, *92*, 6027–6031.

17. Meyers, H. V.; Dilley, G. J.; Durgin, T. L.; Powers, T. S.; Winssinger, N. A.; Zhu, H.; Pavia, M. R. *Mol. Diversity* **1995**, *1*, 13–20.

18. Golf, D. A.; Zuckermann, R. N. *J. Org. Chem.* **1995**, *60*, 5744–5745.

19. Mayer, J. P.; Zhang, J.; Bjergarde, K.; Lenz, D. M.; Gaudino, J. J. *Tetrahedron Lett.* **1996**, *37*, 8081–8084.

20. Kick, E. K.; Ellman, J. A. *J. Med. Chem.* **1995**, *38*, 1427–1430.

21. Bartra, M.; Romea, P.; Urpi, F.; Vilarrasa, J. *Tetrahedron* **1990**, *46*, 587–594.

22. Goff, D. A.; Zuckermann, R. N. *J. Org. Chem.* **1995**, *60*, 5744–5745.

23. Boyd, E. A.; Chan, W. C.; Loh, V. M. *Tetrahedron Lett.* **1996**, *37*, 1647–1650.

24. Chan, W. C.; Mellor, S. L. *J. Chem. Soc., Chem. Commun.* **1995**, 1475–1477.

25. Look, G. C.; Murphy, M. M.; Campbell, D. A.; Gallop, M. A. *Tetrahedron Lett.* **1995**, *36*, 2937–2940.

26. (a) Boyd, E. A.; Regan, A. C.; James, K. *Tetrahedron Lett.* **1992**, *33*, 813–816. (b) Boyd, E. A.; Regan, A. C.; James, K. *Tetrahedron Lett.* **1994**, *35*, 4223–4226.

27. Meutermans, W. D. F.; Alewood, P. F. *Tetrahedron Lett.* **1995**, *36*, 7709–7712.

28. (a) Ley, S. V.; Mynett, D. M.; Koot, W. J. *Synlett* **1995**, 1017–1020. (b) Kurth, M. J.; Ahlberg Randall, L. A.; Chen, C.; Melander, C.; Miller, R. B.; McAlister, K.; Reitz, G.; Kang, R.; Nakatsu, T.; Green, C. *J. Org. Chem.* **1994**, *59*, 5862–5864.

29. Goldwasser, J. M.; Leznoff, C. C. *Can. J. Chem.* **1978**, *56*, 1562–1568.

30. Cuervo, J. H.; Weitl, F.; Ostretch, J. M.; Hamashin, V. T.; Hannah, A. L.; Houghten, R. A. In *Peptides 1994: Proceedings of the 23rd European Peptide Symposium*; Maia, H. L. S., Ed.; Escom: Leiden, 1995; pp 465–466.

31. Ojima, I.; Tsai, C.-Y.; Zhang, Z. *Tetrahedron Lett.* **1994**, *35*, 5785–5788.

32. Rockwell, A.; Melden, M.; Copeland, R. A.; Hardman, K.; Decicco, C. P.; DeGrado, W. F. *J. Am. Chem. Soc.* **1996**, *118*, 10337–10338.

4.7 PREPARATION OF HETEROCYCLIC COMPOUNDS

The solid-phase synthesis of heterocyclic compounds, in particular, has become a focal point of combinatorial research due to the value of heterocyclic libraries in the drug discovery process.[1] Heterocyclic compounds are synonymous with natural products in organic chemistry, including a number of fundamental biologically important molecules such as porphyrins, nucleic acids, aromatic amino acids, and the ergot alkaloids. Many modern pharmaceuticals are totally synthetic compounds, and a large portion are heterocyclic.[2] Familiar examples include the widely prescribed β-lactams, quinolines, and 1,4-benzodiazepines. Consequently, the combinatorial solid-phase synthesis of heterocyclic compounds has become a major focus of drug discovery efforts.

The solid-phase syntheses of 1,4-benzodiazepines and β-lactams are briefly discussed to illustrate some of the differences between solid-phase and solution synthesis of heterocycles. For example, in the classical preparation of 1,4-benzodiazepines in solution, the most acidic site is deprotonated with approximately 1 equiv of a strong base for selective anilide alkylation. In contrast, for the solid-phase alkylation, a large excess of a base (lithiated oxazolidinone, DMSO, pK_a-20.5) was employed because its intermediate pK_a value allows for the selective deprotonation of the anilide in the presence of ester- and amide-protected amino acid side chains.[3]

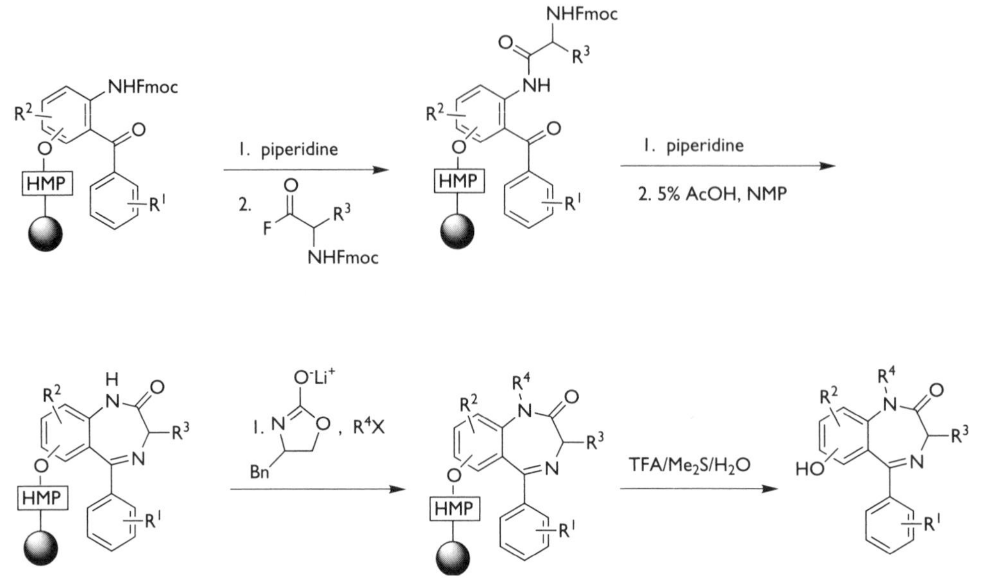

The solid-phase synthesis of 1,4-benzodiazepines. Lithiated oxazolidinone was employed for selective anilide alkylation.

The Staudinger reaction involving highly reactive ketenes was employed in the solid-phase synthesis of β-lactams.[4] The [2 + 2] cyclization was performed with the imine on support and a large excess of ketene in solution. This has the advantage of facile isolation of the product β-lactam from unreacted reagents and reaction byproducts (diketene etc.) by filtration.

In addition to known pharmacophores, novel classes of heterocyclic compounds, such as β-turn mimetics, have also been prepared by solid-phase synthesis.

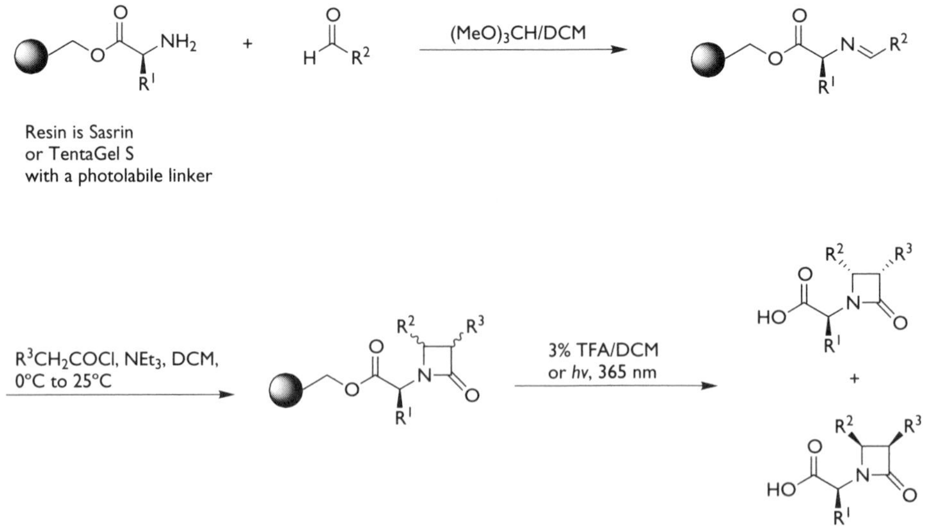

Resin is Sasrin
or TentaGel S
with a photolabile linker

The solid phase synthesis of beta-lactams.

REFERENCES

1. For an overview with leading references on heterocyclic chemistry, see: Gilchrist, T. L. *Hetero-cyclic Chemistry;* Longman Scientific & Technical-John Wiley & Sons: New York, 1985.
2. Grayson, M., Ed. *Antibiotics, Chemotherapeutics, and Antibacterial Agents;* Wiley-Interscience: New York, 1982.
3. Bunin, B. A.; Ellman, J. A. *J. Am. Chem. Soc.* **1992,** *114,* 10997–10998.
4. Ruhland, B.; Bhandari, A.; Gordon, E. M.; Gallop, M. A. *J. Am. Chem. Soc.* **1996,** *118,* 253–254.

4.7.1   Synthesis of 1,4-Benzodiazepines from Support-Bound 2-Aminobenzophenones or Aminoarylstannanes [1,2]

Note: Hydroxyl- and carboxy-substituted 2-*N*-Fmoc-aminobenzophenones were coupled to the [4-(hydroxymethyl)phenoxy]acetic acid (HMP) linker employing solution chemistry. The linker-derivatized aminobenzophenones were then coupled to the solid support by employing standard amide bond-forming methods (the linker may be attached to either ring of the 2-aminobenzophenone).

Points of Interest

1. Derivatives of 1,4-benzodiazepines have widespread biological activities and are one of the most prescribed classes of bioavailable therapeutic agents.[3]

2. The solid-phase synthesis of 1,4-benzodiazepines was the earliest application of combinatorial synthesis to small organic molecules.

3. Three spatially separate 1,4-benzodiazepine libraries of increasing size and diversity were constructed.[4–6]

4. Acylation of the electron-poor aniline was unsuccessful with carbodiimides, pentafluorophenyl esters, or PyBOP; however, acylation was quantitative with Fmoc-amino acid fluorides prepared according to the procedure of Carpino.[7] Acylation of the electron-poor aniline was also successful with mixed carbonic anhydrides using the procedure described by Bodanszky and Bodanszky.[8]

5. Employing lithiated 5-benzyl-2-oxazolidinone as a base allows selective alkylation of the anilide in the presence of protected amino acid functionality such as *tert*-butyl esters and carbamates. No racemization was detected under these alkylation conditions.

Representative Examples

The synthetic methodology provides high yields of fully purified and characterized 1,4-benzodiazepines (average >90% isolated yield relative to 2-aminobenzophenone loading levels). A range of amino acids with protected functionality including amines, carboxylic acids, amides, phenols, indoles, and alcohols were incorporated into 1,4-benzodiazepines. Alkylation is compatible with aromatic and primary aliphatic alkylating agents. The number of commercially available and functionalized 2-aminobenzophenones is limited; therefore, they were initially prepared in solution. The method developed by Frye and co-workers is a particularly efficient and reliable route to functionalized 2-aminobenzophenones.[9] As shown in the following, a solid-phase synthesis of 2-aminobenzophenones and 2-aminoacetophenones was developed.

Literature Procedure

Fmoc-protected 2-aminobenzophenones were derivatized with the (hydroxymethyl)phenoxy (HMP) linker in solution and attached to commercially available (aminomethyl)polystyrene[10] (see p 14) via standard amide bond-forming reaction conditions (DIC, HOBt in DMF). Alternatively, Bpoc-protected 2-aminobenzophenones or 2-aminoacetophenones were prepared on support as described in the following. In either case, after removal of the appropriate protecting group the support-bound aniline was acylated as follows. The support-bound 2-aminobenzophenone (0.17 mmol) was added to a 50-mL round-bottom flask fitted with a stir bar, followed by addition of a DCM solution that was 0.2 M in Fmoc-protected amino acid fluoride and 0.2 M in 2,6-di-*tert*-butyl-4-methylpyridine (at least eightfold excess relative to the molar amount of support-bound 2-aminobenzophenone). After the resulting slurry was stirred for 24 h at room temperature, the solution was removed by filtration, and the support-bound anilide was washed with DCM (3×) and DMF (3×). The Fmoc protecting group was removed with 15 mL of 20% piperidine in DMF for 1 min, the resin was filtered, and the procedure was repeated with a fresh solution of 20% piperidine in DMF for 20–30 min at room temperature. The solvent was removed by filtration and the support was rinsed with DMF (3×) and DCM (3×). The support-bound acyclic intermediate was solvated with 25 mL of 5% acetic acid in DMF, butanol, or NMP and the slurry was stirred at 60°C for 12 h to provide the support-bound unalkylated 1,4-benzodiazepine. The solvent was removed and the support was rinsed with DMF (3×), MeOH (2×), DCM (3×), and THF (3×). The reaction flask was sealed with a fresh rubber septum (to remove residual acetic acid from the cyclization which can quench the deprotonated 1,4-benzodiazepine and lower the conversion during the alkylation step), flushed with nitrogen, and cooled to 0°C. A solution of freshly prepared lithium benzyloxazolidinone in THF (0.1 M, 10 equiv) was transferred by cannula to the solid support with stirring at 0°C. The resulting slurry was stirred for 1.5 h, at which point 15 molar equiv of the appropriate alkylating agent was added, followed by anhydrous DMF to reach a final solvent ratio of ca. 70:30 THF/DMF. The slurry was allowed to warm to room temperature. After 12 h of stirring, the solvent was removed by filtration. The support was then washed with THF, THF/H_2O (2×), MeOH, THF, and DCM (2×). To the fully derivatized 1,4-benzodiazepine on support was added 15 mL of 85:5:10 TFA/H_2O/DMS and the slurry was stirred for 12–24 h. The cleavage solution was removed by filtration and the resin was rinsed with 20 mL of MeOH/DCM (3×). Concentration of the combined filtrates then provided the unpurified product (generally a single compound by TLC), which was purified by silica gel chromatography and fully characterized.

The solid-phase synthesis of 2-aminobenzophenones and 2-aminoacetophenones

Points of Interest

1. The stannane is attached to support via an amide bond formed with a cyanomethyl ester. This prevents destannylation that can occur in the presence of a free acid or HOBt. The support-bound stannane allows a range of acid chlorides to be used as building blocks in solution. The methodology is particularly useful because there are a limited number of commercially available and appropriately functionalized 2-aminobenzophenones.

2. The solid-phase Stille coupling increases the amount of diversity that can be displayed around the 1,4-benzodiazepine scaffold.

3. The support-bound Stille coupling should be done in the presence of Hunig's base to prevent protiodestannylation.

4. The protected aminoarylstannane was prepared in solution; this allows the incorporation of a range of commercially available acid chloride in the Stille coupling. Although multigram quantities have been prepared, the synthesis is relatively lengthy for a building block in a library. The Mpoc aminoarylstannane provides intermediates in solution that are easier to purify than the Bpoc aminoarylstannane initially employed.[11]

5. Benzodiazepines have also been prepared on support with a "traceless linker" using a silicon or germanium linker (see p 66).[12]

Representative Examples

Aliphatic, heteroaromatic, electron-rich, electron-poor, and ortho-substituted acid chlorides were incorporated into 1,4-benzodiazepines.

Literature Procedure

In a dry Schlenk flask under nitrogen a slurry of the protected aminoaryl-stannane cyanomethyl ester (1.50 g, 2.10 mmol), DMAP (256 mg, 2.10 mmol), DIEA (365 μL, 3.15 mmol), and (aminomethyl)polystyrene (2.00 g, 1.2 mmol, 0.60 mequiv/g) in NMP (8 mL) was heated at 65°C for 12 h, at which point the ninhydrin test indicated that the reaction had gone to >90% completion. The resin was transferred to a peptide flask, rinsed with ethyl acetate (3×) and DCM (3×), and then dried under vacuum. The resin was calculated to have a loading of 0.43 mequiv/g assuming the coupling reaction went to completion. The general procedure for the Stille coupling follows. To a dry Schlenk flask under nitrogen were added the derivatized resin (0.40 g, 0.17 mmol), Pd$_2$(dba)$_3$·CHCl$_3$ (60 mg, 0.06 mmol), potassium carbonate (10 mg), THF (4.0 mL), and DIEA (40 μL, 0.20 mmol). The resin was stirred for 3 min to ensure complete solvation, at which point the acid chloride (1.00 mmol) was added slowly. The reaction was allowed to proceed for 1 h, after which the mixture was transferred to a peptide flask and rinsed with DCM (3×), DMSO saturated with KCN (to remove Pd from the resin), MeOH, water, MeOH (3×), and DCM (3×). The Bpoc (or Mpoc) group was then removed by treating the resin with 97:3 DCM/TFA (99:1 is sufficient for cleavage) for 5 min. The resin was rinsed with DCM (2×), and the acid treatment was repeated to afford support-bound 2-amino-benzophenone or 2-aminoacetophenone. The resin was rinsed with DCM (5×) and MeOH (5×) and dried *in vacuo*. The rest of the 1,4-benzodiazepine synthesis—acylation, cyclization, alkylation, and cleavage—proceeded exactly as described above.

4.7.2 Solid-Phase Synthesis of 1,4-Benzodiazepines from Support-Bound Amino Esters [13]

Points of Interest

1. The cleavage by cyclization was designed to enhance the purity of the product, since incomplete products would be expected to stay bound to resin.
2. Three out of the four variable positions around the 1,4-benzodiazepine scaffold were incorporated into the imine building block.
3. In addition to the synthesis, the construction of a Diversomer apparatus for producing 40 reactions simultaneously was described. An eight-unit array is now commercially available, and larger arrays are in the process of becoming commercially available from ChemGlass.[14]
4. The solid-phase synthesis of 40 hydantoins was also described.

Representative Examples

Electron-rich and electron-poor imines were incorporated as well as imines derived from 2-amino-9-fluorenone. Five different amino acids were examined, including valine and tryptophan. A library of 1,4-benzodiazepines incorporating five amino acids and eight 2-aminobenzophenones or 2-amino-9-fluorenone was synthesized. 1,4-Benzodiazepine derivatives were purified by extraction and isolated in crude mass balances of 9–63% (average 34%, 40 compounds) in >90% purity as measured by ^{1}H NMR.

Literature Procedure

Forty discrete benzodiazepines were synthesized by treating each of five amino acid resins with eight 2-aminobenzophenone imines. The two-step synthesis created a 40-unit two-dimensional array. Five Boc-protected amino acid-derivatized Merrifield resins (alanine, glycine, phenylalanine, tryptophan, and valine) were deprotected in bulk (1–5 g) using TFA/DCM (1:1) at room temperature overnight. The resins were washed with dioxane and DCM, dried under vacuum, and used directly as the TFA salts. Approximately 99–107 mg of each of the five amino acid Merrifield resins was measured into 40 pins (the Diversomer apparatus). Each of the resins was swollen with 3 mL of DCM. To form the resin-bound imines, the pins were submerged in reaction wells containing 3 mL of the appropriate 2-aminobenzophenone imines (3–6 equiv) in dichloroethane and heated at 60°C for 24 h. The pin array was then drained and washed by repeatedly dispensing 4-mL portions of DCM through the aperture at the top of each pin 12 times, until the washes were no longer colored. To cyclize and cleave the benzodiazepines from the solid support, the pins were submerged in reaction wells containing 3 mL of TFA and heated for 20 h at 60°C. To isolate and purify the final products, the pins were drained and then extracted by repeatedly dispensing 2-mL portions of DCM through the aperture at the top of each pin three times. The combined washes and reaction filtrates were evaporated under a stream of nitrogen. A simple two-phase extraction procedure was implemented by using a Tecan 5032 processor. The residues from evaporation were dissolved in 3 mL of DCM and mixed with 3 mL of saturated aqueous bicarbonate. The organic phase was withdrawn, and the aqueous layer was extracted twice more with 1.5 mL of DCM. The combined organic extracts were dried with magnesium sulfate, filtered, and concentrated to yield 2–14 mg of the expected 1,4-benzodiazepines in 9–63% (average 34%, 40 compounds) in >90% purity as measured by ^{1}H NMR.

REFERENCES FOR SECTIONS 4.7.1 AND 4.7.2

1. Bunin, B. A.; Ellman, J. A. *J. Am. Chem. Soc.* **1992**, *114*, 10997–10998.
2. Plunkett, M. J.; Ellman, J. A. *J. Am. Chem. Soc.* **1995**, *117*, 3306–3307.
3. Sternbach, L. H. *J. Med. Chem.* **1979**, *22*, 1–11.
4. Bunin, B. A.; Plunkett, M. J.; Ellman, J. A. *Proc. Natl. Acad. Sci. U.S.A.* **1994**, *91* (11), 4708–4712.
5. Bunin, B. A.; Plunkett, M. J.; Ellman, J. A. *New J. Chem.* **1977**, *21*, 125–130.
6. Bunin, B. A.; Plunkett, M. J.; Ellman, J. A. *Methods Enzymol.* **1996**, *276*, 448–465.
7. Carpino, L. A.; Sadat-Aalaee, D.; Chao, H. G.; DeSelms, R. H. *J. Am. Chem. Soc.* **1990**, *112*, 9651–9652.
8. Bodanszky, M.; Bodanszky, A. In *The Practice of Peptide Synthesis*; Springer-Verlag: New York, 1984; pp 109–110.
9. Frye, S. V.; Johnson, M. C.; Valvano, N. L. *J. Org. Chem.* **1991**, *56*, 3750–3753.
10. Zikos, C. C.; Frederigos, N. G. *Tetrahedron Lett.* **1995**, *36*, 3741–3744.
11. Plunkett, M. P.; Ellman, J. A., unpublished results.
12. Plunkett, M. J.; Ellman, J. A. *J. Org. Chem.* **1995**, *60*, 6006–6007.
13. DeWitt, S. H.; Kiely, J. S.; Stankovic, C. J.; Schroeder, M. C.; Cody, D. M. R.; Pavia, M. R. *Proc. Natl. Acad. Sci. U.S.A.* **1993**, *90*, 6909–6913.
14. ChemGlass, Inc., Vineland, NJ 08360.

4.7.3   Preparation of 1,4-Benzodiazepine-2,5-diones from Anthranilic Acids [1]

Points of Interest

1. Amino esters are loaded onto a support-bound aldehyde via a reductive amination. The imines must be reduced immediately upon formation to avoid racemization. Conversely, racemic products can easily be prepared if desired without purchasing L- and D-amino acids.

2. The anthranilic acid building blocks do not require protection due to the reduced nucleophilicity of the aniline relative to the support-bound amine.

3. Deprotonation, cyclization, and alkylation of 1,4-benzodiazepine-2,5-diones are performed in a one-pot reaction. An acyclic tertiary amide precursor was critical to achieve efficient lactamization.

4. Benzodiazepine-2,5-diones have also been prepared in good yield and purity via a cyclic cleavage from solid support.[2]

Representative Examples

Both sterically hindered (Val) and functionalized (Tyr, Lys) amino acids were incorporated into 1,4-benzodiazepine-2,5-diones. Electron-rich and electron-poor anthranilic acids were incorporated, and the incorporation of 7-bromo-anthranilic acid allowed further modification by a Suzuki cross-coupling. Five different aromatic and aliphatic alkylating agents were used. Overall yields of purified 1,4-benzodiazepine-2,5-diones ranged from 62 to 92%.

Literature Procedure

Commercially available Merrifield resin[3] was derivatized with the aldehyde linker under Ar as follows. A solution of 4-hydroxy-2,6-dimethoxybenzalde-hyde[4] (15.3 g, 84.2 mmol) in DMF (800 mL) was degassed with Ar for 30 min. Under positive Ar flow, 95% NaH (1.92 g, 79.9 mmol) was added slowly in 0.5-g aliquots over 15 min. After 30 min, Merrifield resin (31.7 g, 24.1 mmol, loading 0.76 mmol/g) was added and the contents were stirred mechanically for 36 h at 50°C. The DMF suspension was then diluted with MeOH and filtered. The resin was then rinsed with DMF/MeOH (2×), DMF (5×), DCM (7×), and MeOH (4×). The resin was dried *in vacuo* to a constant weight of 33.4 g. A strong aldehyde carbonyl stretch was observed in the IR (KBr) spectrum of this resin: 1690 cm^{-1}.

Amino esters can be loaded onto the resin by reductive amination either with or without racemization under different experimental conditions. For racemization-free loading, an amino methyl ester hydrochloride (2.59 mmol) was dissolved in 1% AcOH in DMF (20 mL) and then NaBH(OAc)$_3$ was added. To this mixture was immediately added 0.700 g (0.518 mmol) of aldehyde-derivatized resin. The reaction was complete at 60 min as determined by the IR spectrum of an aliquot of resin. The bulk of the resin was then rinsed with MeOH (1×), 10% DIEA in DMF (2×), DMF (7×), DCM (7×), and finally MeOH (3×). The resin was dried *in vacuo* to constant weight and characterized by IR spectroscopy. IR (KBr): 1736 cm^{-1}. For racemized benzodiazepine products the aldehyde-derivatized resin was added before the other reagents and 1 equiv of DIEA was added with the amino ester hydrochloride.

The loading of resin-bound amino esters was quantitated by acetylation of the free amine followed by cleavage with subsequent determination of the mass balance of the silica gel-purified product. The loading with different amino esters was consistently 0.36–0.37 mmol/g of resin. 1,4-Benzodiazepine-2,5-diones were synthesized as follows. Amino ester resin (0.500 g, 0.185 mmol) was suspended in NMP (5 mL) and then EDC-HCl (0.425 g, 2.22 mmol) was added. After the solution was saturated in EDC (most will dissolve over the course of 5 min), the anthranilic acid of choice was slowly added (1.85 mmol). It is imperative that the EDC and resin be allowed to mix thoroughly before addition of anthranilic acid and that the anthranilic acid be added slowly. This minimized side reactions that can occur with the unprotected aniline functionality. The reaction mixture was gently stirred for 8–12 h. The resin was then rinsed with DMF (7×), DCM (7×), and finally MeOH (3×). The acylation and rinsing were repeated (2 h is adequate) to ensure reaction completion and the resin was dried to a constant weight.

Cyclization and alkylation were accomplished in one pot: The acylated resin (0.500 g, 0.175 mmol) was placed in a round-bottom flask and purged with Ar for 5 min. The flask was charged with 5 mL of DMF and 5 mL of THF. In a separate flame-dried flask p-methoxyacetanilide or acetaniline (4.44 mmol) was dissolved in THF, the flask was purged with Ar for 5 min, and the solution was cooled to −78°C. n-Butyllithium (1.48 mL, 2.5 M in hexanes, 3.7 mmol) was added dropwise over 10 min and the suspension was mixed for 45 min. Dimethylformamide (5 mL) was then added for solubility, and the mixture was stirred for an additional 15 min and then transferred via cannula into the flask containing the resin. The suspension was stirred gently at room temperature for 30 h under an Ar atmosphere. Alkylating agent (7.40 mmol) was added via syringe and stirring was continued until the suspension no longer turned pH paper dark green or blue (typically 3–6 h). The alkylating solution was removed and the resin was rinsed with DMF (7×), DCM (7×), and MeOH (3×) and dried *in vacuo* to a constant weight. The benzodiazepines were cleaved from support with 20 mL of TFA/Me$_2$S/H$_2$O (90:5:5) for 36 h, and the resin was filtered and rinsed with DCM (4×) and MeOH (2×). The benzodiazepines were purified by silica gel chromatography.

4.7.4   Solid-Phase Synthesis of 1,4-Benzodiazepine-2,5-diones from Support-Bound Peptoids[5]

Points of Interest

1. The methodology developed for solid-phase synthesis of 1,4-benzodiazepine-2,5-diones complements solid-phase peptoid synthesis.

2. Peptoids can be robotically generated. The postmodification strategy is an alternative to preparing monomers in solution through a multistep synthesis. Postmodification of peptoids has been extended to provide isoxazoles and isoxazolines,[6] isoquinolines,[7] and tetrahydroisoquinolines and tetrahydroimidazopyridines.[8]

3. 1,4-Benzodiazepine-2,5-dione derivatives (X = OTf) were further functionalized via support-bound Suzuki couplings (see p 91).[9]

4. 1,4-Benzodiazepine-2,5-dione derivatives (X = NO_2) were also further functionalized by nitro reduction ($SnCl_2$–H_2O, MeOH, reflux 3 h) to give the amino derivative, which was subsequently acylated on resin.

Representative Examples

Amino esters with functional side chains (Tyr, Thr, Ser, Lys, Asp) were incorporated. Anthranilic acids with triflate and nitro functionalities were incorporated; these functional groups expand the diversity of libraries by modification of the aromatic substituents on resin as described above. The yields were modest to good (37–90%), but in many cases the purity of the crude products was excellent.

Literature Procedure

A large batch (3–5 g) of commercially available Rink amide resin[10] (see p 38) was acylated with bromoacetic acid followed by amination with isobutylamine to provide the monopeptoid.[11] Bromoacetylation and displacement with the amino acid methyl or ethyl ester free base in DMSO (room temperature, Ar, on an orbital shaker at 200 rpm for 3 h) gave the support-bound secondary amino ester. The resin was washed with DMF (3×) and DCM (2×) and dried *in vacuo*. The support-bound secondary amino ester was divided into two portions, placed on a Symphony multiple peptide synthesizer (Protein Tech-fjnologies, Inc.), swollen with 1,2-dichloroethane (DCE), drained, and treated 2 × 30 min with a 1.2 M solution of *o*-azidobenzoyl chloride and triethylamine in DCE (4 mL). The azide can be cleaved from support at this stage in the synthesis. Treatment of the support-bound azide for 2 × 30 min with Bu$_3$P in toluene (0.6 M) at room temperature gave the imino phosphorane. This resin (550 mg) was washed with DCE and DCM and then (after the mixture was briefly evacuated) heated to 130°C for 5–7 h in a Schlenk tube with 10 mL of *p*-xylene to give the benzodiazepinedione. The vessels were cooled to room temperature and the resin was filtered and washed with DCM, followed by cleavage with TFA/H_2O (95:5) for 20 min. The resin was filtered and washed with acetic acid. The filtrate was diluted with water and lyophilized twice from glacial acetic acid.

REFERENCES FOR SECTIONS 4.7.3 AND 4.7.4

1. Boojamra, C. G.; Burrow, K. M.; Ellman, J. A. *J. Org. Chem.* **1995**, *60*, 5742–5743.
2. Mayer, J. P.; Zhang, J.; Bjergarde, K.; Lenz, D. M.; Gaudino, J. J. *Tetrahedron Lett.* **1996**, *37*, 8081–8084.
3. Feinberg, R. S.; Merrifield, R. B. *Tetrahedron* **1974**, *30*, 3209–3212.
4. Landi, J.; Ramig, K. *Synth. Commun.* **1991**, *21*, 167–171.
5. Goff, D. A.; Zuckermann, R. N. *J. Org. Chem.* **1995**, *60*, 5744–5745.
6. Pei, Y.; Moos, W. H. *Tetrahedron Lett.* **1994**, *35*, 5825–5828.

7. Goff, D. A.; Zuckermann, R. N. *J. Org. Chem.* **1995**, *60*, 5748–5749.

8. Hutchins, S. M.; Chapman, K. T. *Tetrahedron Lett.* **1996**, *37*, 4865–4868.

9. Deshpande, M. S. *Tetrahedron Lett.* **1994**, *31*, 5613–5614.

10. Rink, H. *Tetrahedron Lett.* **1987**, *28*, 3787–3790.

11. Zuckermann, R. N.; Kerr, J. M.; Kent, S. B. H.; Moos, W. H. *J. Am. Chem. Soc.* **1992**, *114*, 10646–10647.

4.7.5   Solid-Phase Synthesis of β-Turn Mimetics Incorporating Side Chains[1]

p-nitrophenylalanine
derivatized Rink resin

Points of Interest

1. Both $i + 1$ and $i + 2$ side chains were incorporated into nine- and ten-membered cyclic β-turn mimetics on solid support. The flexibility of the turn mimetics and the relative orientations of the side chains can be varied by either the ring size or the stereocenters introduced by the α-halo acid or amino acid.

2. For all of the turn mimetics synthesized, cyclization provided the desired cyclic monomer with no cyclic dimer detected (<5%). Also, <5% of the epimer resulting from racemization at the α-bromo site was observed.

Representative Examples

A range of functionalized and sterically hindered side chains, including phenols, amines, carboxylic acids, and isopropyl groups, were incorporated into β-turn mimetics. β-Turn mimetics were synthesized on Rink-derivatized PEG–PS resin and Chiron Mimotope pins in 59–93% purity by HPLC. *p*-Nitrophenylalanine was introduced in the first step as a UV tag to ensure that all products (and impurities) would be detected by UV-HPLC.

Literature Procedures

The β-turn mimetics were synthesized on a poly(ethylene glycol)–polystyrene (PEG–PS) graft copolymer (Rapp TentaGel) with a loading level of 0.28 mequiv/g. All reaction sequences were performed in peptide flasks except for the final cleavage, which was performed in a round-bottom flask. The synthesis of the β-turn mimetics proceeded from a single batch of resin that was separated into portions of approximately 500 mg of resin per mimetic as the syntheses diverged. All reactions employed at least 2 and typically 3–5 equiv of reagent relative to the resin loading level. To ensure efficient resin mixing during rinses or during reaction steps, nitrogen was gently bubbled through the mixture in a peptide flask. For reaction times greater than 1 h, the peptide flasks were stoppered and shaken at low speed on a Fisher wrist action shaker. A volume of 10 mL of solvent per gram of resin was used for each rinse, which lasted approximately 1 min. Reaction completion was monitored by the Kaiser ninhydrin test[2] for free amines and the Ellman test[3] for free thiols where applicable. Removal of Fmoc protecting groups was generally accomplished with 20% piperidine in DMF (10 mL/g of resin) for 1 min and then 20 min followed by rinses with DMF (4×) and DCM (5×).

β-Turn Synthesis.

The dry support (aminomethyl-PEG–PS) was washed with DCM (2×) before the linker was attached. Addition of a DMF solution that was 0.2 M in acid-labile Rink amide linker, HOBt, and DIC resulted in complete derivatization of the resin. The reagents were then drained, and the resin was rinsed with DMF (4×) and DCM (5×). The Fmoc protecting group on the Rink linker was removed under standard conditions, and a premixed DMF solution that was 0.2 M in both Fmoc-(pNO$_2$)-Phe-Opfp and HOBt was added to introduce the UV tag. After 6 h, the reaction mixture was drained, and the resin was rinsed with DMF (4×) and DCM (5×). The resin was deprotected under standard conditions to yield the free amine. The support-bound amine was treated with a DMF solution that was 0.2 M in bromoacetic acid and 0.22 M (1.1 equiv) in DCC for 8 h. The resin was then rinsed with DMF (4×) and DCM (5×) and an additional DMSO rinse was performed immediately prior to the next step. The backbone element was introduced by treating the support-bound bromide with a DMSO solution 0.25–0.50 M in the appropriate aminothiol $tert$-butyl disulfide and 0.27–0.55 M (1.1 equiv) in 1,1,3,3-tetramethylguanidine (TMG) for 24 h. The resin was then rinsed with DMF (4×), DCM (5×), and MeOH (2×). The Fmoc-protected intermediate was deprotected under standard conditions. Meanwhile, a solution of the symmetric preformed anhydride (corresponding to the side chain at the $i + 1$ position) was prepared in DCM by reacting the α-bromo acid (0.4 M) with 0.5 equiv of DIC. After 15 min, the solution of anhydride in DCM was filtered to remove the precipitated urea followed by dilution with an equal volume of DMF to afford a 0.1 M solution of the anhydride in a 1:1 DCM/DMF mixture. The anhydride solution was then added to the support and the mixture was gently shaken for 8 h. The resin was then rinsed with DMF (4×), DCM (5×), and MeOH (2×). The $tert$-butyl disulfide resin was solvated in a 5:3:2 i-PrOH/DMF/H$_2$O solvent comixture (10 mL/g of resin) that had been purged with Ar for 20 min.

Bu$_3$P (250 μL/10 mL of solvent) was then added. After 1–3 h, the reaction was drained and the resin was washed with DCM (4×) and MeOH (2×). The resin-bound free thiol was solvated in THF (10 mL/g of resin) that had been purged with Ar for 20 min. TMG (250 μL/10 mL of solvent) was then added. After 8 h of gentle shaking, the resin was rinsed with 1% AcOH in DMF, DMF (3×), and MeOH (3×) and then dried *in vacuo*. The β-turn mimetics were liberated from support as primary amides by treatment of the resin with a 90:5:5 mixture of TFA/H$_2$O/Me$_2$S (10 mL/g of resin). The combined cleavage and rinse solutions were concentrated and the crude product was dissolved in 500 μL of DMSO. The purity of the β-turn mimetics was then ascertained by reverse phase HPLC analysis (270 nm). The β-turn mimetics were purified by preparative HPLC prior to analytical characterization.

REFERENCES FOR SECTION 4.7.5

1. Virgilio, A. A.; Ellman, J. A. *J. Am. Chem. Soc.* **1994**, *116*, 11580–11581.
2. Kaiser, E.; Colescott, R. L.; Bossinger, C. D.; Cook, P. I. *Anal. Biochem.* **1970**, *34*, 595–598.
3. Ellman, G. L. *Arch. Biochem. Biophys.* **1959**, *82*, 70–77.

4.7.6   Expedient Solid-Phase Synthesis of Second Generation β-Turn Mimetics Incorporating the *i* + 1, *i* + 2, and *i* + 3 Side Chains[1]

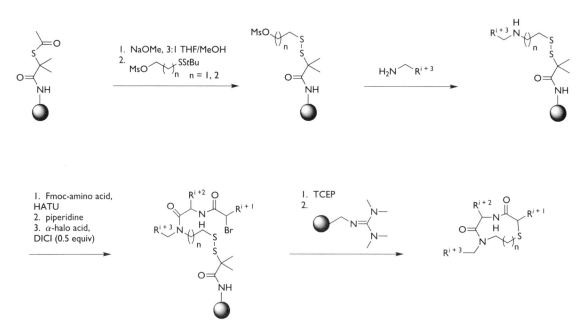

Points of Interest

1. Second-generation mimetics were designed to possess improved binding affinity, solubility, and perhaps bioavailability by incorporating the *i* + 3 side chain and by eliminating the primary amide functionality.

2. The second-generation mimetics involved two major synthetic modifications. First, side chain functionality at the $i + 3$ position was introduced by means of a primary amine. Second, attachment to the solid support during the solid-phase synthesis sequence was accomplished using a disulfide linkage (see p 74) rather than the Rink amide linkage (see p 38). When the final cyclic product is formed, no trace of the linkage to support remains.

Representative Examples

Eight second-generation turn mimetics were prepared to demonstrate the generality of the synthetic sequence. A range of aromatic and aliphatic primary amines were incorporated, in addition to functionalized amino acids and α-bromo acids. High levels of purity ($>90\%$) as determined by HPLC and ^{1}H NMR were observed for each of the crude mimetics obtained directly from the resin. In addition, the average overall yield for the chromatographed and analytically pure material was 55% based upon the expected loading.

Literature Summary

To a slurry of (aminomethyl)polystyrene resin in DMF (10 mL/g of resin) were added S-acetyl 2-mercapto-2-methylpropionic acid (0.2 M), PyBOP (0.2 M), HOBt (0.2 M), and i-Pr$_2$EtN (0.4 M). After 8 h, the resin was filtered, rinsed with DMF ($3\times$), DCM ($4\times$), and MeOH ($2\times$), and then dried *in vacuo*. The derivatized resin was solvated in a 3:1 THF/MeOH comixture (10 mL/g of resin) and then purged with Ar (20 min). NaOMe (0.2 M final concentration) was added, and the reaction vessel was stoppered and shaken for 1 h. The reaction was quenched with excess AcOH. The resin was filtered and rinsed with 3:1 THF/MeOH ($2\times$) and DCM ($3\times$). A solution of benzothiazolyl-activated disulfide mesylate in DCM (0.1 M) was added to the resin, and the reaction vessel was stoppered under an Ar atmosphere. After 12 h, the resin was filtered and rinsed with DCM ($5\times$) and MeOH ($2\times$). To the support-bound mesylate was added a solution of the desired amine in NMP (1.0 M). The resulting mixture was shaken for 16 h at $50°$C. The resin was filtered and rinsed with DMF ($3\times$), DCM ($5\times$), and MeOH ($2\times$). The resin-bound secondary amine was solvated in DMF (10 mL/g of resin), and the desired Fmoc-protected amino acid (0.2 M), HATU (0.2 M), and i-Pr$_2$EtN (0.4 M) were added. After 8 h, the resin was filtered and rinsed with DMF ($3\times$), DCM ($5\times$), and MeOH ($2\times$). A solution of 20% piperidine in DMF was added to the resin (20 min), and it was rinsed as before. The resin was then solvated in DMF (10 mL/g of resin), and the desired α-halo acid (0.1 M), HOAt (0.1 M), and DIC (0.1 M) were added. After 4 h, the resin was filtered and rinsed as before to afford the support-bound acyclic precursor to the β-turn mimetic. The disulfide linkage to the solid support was reduced with a 4.0 mM solution of TCEP in a 9:1 dioxane/H$_2$O solution (1 mL/mmol of resin) that had been purged with Ar (20 min). After 8 h, the solution was transferred via a filtration cannula under Ar pressure to a flask containing support-bound guanidine (30 equiv). Following the disappearance of all the acyclic material (<24 h), the solution was filtered and the filtrate was concentrated *in vacuo* to afford the second-generation cyclic β-turn mimetics.

REFERENCES FOR SECTION 4.7.6

1. Virgilio, A. A.; Schürer, S. C.; Ellman, J. A. *Tetrahedron Lett.* **1996**, *37*, 6961–6964.

4.7.7   Solid-Phase Synthesis of Hydantoins Using a Carbamate Linker and a Novel Cyclization/Cleavage Step [1]

Points of Interest

1. The carbamate linker was formed in nearly quantitative yield as determined by gel-phase ^{13}C NMR (elemental analysis for nitrogen gave a loading capacity of 0.84 mmol/g). The support-bound carbonate can be stored for at least 6 months without loss of activity. Polymer-bound acylimidazole has been reported as an alternative route to carbamate linkers (see p 58).[2]

2. The zwitterionic amino acids were dissolved with gentle heating in DMF using 2.5 equiv of N,O-bis(trimethylsilyl)acetamide (BSA).

3. The base-catalyzed cyclization was performed substituting acetonitrile for methanol but lower yields and product purities were generally observed. Attempts to use other protic solvents such as isopropyl and ethyl alcohol were unsuccessful.

4. The base-catalyzed cyclization is an improvement on previously described methods for the acid-catalyzed cyclization of hydantoins.[3] The conditions for base-catalyzed cyclization reported by Dressman and coworkers provide higher yields of hydantoins under milder reaction conditions.

5. Gel-phase ^{13}C NMR[4] (see p 220) was used for both pilot experiments and quality control in library production.

6. Hydantoins have also been prepared by solid-phase unnatural peptide synthesis[5] (see p 118) followed by the Parke-Davis cyclization methodology.

Representative Examples

Polar, nonpolar, acyclic, cyclic, geminally disubstituted, N-alkyl, and N-aryl amino acids were incorporated; primary amines, including heteroaromatic amines, and hydroxyamines were incorporated as well. From a library of 800 individual hydantoins, a random analytic sampling (15%) by FD MS showed that in 90% of the cases desired product was obtained. Regardless of the mass balance yields the products were obtained in 67–99% purity by HPLC presumably due to the cyclic cleavage.

Literature Summary

p-Nitrophenyl chloroformate (1.31 g, 6.50 mmol, 2 equiv) was added in one portion to a stirring solution of 3.26 g of (hydroxymethyl)polystyrene (1 mmol/g) and NMM (659 mg, 6.50 mmol, 2 equiv) in DCM at 0°C. The reaction mixture was warmed to room temperature, stirred overnight, filtered, and washed with DCM. Drying overnight in a vacuum oven afforded 3.28 g of a light pink resin. ^{13}C NMR (CDCl$_3$) δ 70.96 (broad s, PCH$_2$O). IR (KBr) 1761 cm^{-1}. Gel-phase ^{13}C NMR spectra were obtained using the procedure of Giralt.[6] Amino acids (4 equiv) were dissolved with light heating in DMF using N,O-bis(trimethylsilyl)acetamide (10 equiv) and then added to the resin-bound activated carbonate in the presence of DMAP (2 equiv). The support-bound carboxylic acid was stirred overnight in the presence of excess primary amine (4 equiv), HOBt–H$_2$O (4 equiv), and DCC (4 equiv). Treatment of the support-bound amide with excess triethylamine (14 equiv) in methanol at temperatures between 55 and 90°C for 48 h provided hydantoins.

4.7.8 Solution- and Solid-Phase Synthesis of 5-Alkoxyhydantoins[7]

Points of Interest

1. *N*-(Benzyloxy)-α-amino esters were prepared in one pot and used as precursors for the preparation of 5-alkoxyhydantoins.

2. A library of 50 discrete 5-alkoxyhydantoins with three functional group variations was prepared in solution (see p 246) and six analogs were prepared on Merrifield resin.

Representative Examples

5-Alkoxyhydantoins were prepared on support (average yield 56%) and in solution (average yield 80%). Each of the 50 5-alkoxyhydantoins prepared in solution was fully characterized by ^{1}H NMR, ^{13}C NMR, and mass spectrometric techniques. The authors reported that the 5-alkoxyhydantoins prepared on support exhibited satisfactory ^{1}H and ^{13}C NMR spectra. HPLC analysis of crude products cleaved from the resin showed purities ranging from 89 to 96%.

Literature Summary

The α-hydroxy acid was first loaded onto a chloromethylated polystyrene resin (1% DVB Merrifield resin, 1 mmol/g) via its cesium salt. The support-bound α-hydroxy ester was converted into the *N*-benzyloxyamino ester with trifluoromethanesulfonic acid anhydride in the presence of lutidine in DCM, followed by the addition of *O*-benzylhydroxylamine. The resulting α-amino acid derivative was subsequently condensed with aryl isocyanates in dichloroethane under reflux to give the support-bound urea. Treatment with potassium *tert*-butoxide in alcoholic solution led to cyclization and cleavage from the resin to afford the desired 5-alkoxyhydantoins in 23–72% yield based on the original chloromethyl loading.

REFERENCES FOR SECTIONS 4.7.7 AND 4.7.8

1. Dressman, B. A.; Spangle, L. A.; Kaldor, S. W. *Tetrahedron Lett.* **1996**, *37*, 937–940.
2. Hauske, J. R.; Dorff, P. *Tetrahedron Lett.* **1995**, *36*, 1589–1592.
3. DeWitt, S. H.; Kiely, J. S.; Stankovic, C. J.; Schroeder, M. C.; Cody, D. M. R.; Pavia, M. R. *Proc. Natl. Acad. Sci. U.S.A.* **1993**, *90*, 6909–6913.
4. Giralt, E.; Rizo, J.; Pedroso, E. *Tetrahedron* **1984**, *40*, 4141–4152.
5. O'Donnell, M. J.; Zhou, C.; Scott, W. L. *J. Am. Chem. Soc.* **1996**, *118*, 6070–6071.
6. Giralt, E.; Rizo, J.; Pedroso, E. *Tetrahedron* **1984**, *40*, 4141–4152.
7. Hanessian, S.; Yang, R. Y. *Tetrahedron Lett.* **1996**, *37*, 5835–5838.

4.7.9 Postmodification of Peptoid Side Chains to Isoxazoles and Isoxazolines: [3 + 2] Cycloaddition of Nitrile Oxides with Alkenes and Alkynes on the Solid Phase [1]

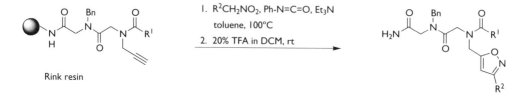

1. $R^2CH_2NO_2$, Ph-N=C=O, Et$_3$N
   toluene, 100°C
2. 20% TFA in DCM, rt

Rink resin

Points of Interest

1. Isoxazoles and isoxazoline were synthesized through [3 + 2] cyclo-addition reactions of highly reactive nitrile oxides with peptoids prepared from propargylamine or allylamine.

2. Peptoids can be robotically generated. The postmodification strategy is an alternative to preparing monomers in solution through a multistep synthesis. Postmodification of peptoids has been extended to provide 1,4-benzodiazepine-2,5-diones,[2] isoquinolines,[3] and tetrahydroisoquinolines and tetrahydroimidaz-opyridines.[4]

3. Leznoff prepared isoxazoles while examining the regioselectivity of cycloadditions on solid support.[5]

4. Isoquinoline- and isoxazoline-containing heterocycles have been prepared on a "traceless" Reissert resin (see p 71).[6] Polyisoxazolines have been prepared by an iterative solid-phase synthesis alternating between [3 + 2] cycloadditions and oxidation/elimination reactions.[7]

Representative Examples

All the products contained an acylated peptoid dimer. Postmodification of peptoid side chains afforded heterocycles with additional aliphatic, aromatic, alcohol, and ester substituents.

Literature Summary

Peptoids prepared on Rink amide resin (acid labile, 0.55 mmol/g) through the submonomer method were acylated with acetic anhydride or with nicotinic acid and isonicotinic acid. Acylation with the pyridine derivatives was achieved using PyBop/HOBt/DIEA. Cycloaddition reactions were carried out either in toluene at 100°C or in DCM/water with NaOCl at room temperature, depending on the precursors of the nitrile oxides. Nitrile oxides were generated *in situ* by the treatment of nitroalkyl compounds with phenyl isocyanate and triethylamine or through the oxidation of oximes with sodium hypochlorite in the presence of triethylamine. The following is a typical experimental procedure for the [2 + 3] cycloaddition reaction of nitrile oxides on the solid phase. To a suspension of dried resin-bound, acylated peptoid incorporating an alkyne (100 mg, 0.055 mmol) in toluene (5 mL) under Ar were added nitroethane (0.040 mL, 0.55 mmol), phenyl isocyanate (0.12 mL, 1.10 mmol) and triethylamine (0.153 mL, 1.10 mmol). The resulting slurry was stirred at 100°C for 5 h. The resin was filtered, washed with DMF (3×) and DCM (3×), and air-dried for 15 min. The product was cleaved from resin with 20% TFA in DCM at room temperature for 30 min. The resin was removed by filtration and the filtrate was collected and concentrated under an argon stream. The residue was lyophilized from acetic acid to afford a hygroscopic white powder (72% crude yield).

REFERENCES FOR SECTION 4.7.9

1. Pei, Y.; Moos, W. H. *Tetrahedron Lett.* **1994,** *35,* 5825–5828.
2. Goff, D. A.; Zuckermann, R. N. *J. Org. Chem.* **1995,** *60,* 5744–5745.
3. Goff, D. A.; Zuckermann, R. N. *J. Org. Chem.* **1995,** *60,* 5748–5749.
4. Hutchins, S. M.; Chapman, K. T. *Tetrahedron Lett.* **1996,** *37,* 4865–4868.

5. Yedidia, V.; Leznoff, C. C. *Can. J. Chem.* **1980**, *58*, 1144–1150.
6. Lorsbach, B. A.; Miller, R. B.; Kurth, M. J. *J. Org. Chem.* **1996**, *61*, 8716–8717.
7. Kurth, M. J.; Ahlberg Randall, L. A.; Takenouchi, K. *J. Org. Chem.* **1996**, *61*, 8755–8761.

4.7.10 Solid-Phase Synthesis of Pyrazoles and Isoxazoles[1]

Rink amide linker

Ar[1] = aromatic spacer

Y = NR[4], O

Points of Interest

1. For the alkylation reaction best results were obtained with TBAF as base to avoid O-alkylation and increase nucleophilicity.
2. Mixtures of regioisomers were obtained from cyclization.
3. The yields for individual steps were high (60–100%).

Representative Examples

Ar[1] = 4-Ph, R[2] = aliphatic, aromatic, and heteroaromatic alkyl esters, R[3] = aliphatic, allyl, and α-carbonyl alkylating agents, Y = NMe, NPh.

Literature Summary

Commercially available Fmoc-protected Rink amide resin[2] (see p 38) was deprotected with a 20% piperidine/DMF, rinsed, and acylated with a 0.3 M solution of acetyl carboxylic acid (3 equiv) at room temperature (the carboxylic acid was preactivated for 40 min with 3.3 equiv DIC/HOBt) until the Kaiser ninhydrin test (see p 214) was negative.[3] Rinses between reactions were not given (conservatively use 12 rinses between reactions). To the resin (50 mg, 22.5 μmol) was added a solution of 675 μmol of carboxylic ester in 670 μl of DMA. Under inert gas 18 mg (0.45 mmol) of sodium hydride (60%) was added and the reaction mixture was shaken well at 90°C for 1 h. The resulting resin was filtered, washed (30% v/v acetic acid/water, DMA, DMSO, and *i*-PrOH), and then dried under reduced pressure. The resin (20 mg, 8.6 μmol) was treated with 86 μl of 1 M TBAF in THF for 2 h. After the addition of 150 μl of a 2.5 M solution of the appropriate alkylating agent, the reaction was continued for 2 h.

The resin was filtered and rinsed with DCM and THF. Cyclization was performed in 0.50 mL of a 2.5 M solution of hydrazines or hydroxylamines (HCl was neutralized by triethylamine) in DMA at 80°C for 24 h. Final cleavage from support was performed as described by Rink (see p. 38).[4]

4.7.11 Solid-Phase Synthesis of Structurally Diverse 1-Phenylpyrazolone Derivatives [5]

hydroxypropylether linkage

Points of Interest

1. The dianion of support-bound β-keto ester was exclusively alkylated at the primary carbon.

2. Cyclic cleavage was performed during the final synthetic step with excess phenylhydrazine (20 equiv).

Representative Examples

Acetoacetate, both activated and unactivated alkylating agents, and phenylhydrazine were incorporated into 1-phenylpyrazolones. Phenylpyrazolones were prepared in 40–76% yield (average 90% purity).

Literature Procedure

The support-bound β-keto ester was obtained by a transacetoacetylation through heating the hydroxypropyl ether-modified (toluene swollen) polystyrene resin with a 10-fold excess of *tert*-butyl acetoacetate (100°C, 3 h). Treatment of the derivatized resin (400 mg, 0.3 mmol, swollen in 5 mL of THF) with LDA (6 equiv, THF, 0°C, 60 min) afforded its dianion, which after removal of the excess base was alkylated with a variety of haloalkanes (2–5 equiv, THF, 0°C to room temperature, 1–12 h) to selectively afford the corresponding polymer-bound γ-alkylated β-keto ester monoanion. Subsequent treatment with 2 N HCl/THF (1:1) afforded the corresponding β-keto ester. The color of the resin changed during the generation of the dianion from light yellow to a deep red. During alkylation the color fades, serving as a convenient indicator of the completeness of the reaction. Additional treatment of the monoanion with *n*-BuLi (5 equiv, 0°C, 30 min) allowed the re-formation of the red dianion, which could be alkylated with different haloalkanes to yield the γ-dialkylated β-keto ester exclusively after addition of 2 N HCl/THF (1:1). Subsequent formation of the hydrazones was carried out by addition of an excess of phenylhydrazine (20 equiv, 3 h, THF) to the resin-bound, dialkylated β-keto ester.

Heating the hydrazones to 100°C in toluene for 5 h resulted in cyclization and concomitant release of final 1-phenylpyrazolones from the support in 40–76% yield (average 90% purity) relative to the number of initial hydroxyl groups.

REFERENCES FOR SECTIONS 4.7.10 AND 4.7.11

1. Marzinzik, A. L.; Felder, E. R. *Tetrahedron Lett.* **1996**, *37*, 1003–1006.
2. Rink, H. *Tetrahedron Lett.* **1987**, *28*, 3787–3790.
3. Kaiser, E.; Colescott, R. L.; Bossinger, C. D.; Cook, P. I. *Anal. Biochem.* **1970**, *34*, 595–598.
4. Rink, H. *Tetrahedron Lett.* **1987**, *28*, 3787–3790.
5. Tietze, L. F.; Steinmetz, A. *Synlett* **1996**, 667–668.

4.7.12   Imidazole Libraries on Solid Support [1]

Points of Interest

1. The support-bound imidazoles are prepared via either a three-component or a four-component process in a one-pot procedure.

2. The imidazoles can be formed with either the amine or aldehyde component to the support.

3. Carboxybenzaldehyde and Fmoc-aminohexanoic acid were linked to Wang resin in 75 and 99% yields, respectively, via standard DIC–DMAP solid-phase esterification conditions.[2] The phenol was linked to Wang resin via a modified Mitsunobu coupling (see p 106) with sonication in 57% yield.[3]

4. The Wang resins were used due to the lability of the Rink amide linker during imidazole formation under acidic conditions at high temperature.

5. Unsymmetrical 1,4-bisimidazoles were prepared using 1,4-bisbenzil in high purity as well:

unsymmetrical 1,4-*bis*-imidazole

Representative Examples

Aromatic, ortho-substituted, and heteroaromatic diones as well as non-functionalized aromatic and aliphatic aldehydes and amines were incorporated. TFA cleavage afforded crude imidazoles in 90–95% purity; yields after chromatography were also reported to be good.

Literature Summary

To 0.07 mmol (1 equiv) of support-bound aldehyde weres added 1.4 mmol (20 equiv) of 1,2-dione, 1.4 mmol (20 equiv) of the primary amine, 0.1 mmol (1.4 equiv) of NH_4OAc, and 1.2 mL of AcOH. The mixture was stirred for 4 h at 100°C and then the resin was filtered and washed with DCM (40 mL), MeOH (40 mL), DCM (40 mL), and MeOH (40 mL). The product was cleaved from the resin with 5 mL of a 20% solution of TFA in DCM for 20 min at room temperature. The solvent was removed and residual TFA was azeotroped with heptane (2 × 100 mL) to afford the imidazole as the TFA salt. The trisubstituted imidazoles were formed in a similar manner.

To 0.07 mmol (1 equiv) of support-bound amine were added 1.4 mmol (20 equiv) of 1,2-dione, 1.4 mmol (20 equiv) of the aldehyde, 2.8 mmol (40 equiv) of NH_4OAc, and 1.2 mL of AcOH. The mixture was stirred for 4 h at 100°C and then the resin was filtered and washed with DCM (40 mL), MeOH (40 mL), DCM (40 mL), and MeOH (40 mL). The product was cleaved from the resin with 5 mL of a 20% solution of TFA in DCM for 20 min at room temperature. The solvent was removed and residual TFA was azeotroped with heptane (2 × 100 mL) to afford the imidazole as the TFA salt.

4.7.13 Tetrasubstituted Imidazoles via α-N-Acyl-N-alkylamino-β-ketoamides[4]

Points of Interest

1. The isocyanide component was attached to Wang resin; this is advantageous in light of the limited number of commercially available isocyanides.

2. The Ugi four-component coupling and cyclization on support provide the imidazoles with few, and rather innocuous, byproducts for biological screening.

Representative Examples

Two different arylglyoxals were reported; aliphatic and aromatic acids and primary amines (but not anilines) could be incorporated. Yields ranged from 16 to 56% following multistep solid-phase synthesis.

Literature Summary

Formylated aliphatic amino acids[5] were attached to Wang resin (DIC/DMAP) and dehydrated (Ph₃P–EtN₃–CCl₄)[6] to provide support-bound isocyanates. Ugi-4CC products were obtained from the reaction of support-bound isocyanates with aldehyde, amine, and carboxylic acid in 1:1:1 $CHCl_3$/MeOH/pyridine for 3 days at 65°C. Treatment with 60 equiv of NH_4OAc in AcOH (100°C for 20 h) followed by 10% TFA/DCM (room temperature for 20 min) provided imidazoles in 45% yield after purification.

REFERENCES FOR SECTIONS 4.7.12 AND 4.7.13

1. Sarshar, S.; Mjalli, A. M. M.; Siev, D. *Tetrahedron Lett.* **1996**, *37*, 835–838.
2. For a review, see: Fields, G. B.; Noble, R. L. *Int. J. Pept. Protein Res.* **1990**, *35*, 161–214.
3. Richter, L. S.; Gadek, T. R. *Tetrahedron Lett.* **1994**, *35*, 4705–4706.
4. Zhang, C.; Moran, E. J.; Woiwode, T. F.; Short, K. M.; Mjalli, A. M. M. *Tetrahedron Lett.* **1996**, *37*, 751–754.
5. (a) Lash, T. D.; Bellettini, J. R.; Bastian, J. A.; Couch, K. B. *Synthesis* **1994**, 170. (b) Solladie-Cavallo, A.; Quazzotti, S. *Tetrahedron Asymmetry* **1992**, *3*, 39.
6. Appel, R.; Kleinstuck, R.; Ziehn, K. *Angew. Chem.* **1971**, *83*, 143.

4.7.14 Pyrroles Derived from a Four-Component Condensation[1]

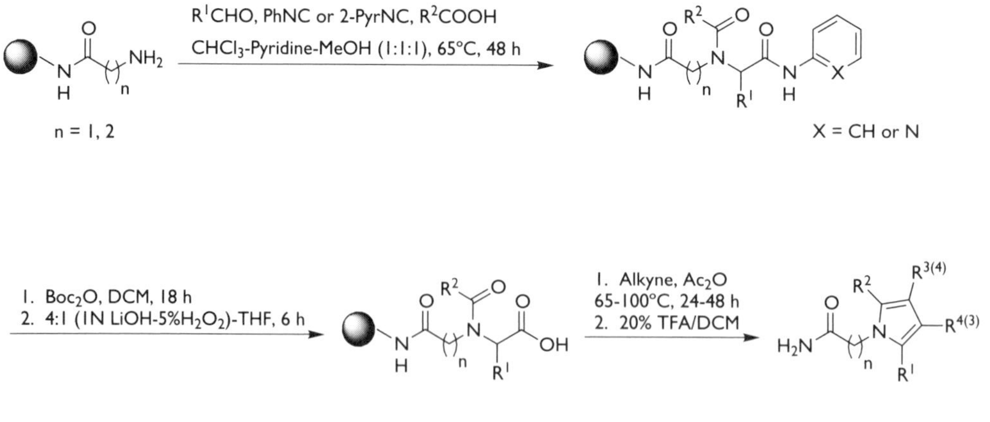

Points of Interest

1. Tetra- and pentasubstituted pyrroles were prepared via a 1,3-dipolar cycloaddition of alkynes to polymer-bound münchnones.

2. The authors mention that the Ugi-4CC (with commercially available primary amines, aldehydes, and carboxylic acids) provides access to a larger number of *N*-acyl-*N*-alkylamino amides than the corresponding reductive amination of a support-bound amino acid derivative followed by acylation of the secondary amine.

3. Specialized secondary amides such as 1-cyclohexenyl, phenyl, and 2-pyridyl amides can be converted to their corresponding acids and/or esters.[2]

4. The pyrroles were prepared in 26–72% yield.

Representative Examples

Esters, amides, and methyl ethers were incorporated. Aliphatic aldehydes and aromatic carboxylic acids were used as building blocks in all examples.

Literature Summary

After the Ugi-4CC with PhNC the amide was acylated and hydrolyzed in one pot as shown in the reaction scheme. Reaction with alkynes either in neat Ac_2O or with isobutyl chloroformate–TEA–toluene (65–100°C, 24–48 h) followed by cleavage from the solid support (20% TFA/DCM) yielded pyrroles.

REFERENCES FOR SECTION 4.7.14

1. Mjalli, A. M. M.; Sarshar, S.; Baiga, T. J. *Tetrahedron Lett.* **1996**, *37*, 2943–2946.
2. Keating, T. A.; Armstrong, R. W. *J. Am. Chem. Soc.* **1995**, *117*, 7842–7843.

4.7.15   Reductive Alkylation on a Solid Phase: Synthesis of a Piperazinedione Combinatorial Library[1]

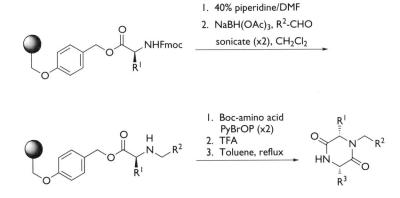

Points of Interest

1. Reductive alkylation was performed twice with sonication to give 85–95% conversions except for sterically hindered amino acids and/or electronically deactivated aldehydes. After reductive alkylation 10% racemization of the resulting secondary amino acids was observed by chiral HPLC.

2. A double-coupling protocol with PyBrOP was necessary to achieve >90% acylation of the secondary amines with Boc-protected amino acids. Diketopiperazines were formed in refluxing toluene (no DKP formation was observed without thermolysis).

Representative Examples

Two purified diketopiperazines were prepared in 24% yield (DKP from Val, 2,4,6-trimethoxybenzaldehyde, and Leu) and 42% yield (DKP from Phe, *p*-methoxybenzaldehyde, and Ala). A library of 1000 diketopiperazines was prepared with the split and mix method.

Literature Summary

To a flask containing amino acid-derived Wang resin suspended in DCM (0.5 mL) was added an aldehyde (0.24 mmol) in DCM (0.5 mL). The reaction mixture was sonicated in an ultrasound bath for 20 min followed by addition of a presonicated solution of sodium triacetoxyborohydride (0.28 mmol) in DCM (0.5 mL). The mixture was sonicated for 5 min and then stirred vigorously for 16 h. The resin was filtered, washed (water, aqueous bisulfate, and water/THF, 3×2 mL each), and dried, and the reductive alkylation procedure was repeated. Samples (10% by weight) of the dry resin-bound secondary amine intermediate were retained for future analysis and iterative followup. All remaining resin was suspended in DCM (3 mL) and treated with a solution of the Boc-amino acid components (0.2 mmol) in DCM (0.5 mL and DMF, if necessary, to achieve complete solubility), followed by a solution of PyBrOP (0.2 mmol) in DCM (0.5 mL) and finally DIEA (0.4 mmol). The reaction mixture was stirred for 24 h, filtered, washed (DMF, water, THF), and dried, and the coupling process was repeated. The resin was treated with TFA (1 mL) for 3 h with occasional agitation, filtered, and washed with DCM. The filtrate was concentrated, and the residue was dissolved in toluene and concentrated once more. The residue was dissolved in toluene (10 mL), stirred under reflux for 5 h, and then evaporated to dryness. Mixtures were evaluated by LC/MS.

4.7.16   Solid-Phase Synthesis of Diketopiperazines and Diketomorpholines[2]

Points of Interest

1. The diketopiperazines (DKP) and diketomorpholines (DKM) for this study were prepared from amines and α-bromo carboxylic acids. Amines purchased as salts were treated with either an anion-exchange resin (AG-1-X8 resin, hydroxide form; Bio-Rad, Richmond, CA) or Na_2CO_3 extraction prior to use.

2. Amine displacement of the support-bound bromide is sterically controlled. The reaction rate is dependent upon the substituents on the α-carbons (R^1 or R^3). Thus, prolonged reaction time (>12 h), elevated temperatures ($50-70°C$), and a large excess of amine ($1-2$ M or ca. 40 equiv) were required to drive the reactions to completion. Aminolysis of the ester linkage was not observed under these reaction conditions.

3. Best results for acylation of the resin-bound, sterically hindered secondary amines were achieved by using THF as the solvent and PyBrOP as the activating agent in the presence of DIEA at $50°C$.

4. Intramolecular cyclization could occur either during the second amine displacement or with TFA, which is also necessary to remove protecting groups. During the library synthesis products from both modes of cleavage were obtained by subjecting the amine displacement filtrates with excess cation-exchange resin (AG50W-X8, hydrogen form), followed by recombination with the filtrates from the resin treated with TFA/H_2O ($95:5$) to afford diketopiperazines.

5. Substituted piperazine libraries have also been used to discover bradykinin antagonists and other G-protein-coupled receptor ligands.[3]

Representative Examples

A mass balance yield of 90% after extraction was reported for a DKP with $R^1 = H$, $R^2 =$ benzyl, $R^3 = H$, $R^4 =$ isobutyl. Individual reactions were validated by HPLC and GC–MS with a range of commercially available building blocks. Anilines produced variable results. Building blocks with acid-labile functional groups were discussed but not disclosed.

Representative Literature Procedure

To a slurry of hydroxymethyl resin (5.0 g, having a loading of 0.50 mmol/g, 2.50 mmol) in 50 mL of DMF were added an α-bromo acid (7.50 mmol) and catalytic DMAP (30 mg, 0.25 mmol). DIC (1.17 mL, 7.50 mmol) was added in one portion to the reaction mixture, which was agitated at room temperature for 30 min, after which the resin was drained and the acylation repeated. The resin was filtered and washed with DMF (2 × 10 mL) and DCM (2 × 10 mL). This resin was treated with a solution of an amine in DMSO (50 mL of a 2 M solution) at 50°C for 22 h. The resin was filtered and washed with DCM (2 × 10 mL), MeOH (2 × 10 mL), and DCM (2 × 10 mL). A ninhydrin test confirmed the presence of amine on the resin. To a slurry of a portion of this resin (i.e., R^1 = propyl, R^2 = benzyl; 0.20 g, 0.10 mmol) in 2 mL of THF were added another α-bromo acid (1.0 mmol) and DIEA (348 μL, 2.0 mmol). PyBrOP (446 mg, 1.0 mmol) was added in one portion to the reaction mixture, which was subsequently agitated at 50°C until a ninhydrin test confirmed that the acylation was complete (18 h). The resin was filtered and washed with DMF (2 × 10 mL) and DCM (2 × 10 mL). The resin was finally treated with another amine (R^4) in DMSO (1 mL of a 2 M solution) for 96 h at 70°C. The product was either cleaved off of support during this step or cleaved off later with TFA/H_2O depending on the substitution pattern. The filtrates from both aminolysis and acidic cleavage were combined for library synthesis. The resin (or mixture of resins) was washed with DMSO and DCM (2×). These filtrates were combined, evaporated, and then added to a previously prepared cation-exchange resin (AG50W-X8 resin, hydrogen form; 3 molar equiv) for 1 h. The cation-exchange resin was drained and washed with two column volumes of DCM, MeOH, DCM, and MeOH. The eluents were then concentrated under vacuum to furnish partial DKP filtrates. The solid-phase resin was treated separately with TFA/H_2O (95:5). The cleavage mixture was filtered into the partial DKP filtrates and washed once with an additional cleavage cocktail (3 mL). The resulting solution was allowed to stand at room temperature for 2 h and evaporated. Mixtures and individual compounds were analyzed by HPLC or GC–MS.

4.7.17   Solid-Phase Synthesis of 2-Oxopiperazines by Intramolecular Michael Addition[4]

Points of Interest

1. Attempted cyclopropanation of unsaturated peptoids on solid support led to the discovery of a facile method for generating libraries of constrained cyclic peptoids. Diastereomeric monoketopiperazines were prepared under standard conditions for Fmoc deprotection of the appropriate precursor.
2. Broad NMR resonances precluded stereochemical assignment.
3. The simplicity of the synthesis makes it readily adaptable to robotics.

Representative Examples

A variety of Fmoc-amino acids were used to produce monoketopiperazines. Aromatic amino acids (tryptophan and phenylalanine) as well as hindered aliphatic amino acids (valine and proline) all gave products of greater than 80% purity by HPLC. The monoketopiperazines were further derivatized with isocyanate and carboxylic acids. Anilines or hindered amines did not work as NR^2 under the reported conditions.

Literature Summary

Rink amide resin (5.7 g, 0.51 mmol/g) was swollen in DMF, filtered, and treated with 50 mL of 20% piperidine in DMF (5 min, then 30 min). The resin was washed with DMF and then treated 2 × 30 min with a solution of bromoacetic acid (30 mmol) and DIC (30 mmol) in DMF (50 mL). The resin was then washed with DMF and treated for 2 h with a 2.0 M solution of isobutylamine in DMSO (50 mL). The resin was again washed with DMF and treated 2 × 30 min with 4-bromo-2-butenoic acid (30 mmol) and DIC (30 mmol) in DMF (50 mL). The resin was then washed with DMF, treated for 2 h with a 2 M solution of benzylamine in DMSO (50 mL), washed with DMF and DCM, and dried overnight *in vacuo*. All of the synthesis operations were performed under Ar. A 1.5-g portion of this unsaturated dipeptoid was then swollen in DMF and drained. For 1 h the resin was treated with a 0.6 M solution of Fmoc-L-phenylalanine/HOBt/DIC (1:1:1) in DMF (25 mL) and then it was washed and dried as before. A 190-mg portion of the capped dipeptoid on resin was allowed to expand in DMF for 5 min followed by treatment with 20% piperidine in DMF (1 × 5 min, 1 × 30 min, 5 mL). After the resin was washed thoroughly with DMF and then dichloroethane (DCE), it was acylated with 2 mL of 1.0 M benzoyl chloride in DCE and 2 mL of 1.0 M Et_3N in DCE (2 × 30 min). The resin was then washed thoroughly with DMF and DCM and cleaved with TFA/H_2O (95:5) for 20 min to give the crude monoketopiperazine. The crude products were characterized by HPLC (214 nm) and electrospray MS. Proton NMR spectra were obtained for all compounds.

REFERENCES FOR SECTIONS 4.7.15 TO 4.7.17

1. Gordon, D. W.; Steele, J. *BioMed. Chem. Lett.* **1995**, *5*, 47–50.
2. Scott, B. O.; Siegmund, A. C.; Marlowe, C. K.; Pei, Y.; Spear, K. L. *Mol. Diversity* **1995**, *1*, 125–134.
3. Goodfellow, V. S.; Laudeman, C. P.; Gerrity, J. I.; Burkard, M.; Strobel, E.; Zuzack, J. S.; McLeod, D. A. *Mol. Diversity* **1996**, *2*, 97–102.
4. Goff, D. A.; Zuckermann, R. N. *Tetrahedron Lett.* **1996**, *37*, 6247–6250.

4.7.18   Solid-Phase Synthesis of 1,4-Dihydropyridines[1]

Points of Interest

1. Numerous DHPs (dihydropyridines) have found commercial utility as calcium channel blockers, as exemplified by therapeutic agents such as Nifedipine, Nitrendipine, and Nimodipine. Gordeev and coworkers reference over 20 examples.[1]

2. Pyridine is essential for the two- or three-component cyclocondensation of enamino esters with 2-arylidene β-keto esters or β-keto esters and aldehydes.

3. An NMR study with ^{13}C-labeled building blocks suggested that the acyclic adduct is cleaved from support prior to intramolecular cyclization. In addition, the IR spectrum of a Rink-supported intermediate prior to acid treatment displayed a signal of the unconjugated ester group at 1735 cm^{-1}. By contrast, the IR spectrum of the corresponding 1,4-dihydropyridine displayed only the expected signal of a conjugated ester carbonyl at 1705 cm^{-1}.

Representative Examples

Aliphatic β-keto esters, 1,3-diketones, and aromatic aldehydes were incorporated. Nine dihydropyridines were prepared in 65–78% yield prior to chromatography.

Literature Procedure

N-Immobilized enamino esters and 1,4-dihydropyridines were prepared from commercially available Rink or PAL resin as follows. An appropriate Fmoc-protected resin (0.23 mmol) was vortexed with 10% piperidine in DMF (5 mL) for 40 min. The resulting amine resin was filtered, washed sequentially with DMF (4×), MeOH (3×), CHCl₃ (3×), and ether, and dried in a vacuum desiccator (room temperature, 0.5 Torr). The resin was then vortexed with an appropriate β-keto ester (6.9 mmol) and 4-Å molecular sieves (1 g) in DCM (4 mL) for 3 days at room temperature. The resin was then filtered, washed sequentially with CHCl₃ (3×), ethyl acetate (3×), and ether, and dried in a vacuum desiccator (room temperature, 0.5 Torr). The appropriate resin N-immobilized enamino ester (0.023 mmol), β-keto ester (1 mmol) (or acetylacetone in one example), and aromatic aldehyde (1 mmol) with 4-Å molecular sieves (250 mg) in dry pyridine (0.75 mL) were stirred at 45°C under argon in a sealed amber vial for 24 h. Note: a preformed benzylidene β-keto ester (1 mmol) can be used instead of a mixture of β-keto ester and aldehyde for reaction with resin N-immobilized enamino ester. The resin was filtered, washed sequentially with MeOH (4×) and ethyl acetate (4×), and dried in a vacuum desiccator (room temperature, 0.5 Torr). The resulting resin was stirred under argon with 3% TFA in DCM (1 mL, 45 min; for cleavage from Rink resin) or 95% TFA in THF (1 mL, 1.5 h; for cleavage from PAL resin). Degassed acetonitrile (4 mL) was added, and the supernatant layer was separated and quickly evaporated *in vacuo* with the addition of toluene to remove TFA.

4.7.19 Solid-Phase Synthesis of Pyridines and Pyrido[2,3-*d*]pyrimidines[2]

Points of Interest

1. The pyridine nucleus is a key feature of various drugs, including numerous antihistamines, as well as antiseptic, antiarrhythmic, antirheumatic, and other pharmaceuticals.

2. Initially, a model study was performed employing ^{13}C-labeled benzaldehyde to monitor the course of reaction on solid support by fast gel-phase ^{13}C NMR spectroscopy (see p 227).[3]

3. The β-keto ester can be analyzed by cleavage/heterocyclization into 3-methyl-3-pyrazolin-5-one (5% hydrazine in EtOH, 30 min, room temperature). A completion of the Knoevenagel condensation step is ensured when a negative pyrazoline test is observed with ethanolic hydrazine (by TLC).

Representative Examples

A range of diverse substituents were incorporated into the scaffolds in 70–100% HPLC purity. Aliphatic, heteroaromatic, and electron-rich and electron-poor aldehydes were incorporated.

Literature Summary

Wang or SASRIN resin was acetoacetylated with diketene (DMAP catalyst, DCM, −50°C to room temperature). The resulting β-keto ester was cleaved/cyclized into 3-methyl-3-pyrazolin-5-one (5% hydrazine in EtOH, 30 min, room temperature). The β-keto ester was observed by this method with nearly quantitative conversion. Knoevenagel condensation with benzaldehyde (*i*-PrOH/benzene, 60°C, 4-Å molecular sieves or HC(OMe)$_3$) afforded the corresponding benzylidene resin (δ [13]CH ca. 140 ppm). Next, a Hantzsch-type heterocyclization with methyl aminocrotonate (DMF, 80°C, 4-Å molecular sieves or HC(OMe)$_3$, 10 h) proceeded quantitatively by ^{13}C NMR to generate the expected 1,4-dihydropyridine derivative. The support-bound 1,4-dihydropyridine was oxidized with cerric ammonium nitrate (CAN) in dimethylacetamide (15 min, room temperature) to afford the immobilized pyridine (this reaction was monitored by fast gel-phase ^{13}C NMR). Pyridines and pyrido[2,3-*d*]pyrimidines were obtained in greater than 70% purity by HPLC after cleavage from resin.

REFERENCES FOR SECTIONS 4.7.18 AND 4.7.19

1. Gordeev, M. F.; Patel, D. V.; Gordon, E. M. *J. Org. Chem.* **1996**, *61*, 924–928.
2. Gordeev, M. F.; Patel, D. V.; Wu, J.; Gordon, E. M. *Tetrahedron Lett.* **1996**, *37*, 4643–4646.
3. Look, G. C.; Holmes, C. P.; Chinn, J. P.; Gallop, M. A. *J. Org. Chem.* **1994**, *59*, 7588–7590.

4.7.20 Solid-Phase Protocol of the Biginelli Dihydropyrimidine Synthesis[1]

Points of Interest

1. Dihydropyrimidines were prepared on support in one step from a Biginelli condensation of β-keto esters, aldehydes, and ureas.

2. All dihydropyrimidines were prepared with GABA (γ-aminobutyric acid) as an auxiliary functional group for attachment to resin. The carboxylic acids obtained upon cleavage from resin were alkylated with benzyl bromide prior to full characterization.

3. The three-component coupling was compared on support versus in solution. The solid-phase protocol provides the dihydropyrimidines in higher yield and superior purity. The solid-phase protocol is advantageous because the Biginelli reaction is typically performed with an excess of the β-keto ester and aldehyde, which can be rinsed away on resin.

Representative Examples

Aliphatic β-keto esters and aromatic aldehydes were incorporated into dihydropyrimidines. After cleavage, the Biginelli products were isolated in 95% purity by NMR as carboxylic acids that were then converted into the corresponding benzyl esters prior to full characterization.

Literature Summary

To 1.0 g (0.96 mmol) of Wang resin (0.96 mmol/g) was added 15 mL of dry DMF followed by 0.56 g (3.84 mmol) of 4-ureidobutyric acid,[2] 0.738 g (3.84 mmol) of 1-(3-(dimethylamino)propyl)-3-ethylcarbodiimide hydrochloride, and 0.035 g (0.288 mmol) of DMAP. The reaction was stirred for 24 h and the resin was then washed with water (4×) and dried under vacuum. 4-Ureidobutyric acid was cleaved from a portion of the resin in 98% yield to establish the loading level. After the loading was determined, the three-component coupling was performed. A suspension of 50 mg (0.048 mmol) of the support-bound urea in 1.5 mL of THF was treated with 4 equiv (0.192 mmol) of aldehyde, 4 equiv (0.192 mmol) of β-keto ester, and 50 μL of a 4:1 THF/concentrated HCl solution. The reaction mixture was stirred at 55°C for 36 h, and the resin was filtered and washed with THF (3×), hexanes (3×), MeOH (3×), and DCM (3×). The Biginelli product was cleaved with TFA/DCM (1:1) followed by rinses with DCM. The filtrate was concentrated *in vacuo*, and the crude residue was examined by ^{1}H NMR and then converted to the benzyl esters prior to full characterization.

4.7.21 Solid-Phase Synthesis of 5,6-Dihydropyrimidine-2,4-diones[3]

Points of Interest

1. Cyclization of β-ureido esters occurs with concomitant cleavage from resin and formation of pyrimidinediones. A similar cyclization–cleavage strategy has been reported for the formation of hydantoins from resin-bound α-ureido esters and benzodiazepines from resin–bound imines.[4]

2. To compensate for the limited number of commercially available N-protected β-amino acids with nitrogen substituents, N-protected β-amino acids were prepared by Michael addition of primary amines to a polymer-bound acrylate.

3. The polymer-bound acrylate was characterized by IR microspectroscopy that showed complete disappearance of the hydroxyl OH stretch and the appearance of a C=O stretch at 1725 cm^{-1}.

4. Cleavage with TFA/water afforded β-ureido acids or a mixture of the urea and pyrimidinedione. Cleavage with HCl in toluene at high temperatures afforded pyrimidinediones after chromatography.

Representative Examples

Benzyl-, methyl-, isopropyl-, and isobutylamines and an amino ester were incorporated into pyrimidinediones. Aliphatic, aromatic, and sterically hindered isocyanates were also incorporated into pyrimidinediones. Crude products were greater than 95% pure by HPLC. Cleavage with HCl in toluene at high temperatures afforded predominantly pyrimidinediones; however, slight contamination by the HCl salt of the corresponding β-amino acid was observed. The HCl salts were easily removed from the crude product by filtration through silica using $1:1$ ethyl ether/DCM, and the pyrimidinediones were isolated in $13–69\%$ yield.

Literature Summary

A slurry of 0.5 g of Wang's resin (0.88 mequiv/g) in 4 mL of DCM was treated twice with 200 μL of NEt$_3$ followed by 100 μL of acryloyl chloride and stirred for 2 h. After filtration, the resin was washed three times each with DCM, MeOH, DMF, MeOH, and DMSO. The support-bound acrylate was treated with 2 mL of DMSO and 0.28 g (2.6 mmol) of benzylamine and stirred for 24 h. Reactions with α-branched amines required longer reaction times or higher temperatures. Washing three times each with DMSO, MeOH, and DCM and drying *in vacuo* afforded 0.54 g (0.74 mmol) of derivatized resin. The resin (0.42 g, 0.31 mmol) was treated with 3 mL of DCM followed by 0.10 g (0.8 mmol) of phenyl isocyanate for 4 h and then washed three times each with DCM, MeOH, DCM, and diethyl ether. Cleavage and cyclization of the support-bound β-ureido ester was performed in a glass vial with 4 mL of a saturated solution of dry HCl in toluene, and the vessel was capped and heated to 95°C for 4 h. After cooling, the resin was filtered and washed three times with MeOH and DCM. The combined filtrates were concentrated and the crude product was purified by silica chromatography, eluting with 1 : 1 DCM/diethyl ether to provide 39.7 mg (46% yield) of 5,6-dihydropyrimidine-2,4-dione.

REFERENCES FOR SECTIONS 4.7.20 AND 4.7.21

1. Wipf, P.; Cunningham, A. *Tetrahedron Lett.* **1995**, *36*, 7819–7822.
2. GABA urea was prepared by treatment of GABA with KOCN in H$_2$O/THF according to: Viret, J.; Gabard, J.; Collet, A. *Tetrahedron* **1987**, *43*, 891.
3. Kolodziej, S. A.; Hamper, B. C. *Tetrahedron Lett.* **1996**, *37*, 5277–5280.
4. DeWitt, S. H.; Kiely, J. S.; Stankovic, C. J.; Schroeder, M. C.; Cody, D. M. R.; Pavia, M. R. *Proc. Natl. Acad. Sci. U.S.A.* **1993**, *90*, 6909–6913.

4.7.22   Solid-Phase Synthesis of Dihydro- and Tetrahydroisoquinolines[1]

Points of Interest

1. The Bischler–Naperalski reaction was applied to the solid-phase synthesis of isoquinolines. The crucial step was the treatment of N-acylated dimethoxyphenylalanine with POCl₃ on resin.

2. Four derivatives were synthesized and cleaved from support with HF. All four products were obtained in high purity by HPLC, although the isolated yields were 25–40%.

3. Equimolar mixtures of eight different dihydro- and tetrahydroisoquinolines were prepared and identified by ion spray MS of the mixture.

Representative Examples

Support-bound dimethoxyphenylalanine was acylated with eight aliphatic and aromatic acetic acids.

Literature Summary

Boc-protected 3,4-dimethoxyphenylalanine[2] was linked to chloromethylated polystyrene resin (1% DVB) via the cesium salt. The cesium salt was prepared from 0.5 equiv of cesium carbonate in ethanol/water (1:1). After evaporation of the ethanol/water, the residue was dissolved in DMF, added to the resin (0.75 mequiv/g; 1.5 Cs salt to 1 Cl on Merrifield resin), and heated overnight to 50°C. The loading was 0.31 mmol/g as determined by both cleavage and isolation as well as by amino acid analysis of the resin. The resin was treated with 40% TFA in DCM (1 min, 10 min). The resin was then rinsed with DCM, DCM/MeOH, and DMF. The free amine was acylated with acetic acid derivatives. Amide formation was accomplished with 4 equiv of the acid, 4 equiv of HBTU (0.5 M), and 5 equiv of DIEA in DMF (10 min). The N-acylated dimethoxyphenylalanine was then treated with freshly distilled POCl₃ (30 equiv) for 8 h at 80°C in toluene. Initial experiments demonstrated that at room temperature cyclization was incomplete and that partial demethylation occurs at elevated temperatures. After cleavage with HF/p-cresol (9:1) (*Caution:* HF is toxic!) at 0°C for 1 h, the dihydroisoquinoline was isolated in 40% yield after HPLC purification. Attempts to convert the imine to the tetrahydroisoquinoline with NaBH₄ resulted in reduction of the ester linkage and premature cleavage from the resin. However, the dihydroisoquinoline was selectively reduced by treatment of 2 g of the resin with NaBH₃CN (5 equiv, 140 mg) in 10 mL of MeOH. When gas formation ceased, a few drops of MeOH/HCl (anhydrous) was added until no more gas evolved. The process was repeated twice to ensure complete reduction. Tetrahydroisoquinolines were isolated after HPLC purification in 25–30% yield as a 6:1 ratio of (1,3)-*cis:trans* isomers.

4.7.23   Solid-Phase Synthesis of Highly Substituted Peptoid 1(2*H*)-Isoquinolines[3]

Points of Interest

1. *trans*-4-Bromo-2-butenoic acid (bromocrotonic acid) was used as a scaffold for the preparation of isoquinolines derived from the submonomer method of peptoid synthesis.[4]

2. The Pd-catalyzed intramolecular Heck reaction produced predominantly the endocyclic aromatic alkene. When a 2-iodobenzoyl chloride derivative with a substituent at the 3-position was incorporated, a mixture of endocyclic and exocyclic alkenes was obtained. Further experiments led the authors to conclude that the exocyclic alkene is produced first, followed by equilibration to the isoquinoline under the reaction conditions.

3. Isoquinoline- and isoxazoline-containing heterocycles have been prepared on a "traceless" Reissert resin (see p 73).[5]

Representative Examples

Substituted *o*-iodo or *o*-bromo carboxylic acids prepared from commercially available anthranilic acids as well as from heterocycles such as pyridine or pyrazinecarboxylic acids were incorporated into isoquinolinones in 65–92% yield. Aliphatic, aromatic, and heteroaromatic amines were also incorporated. The process works equally well when a support-bound dipeptoid is used, producing an isoquinoline with an extended side chain.

Literature Procedure

Rink amide resin (3.0 g, Advanced Chemtech, 0.51 mmol/g) was placed into a 250-mL-capacity silanized glass reaction vessel and swollen in DMF (ca. 50 mL) for 5 min. The solvent was drained and the resin was mixed with 30 mL of 20% piperidine in DMF (2 × 20 min) on an orbital shaker at 200 rpm. The solvent was drained and the resin was washed with DMF (6×). The deprotected resin was then treated with a solution of 4-bromo-2-butenoic acid (2.46 g, 15 mmol) and DIEA (2.35 mL, 15 mmol) in DMF (25 mL). The acylated resin was then washed with DMF (3×). Isobutylamine (4.39 g, 60 mmol) in DMSO (30 mL) was added. After 2 h of mixing under Ar, the vessel was drained and the resin was washed well with DMF and then DCM, followed by drying *in vacuo* overnight at room temperature. A 140-mg portion of this resin was loaded into a reaction vessel and placed on a Symphony Multiple Peptide Synthesizer (Protein Technologies, Inc.) and treated first with a solution of freshly prepared 2-iodo-6-fluorobenzoyl chloride (2.4 mmol) in 1,2-dichloroethane (DCE) (2 mL) followed by treatment with a solution of triethylamine (2.4 mmol) in DCE (2 mL). After 30 min of mixing, the vessel was drained and the acylation was repeated. The resin was washed well with DMF and DCE, and an aliquot was treated with 95:5 TFA/H$_2$O (v/v) for 20 min at room

temperature for HPLC analysis. The remainder of the resin was loaded into a
10-mL Schlenk tube and treated with $Pd(PPh_3)_4$, anhydrous NaOAc (75 mg),
Ph_3P (35 mg), and anhydrous dimethylacetamide (8 mL). A gentle vacuum was
pulled on the tube for 2 min before Ar was introduced. The sealed tube was
then placed in a preheated 90°C block heater and mixed at 200 rpm on an
orbital shaker for 6 h. The resin was filtered off, washed with DMF, water, and
DMF, and stirred in a solution of sodium diethyl dithiocarbamate (100 mg) in
DMF (5 mL) for ca. 10 min to remove residual Pd. The resin was again filtered,
washed with DMF, THF, and DCM, and then cleaved with $95:5$ TFA/H_2O for
20 min. The cleavage mixture was diluted with HOAc and H_2O and analyzed
by HPLC. The mixture was lyophilized twice to give 16.5 mg of the desired
product.

4.7.24 Solid-Phase Synthesis of 2,6-Disubstituted Quinoline Derivatives[6]

Points of Interest

1. 2,6-Disubstituted quinoline derivatives were generated by parallel synthesis in seven steps, the last five of which proceed on a polymeric support. Variable building blocks were selected from three commercially available classes of compounds: ω-functionalized fatty acids, aryl methyl ketones, and primary amines.

2. Hydroxy-2-nitrobenzaldehyde was the key component of the benzo ring of the quinoline nucleus. This key component displays excellent formyl group reactivity toward nucleophiles and has a phenolic hydroxyl group to attach the nucleus to the resin.

3. Reductive cyclization of the nitro functionality (with tin(II) chloride) to the quinoline N-oxide followed by further reduction to the quinoline with titanium(III) chloride was demonstrated on support.

Representative Examples

A total of 12 separate quinoline derivatives were prepared with 6 distinct aryl methyl ketones (including a variety of heteroaromatic ketones) for the aldol addition step and 2 primary amines for the final cleavage by aminolysis with trimethylaluminum. Purification by chromatography on silica gel led to a set of spectroscopically homogeneous, crystalline compounds in overall yields ranging from 20 to 65%. The authors mentioned that low recovery of the desired products correlated with the presence of a sensitive indole/furan at the C-2 position, thereby revealing some drawbacks in the protocol.

Literature Summary

5-Hydroxy-2-nitrobenzaldehyde functionalized with an ω-fatty acid was loaded via esterification with (hydroxymethyl)polystyrene employing DIC/DMAP at room temperature for 24 h. All the crucial aldol additions were performed on the resin in DCM/THF (1:4) with the appropriate aryl methyl ketones and potassium carbonate for 2 days at reflux. All transformations were complete as judged by disappearance of the support-bound aldehyde by IR spectroscopy. The resin was filtered and washed thoroughly with tetrahydrofuran and water, ethanol, acetone, ether, and DCM. The resin was then dried to constant weight. The resin samples were preswelled in DCM, and reductive cyclizations were performed with ethanolic tin(II) chloride under reflux for 4 days. The derivatized resin was filtered off from the hot reaction mixtures and washed again with ethanol, water, ethanol, acetone, ether, and DCM. A second reductive step with titanium(III) chloride in toluene/DCM at room temperature overnight afforded the N-oxide-free quinoline. After further rinses, the quinoline derivative was cleaved from support by amination. The resin was treated with iso-butylamine/Me$_3$Al or N-(2-aminoethyl)morpholine/Me$_3$Al for 20 h at room temperature. Aqueous workup delivered 12 discrete products, the majority of which were in a state of purity close to 80% (HPLC).

4.7.25 Solid-Phase Synthesis of 1,3-Dialkylquinazoline-2,4-diones[7]

Points of Interest

1. In recent years, quinazoline (benzopyrimidine) derivatives have attracted attention as pharmacophores, especially as inhibitors of protein tyrosine kinases.

2. Quinazoline derivatives were alkylated under reaction conditions previously employed for the alkylation of 1,4-benzodiazepines (see p 115). In addition, the anthranilate scaffold was prepared in solution and loaded onto resin under reaction conditions previously employed for loading 2-aminobenzophenones onto resin (see p 148).[8]

Representative Examples

Alkyl, alkenyl, halo- and alkylaryl, and heteroaryl substituents were incorporated into 1,3-dialkylquinazoline-2,4-diones in greater than 82% yield by HPLC.

Literature Summary

TentaGel S-NH$_2$-supported anthranilic acid derivative (0.20 g, 0.06 mmol) was slurried in 2 mL of DMF. Piperidine (0.5 mL) was added and the mixture was shaken for 1 h to afford the free aniline. The resin was rinsed with DMF and DCM. The resin was solvated in DCM (2 mL), the isocyanate (1.16 mmol, 20 equiv) was added, and the resulting mixture was shaken for 18 h to form the urea (Method A). Alternatively, the support-bound aniline was added to *p*-nitrophenyl chloroformate (0.5 M) and triethylamine (0.5 M) in 1:1 THF/DCM (2 mL), and the resulting solution was shaken for 18 h to form the support-bound carbamate. The resin was washed with DCM, a primary amine (2 mL of 0.5 M in DCM) was added, and the reaction mixture was shaken for 12 h to form urea (Method B). The support-bound urea was washed with DCM and EtOH. KOH (1 M) in EtOH (2 mL) was added, the mixture was shaken for 1 h to afford monoalkylquinazoline, the resin was then rinsed with EtOH and THF. THF (1 mL) was added, followed by lithiated benzyloxazolidinone (3 mL, 0.3 M in THF, 0.90 mmol, 15.5 equiv). The resulting slurry was shaken for 1.5 h. Alkylating agent (2.32 mmol, 40 equiv) was added followed by DMF (1 mL), and the reaction mixture was shaken for 18 h. The resin was filtered and the alkylation procedure was repeated to afford the support-bound dialkylquinazoline. The resin was rinsed with THF, THF/water, and THF, TFA/water (95:5) was added to the resin, and the mixture was shaken for 1 h. The resulting solution was filtered from the resin, diluted with water, and lyophilized to afford quinazolines.

4.7.26 Solid-Phase Synthesis of 1,3-Disubstituted 2,4(1*H*,3*H*)-Quinazolinediones[9]

Points of Interest

1. Custom-made chloroformate-poly(ethylene glycol)-derivatized polystyrene resin was prepared with triethylene glycol bis(chloroformate) immediately prior to use.

2. Anthranilic acids were linked to a chloroformate resin through the nitrogen. Amines were coupled to the free carboxylic acid, and thermal cyclization led to heterocycle formation and concomitant resin release, resulting in a traceless linkage.

3. In addition to the approximately 50 commercially available anthranilic acids, a modified literature procedure[10] was employed in parallel for the facile synthesis of anthranilic acids from o-chlorobenzoic acids and amines.

4. The products obtained from cyclization were usually of >95% purity and a uniform quantity of 0.2 mmol/g. Substantially reduced yields (0.02 mmol/g) were observed for secondary alkyl amines under the standard cyclization conditions (125°C, 16 h).

Representative Examples

Structurally diverse primary and secondary anthranilic acids were incorporated into quinazolinediones. A range of primary alkyl, aryl, and heteroaryl amines were also incorporated.

Literature Summary

A chloroformate-PEG–PS resin was prepared immediately prior to use by reaction of (aminomethyl)polystyrene resin (1.0 mmol/g) with triethylene glycol bis(chloroformate) (3 equiv) in the presence of Hunig's base (3 equiv) in DCM for 1 h at 20°C. Loading was estimated gravimetrically, with the difference in loadings being attributed to cross-linking. The resulting resin has good solvation properties in the solvents used, while avoiding PEG-contaminated products observed with chloroformate-functionalized commercial PEG-grafted polystyrene resins. The chloroformate-functionalized polystyrene resin (loading: ca. 0.30 mmol/g) was treated individually with a wide range of substituted anthranilic acids (3 equiv) and Hunig's base in DCM at 20°C for 1 h to afford the urethane-derivatized resins. Structurally diverse primary amines (2 × 3 equiv) were coupled twice to the free carboxylic acid using standard PyBOP conditions (2 × 1 h) to afford the resin-bound anthranilamides. Cyclization in DMF at 125°C for 16 h generated the 1,3-disubstituted quinazolinediones and simultaneously liberated the molecules (in high purity) from the resin into solution.

4.7.27   Solid-Phase Synthesis of Tetrahydroisoquinolines and Tetrahydroimidazopyridines[11]

or

Points of Interest

1. Support-bound peptoids[12] incorporating terminal *m*-tyramines or histamines were condensed with aromatic, aliphatic, and heterocyclic aldehydes.

2. Substituted *m*-tyramines were also incorporated into tetrahydroisoquinolines.

Representative Examples

Branched aliphatic, ortho-substituted aromatic, and heteroaromatic aldehydes were incorporated into tetrahydroisoquinolines and tetrahydroimidazopyridines. In all cases reported, the desired product was obtained in greater than 74% yield by HPLC and EIMS.

Literature Summary

Support-bound peptoids with terminal *m*-tyramines or histamines were prepared as described by Zuckermann and co-workers.[12] The derivatized resins (100 mg) were rinsed with pyridine (4×), 1 mL of the desired aldehyde (0.5 M) in pyridine was added, and the resulting mixture was heated to 100°C. After 14 h of mixing at 100°C, the resin was cooled to room temperature and rinsed with pyridine (4×), DMA (4×), THF/DCM (4×), DCM (4×), and AcOH (4×). Cleavage was performed in duplicate with 1 mL of TFA/H_2O (9:1) for 20 min. The solution was concentrated to dryness and analyzed by HPLC and EIMS.

4.7.28   A Solid-Phase Approach to Quinolones[13]

1. (CH$_3$O)$_2$CHN(CH$_3$)$_2$, THF, 18 h
2. NH$_2$C$_3$H$_5$, THF, 72 h

TMG, DCM, 55°C, 18 h

1. piperazine, NMP, 110°C, 4 h
2. TFA/DCM, 1 h

Points of Interest

1. Quinolones are highly potent, broad-spectrum antibiotics. Their mode of action is believed to involve inhibition of DNA gyrase.
2. Reactions on support were monitored by gel-phase ^{13}C NMR.

Representative Examples

Cyclopropylamine and *p*-fluoroaniline were incorporated into resin-bound enamides. Seven piperazines were incorporated into quinolones. After cleavage, quinolones were isolated in 4–24% yield (76–96% purity).

Literature Summary

Transesterification of Wang resin with 2,4,5-trifluorobenzoylacetic acid ethyl ester was achieved by heating the mixture in toluene with catalytic DMAP at 100°C for 72 h. Preparation of the resin-bound enamide was achieved by activation of the resin-bound β-keto ester with dimethylformamide dimethyl acetal for 18 h followed by *in situ* addition of cyclopropylamine at room temperature for 72 h. Cyclization with TMG in DCM at 55°C for 18 h afforded the corresponding resin-bound quinolone. Quinolones were detected on support by gel-phase ^{13}C NMR or cleaved from resin. Nucleophilic substitution with piperazines in NMP at 110°C provided fully derivatized quinolones that were then cleaved from resin with 40% TFA in DCM.

REFERENCES FOR SECTIONS 4.7.22 TO 4.7.28

1. Meutermans, W. D. F.; Alewood, P. F. *Tetrahedron Lett.* **1995**, *36*, 7709–7712.
2. Gensler, J.; Bluhm, A. L. *J. Org. Chem.* **1956**, *21*, 336.
3. Goff, D. A.; Zuckermann, R. N. *J. Org. Chem.* **1995**, *60*, 5748–5749.
4. Zuckermann, R. N.; Kerr, J. M.; Kent, S. B. H.; Moos, W. H. *J. Am. Chem. Soc.* **1992**, *114*, 10646–10647.
5. Lorsbach, B. A.; Miller, R. B.; Kurth, M. J. *J. Org. Chem.* **1996**, *61*, 8716–8717.

6. Ruhland, T.; Kunzer, H. *Tetrahedron Lett.* **1996**, *37*, 2757–2760.

7. Buckman, B. O.; Mohan, R. *Tetrahedron Lett.* **1996**, *37*, 4439–4442.

8. Bunin, B. A.; Ellman, J. A. *J. Am. Chem. Soc.* **1992**, *114*, 10997–10998.

9. Smith, A. L.; Thomson, C. G.; Leeson, P. D. *Bioorg. Med. Chem. Lett.* **1996**, *6*, 1483–1486.

10. Brennan, S. T.; Colbry, N. L.; Leeds, R. L.; Leja, B.; Priebe, S. R.; Reily, M. D.; Showalter, H. D. H.; Uhlendorf, S. E.; Atwell, G. J.; Denny, W. A. *J. Heterocycl. Chem.* **1989**, *26*, 1469–1476.

11. Hutchins, S. M.; Chapman, K. T. *Tetrahedron Lett.* **1996**, *37*, 4865–4868.

12. Zuckermann, R. N.; Kerr, J. M.; Kent, S. B. H.; Moos, W. H. *J. Am. Chem. Soc.* **1992**, *114*, 10646–10647.

13. MacDonald, A. A.; DeWitt, S. H.; Hogan, E. M.; Ramage, R. *Tetrahedron Lett.* **1996**, *37*, 4815–4818.

4.7.29 Solution and Polymer-Supported Synthesis of 4-Thiazolidinones and 4-Metathiazanones [1]

Points of Interest

1. The three-component condensations of amino esters, aldehydes, and α-mercapto acids were performed both on support and in solution. Solution studies afforded moderate to high yields of 4-thiazolidinones and low yields of 4-metathiazanones. No further attempts were made to improve yields of the solution-phase reaction, as the authors' primary goal was to exploit the advantages of solid-phase organic synthesis to drive the reaction to completion. Polymer-supported synthesis gave high yields and high purities of 4-thiazolidinones and lower yields of 4-metathiazanones.

2. For the solution-phase condensation, a relative stoichiometry of 1:2:3 of amine:aldehyde:mercapto acid was optimal. The use of fewer equivalents of aldehyde and mercaptoacetic acid gave dramatically lower yields. For solid-phase synthesis, 15–25 equiv of solution reactants was routinely employed.

3. The polymer-supported[2] synthesis was performed on commercially available solid supports bearing an Fmoc-protected amino acid attached to either an acid-cleavable or photocleavable linker.[3]

4. Diastereomeric products are observed from the two new asymmetric centers formed in the reaction. Roughly equal mixtures of *cis* and *trans* isomers were observed when no chiral center was present in the amine component.

Representative Examples

Support-bound aliphatic amino esters (including the sterically hindered valine derivative), aromatic aldehydes, and mercapto carboxylic acids were incorporated into 4-thiazolidinones in high overall yields, although analogous solid-phase condensations to generate 4-metathiazanones were less successful.

Literature Procedures

For Solution-Phase Condensation. A mixture of the amine hydrochloride salt (12 mmol), aldehyde (24 mmol), mercaptoacetic acid (36 mmol), and DIEA (15 mmol) in 50 mL of benzene was heated to reflux with a Dean–Stark trap for 18 h during which time about 0.5 mL of water collected in the trap. The reaction mixture was cooled to room temperature and diluted with EtOAc. The organic phase was washed (saturated $NaHCO_3$, 1 N HCl, and saturated NaCl), dried ($MgSO_4$), and concentrated to give a colorless oil. Chromatography on silica gel (100% DCM to 3% acetone/DCM) afforded pure product as a colorless oil. The ester was taken on to the acid (for direct comparison to material obtained from support) without further characterization. To a solution of the ester (5 mmol) in 35 mL of MeOH was added 1 N NaOH (7.5 mL, 7.5 mmol). After 2 h of stirring at room temperature, the reaction mixture was partitioned between EtOAc and 1 N HCl. The organic phase was dried ($MgSO_4$) and concentrated to give a white semisolid. Chromatography on silica gel (1% HOAc/10% MeOH/89% $CHCl_3$) gave the pure product as a white solid.

For Solid-Phase Reaction. Commercially available Fmoc-amino resins containing an acid-cleavable linker were used. The terminal Fmoc group was deprotected by immersing the resin in 30% piperidine/DMF for 30 min in a standard peptide synthesis vessel followed by washing with DMF, DCM, MeOH, and Et_2O. The resin was dried under reduced pressure and transferred to glass vials equipped with Teflon caps ($\approx$50 mg of resin in a 4-mL vial), and 2 mL of a solution of 0.25 M aldehyde and 0.5 M mercapto carboxylic acid in THF was added along with 20–30 pellets of 3-Å molecular sieves. The vial was capped and heated to 70°C for 2 h with occasional shaking. The reaction mixture was cooled to room temperature and transferred to a disposable filter tube. The resin was washed (DCM, DMF, MeOH, and Et_2O), dried under reduced pressure, and treated with a solution of TFA/DCM (1:1) for 30 min to release the polymer-bound heterocycle. Removal of the solvent under reduced pressure followed by preparative HPLC afforded the pure 4-thiazolidinone or 4-metathiazanone. Alternatively, in a stepwise procedure the deprotected resin was treated with a 0.25 M solution of aldehyde in THF for 2 h at 70°C. The resin was washed (THF, MeOH, Et_2O) and subsequently treated with a 0.5 M (a 0.5m) solution of mercapto carboxylic acid for 2 h at 70°C. Washing and cleavage in the foregoing afforded the 4-thiazolidinones or 4-metathiazanones.

4.7.30 Solid-Phase Synthesis of Thiazolidines[4]

Points of Interest

1. The authors reference derivatives of the thiazolidine skeleton that exhibit significant antimicrobial, antihypertensive, and antitumor activity.

2. Racemic *N*-acylthiazolidines were obtained as a single diastereomer since epimerization occurs upon hydrolysis from PEG–PS resin.

3. The *N*-acylthiazolidines prepared from electron-rich aromatic aldehydes were unstable to TFA (which is commonly used for side-chain deprotection). However, oxidation of the thioether to the sulfoxide was accomplished in good yield using MCPBA, providing acid-stable products. The analogous desired sulfones were not obtained under a variety of conditions.

Representative Examples

Formaldehyde, aliphatic, and electron-rich aromatic aldehydes were incorporated into support-bound thiazolidines. Thiazolidines were acylated with acetic anhydride, benzoyl chloride, or an isocyanate.

Literature Summary

Fmoc-Cys-(Trt)-OH was attached directly to TentaGel S-OH resin with DIC/HOBt/NMM for 18h. The Fmoc and trityl protecting groups were removed by treatment with 20% piperidine/DMF (20 min) and TFA/5% *i*-Bu$_3$SiH/DCM (3 × 30 min). Reaction with an electron-rich aldehyde (*p*-anisaldehyde) in a mixture of toluene/acetonitrile/acetic acid (45:45:10) afforded 2-(4-methoxyphenyl)thiazolidinecarboxylic acid (the intermediate was not isolated), which after reaction with 5 equiv of Ac$_2$O/pyridine in DMF and final hydrolytic cleavage from the resin (0.5% NaOH) afforded the corresponding acylthiazolidine in >97% purity (HPLC) following extractive workup.

REFERENCES FOR SECTIONS 4.7.29 AND 4.7.30

1. Holmes, C. P.; Chinn, J. P.; Look, G. C.; Gordon, E. M.; Gallop, M. A. *J. Org. Chem.* **1995**, *60*, 7328–7333.

2. SASRIN resin (polystyrene) from Bachem BioScience Inc, TentaGel (PEG–PS) from Rapp Poly-mere, and Novasyn PR 500 (poly(dimethylacrylamide)/polyhipe) from Novabiochem were all used with equal success.

3. For the photocleavable linker, see: Homes, C. P.; Jones, D. G. *J. Org. Chem.* **1995**, *60*, 2318–2319.

4. Patek, M.; Drake, B.; Lebl, M. *Tetrahedron Lett.* **1995**, *36*, 2227–2230.

4.7.31　Combinatorial Synthesis of Highly Functionalized Pyrrolidines [1]

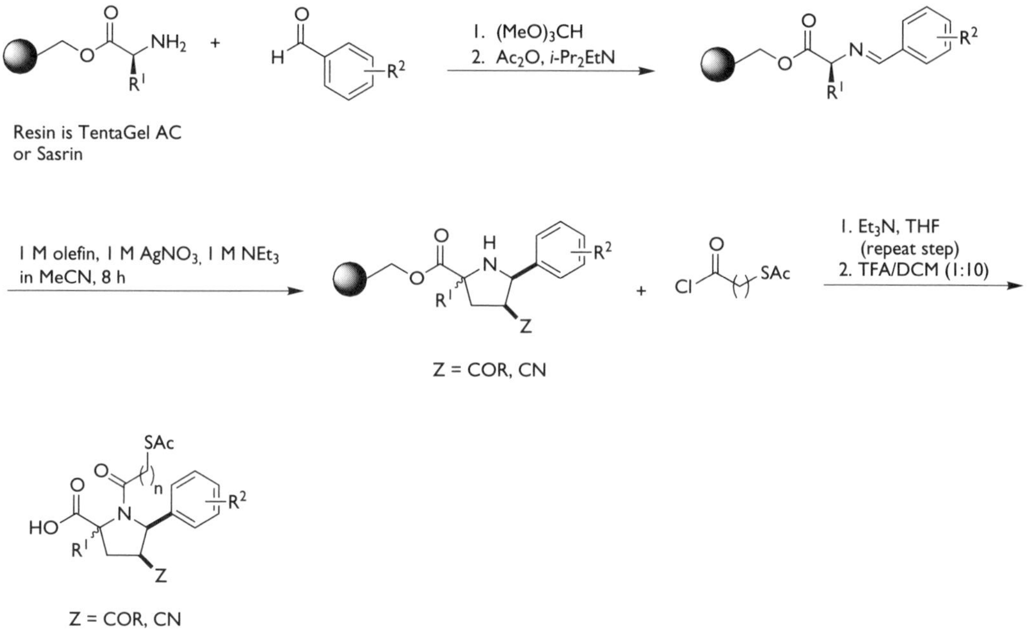

Resin is TentaGel AC
or Sasrin

Z = COR, CN

Z = COR, CN

Points of Interest

1. A combinatorial library of ≈500 mercaptoacyl prolines was prepared. A potent inhibitor, approximately 3-fold more active than captopril against angiotensin converting enzyme (ACE), was identified from the library.

2. The authors have previously reported dehydrating conditions for imine formation on support employing either neat trimethyl orthoformate or trimethyl orthoformate/DCM.[2]

3. The 1,3-dipolar cycloaddition was performed with the azomethine ylide derived from the support-bound Schiff base and an acrylate or acrylonitrile as the dipolarophile under Lewis acid-mediated conditions that are analogous to those initially reported by Grigg and Tsuge for the corresponding reaction in solution.[3]

4. Six representative racemic substituted ω-(acylthio) proline derivatives were isolated in 50–80% yield and the diastereoselectivities ranged from 2.5:1 to 10:1 following cleavage from support.

5. Reaction optimization was performed with the aid of gel-phase ^{13}C NMR using ^{13}C-enriched building blocks in the synthesis (see p 220).[4]

6. Pyrrolidines have been prepared via 1,3-dipolar cycloadditions between support-bound enones and an azomethine ylide derived from glycine methyl ester in the presence of DBU and LiBr.[5]

Representative Examples

A library was prepared from the amino acids Gly, Ala, Leu, and Phe, electron-rich aromatic aldehydes and benzaldehyde, electron-poor alkenes, and thioacyl chlorides.

Literature Procedure

TentaGel-AC preloaded with an Fmoc-protected amino acid (0.50 g/loading, 0.20–0.26 mmol) was added to a solution of 20% piperidine in DMF (3 mL) and gently vortexed every 5 min for 20 min to remove the Fmoc protecting group. The resin was filtered using a fine Büchner funnel and washed with DMF and DCM (2×). The resin was added to a 1.0 M solution of aromatic aldehyde in trimethyl orthoformate (4 mL), vortexed gently, and left for 4 h. The resin was again filtered using a fine Büchner funnel and washed with DCM (2×). The resin was then added to a solution of silver(I) nitrate and the appropriate olefin, each at 1.0 M. To the solution was added a 1 molar equivalence of triethylamine, and the resulting solution was gently vortexed and left for 4–8 h. The solution turned black after 5–10 min, with plating of silver upon the walls of the vessel occurring after 2 h. The resin was filtered using a fine Büchner funnel and washed with saturated ammonium chloride (2×), MeOH (2×), and DCM (2×). The product was cleaved from the resin by treatment with a 10% solution of trifluoroacetic acid in DCM (2 mL) for 30 min. The solution was filtered and evaporated to dryness, leaving 10–30 mg of crude product. After semipreparative HPLC two diastereomeric products (each product is a racemic mixture) were isolated.

4.7.32   Solid-Phase Synthesis of Proline Analogs via a Three-Component 1,3-Dipolar Cycloaddition[6]

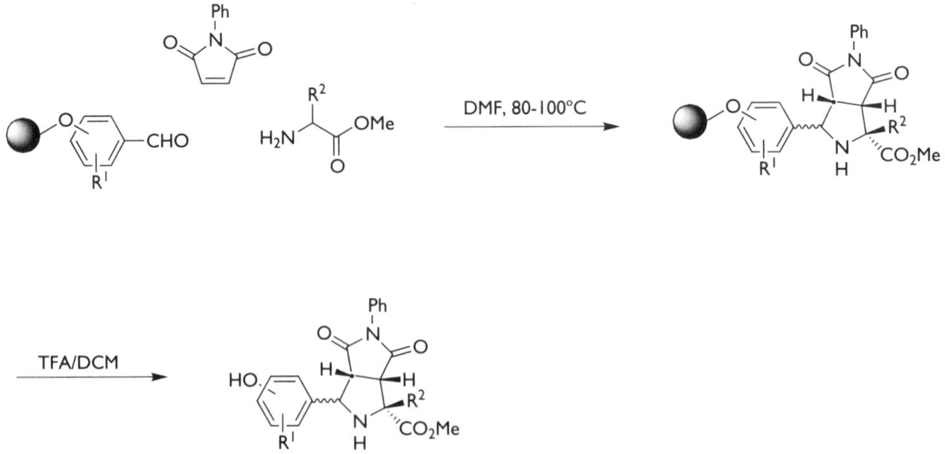

Points of Interest

1. The 1,3-dipolar cycloaddition is similar to the method described by Murphy and coworkers although the pyrrolidines were synthesized in a one-pot, three-component cycloaddition instead of by a two-step process with a Lewis acid.

2. Bicyclic pyrrolidines were formed with maleimide as the activated dipolarophile.

3. Hydroxybenzaldehydes were attached to Wang resin via a Mitsunobu coupling (see p 106). Nine substituted benzaldehydes were attached to Wang resin with loadings ranging from 0.24 to 0.89 mequiv/g.

Representative Examples

Substituted hydroxybenzaldehydes were attached to support for a three-component coupling with an amino ester and N-phenylmaleimide. Yields were not reported, although a mixture of products was analyzed by LC-MS from a mixture of different aldehyde resins.

Literature Summary

Wang resin (15 g, 15 mmol) was rinsed with DCM and anhydrous THF (3×). The resin was then treated with 100 mL of anhydrous THF, 5.90 g (22.5 mmol) of triphenylphosphine, and 6.11 g (50 mmol) of 4-hydroxybenzaldehyde. Diethyl azodicarboxylate (3.95 mL, 25 mmol) was added dropwise to the stirred slurry over a period of 15 min. After the mixture was allowed to stir overnight, the reaction solvent was removed and the resin washed three times each with THF, DMF, MeOH, DCM, and ether. The resin was collected and dried *in vacuo* to afford 15.25 g of resin (0.70 mequiv/g).

In a dry 15-mL vial containing a magnetic stirrer (stirring was important because some reagents were sparingly soluble) were placed 0.25 g (0.13 mmol) of support-bound aldehyde, 75 mg (0.42 mmol) of leucine methyl ester hydrochloride, 71 mg (0.42 mmol) of N-phenylmaleimide, and 8 mL of dry DMF. The mixture was treated with 60 μL (0.2 mmol) of TEA followed by the addition of two drops of acetic acid. The vial was capped, magnetically stirred, and treated at 100°C for 18 h. After the mixture cooled, the resin was filtered, washed three times each with DMF, MeOH, DCM, and ether, and dried *in vacuo*. The product was obtained as a mixture of isomers by cleavage of the resin twice with TFA/DCM (1:1) for 1 h.

REFERENCES FOR SECTIONS 4.7.31 AND 4.7.32

1. Murphy, M. M.; Schullek, J. R.; Gordon, E. M.; Gallop, M. A. *J. Am. Chem. Soc.* **1995**, *117*, 7029–7030.

2. Look, G. C.; Murphy, M. M.; Campbell, D. A.; Gallop, M. A. *Tetrahedron Lett.* **1995**, *36*, 2937–2940.

3. (a) Barr, D. A.; Grigg, R.; Gunaratne, H. Q. N.; Kemp, J.; McMeekin, P.; Sridharan, V. *Tetrahedron* **1988**, *44*, 557–570. (b) Tsuge, O.; Kanemasa, S.; Yoshioka, M. *J. Org. Chem.* **1988**, *53*, 1384–1391.

4. Look, G. C.; Holmes, C. P.; Chinn, J. P.; Gallop, M. A. *J. Org. Chem.* **1994**, *59*, 7588–7590.

5. Hollinshead, S. P. *Tetrahedron Lett.* **1996**, *37*, 9157–9160.

6. Hamper, B. C.; Dukesherer, D. R.; South, M. S. *Tetrahedron Lett.* **1996**, *37*, 3671–3674.

4.7.33 Solid-Supported Combinatorial Synthesis of β-Lactams [1]

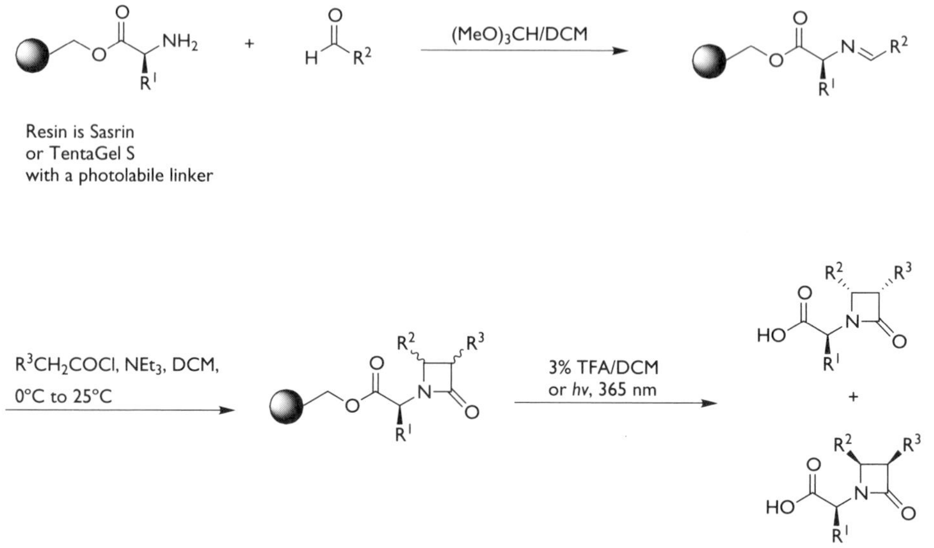

Resin is Sasrin
or TentaGel S
with a photolabile linker

Points of Interest

1. The Staudinger reaction was performed between a resin-bound imine [2] and an excess of ketene in solution. This has the advantage of facile isolation of the product β-lactam from unreacted reagents and reaction byproducts (diketene etc.) by filtration.

2. Cycloaddition reactions of achiral ketenes with resin-bound imines derived from homochiral amino acids and either aromatic or α,β-unsaturated aldehydes were highly *cis* selective but proceeded with only modest levels of stereoinduction from the asymmetric center of the amino acid (in 1:1 to 3:1 ratios).

3. β-Lactams were prepared on both acid-labile and photolabile linkers.

4. α-Amino-β-lactams were prepared with the ketene derived from phthalimidoacetyl chloride. Treatment of resin-bound α-phthalimido-β-lactam derivatives with N-methylhydrazine afforded the corresponding 3-amino-2-azetidinones. The 3-amino functionality was further derivatized under standard solid-phase peptide bond-forming conditions with Fmoc-valine.

5. A small combinatorial β-lactam library was prepared from valine-SASRIN resin, five aldehydes, and ketenes derived from three acid chlorides.

6. The [2 + 2] cycloaddition could be monitored by magic-angle-spinning (MAS) NMR spectroscopy on support. [3]

Representative Examples

[2 + 2] cycloaddition reactions were performed with a number of different ketenes, including N-protected amino ketenes, O-protected hydroxy ketenes, and vinyl ketenes. Using a large molar excess of ketene at high concentrations, even cycloadditions to sterically hindered amino acids could be driven to completion. Aliphatic, aromatic, and heteroaromatic aldehydes as well as enals were incorporated into support-bound imine precursors.

Literature Procedure

SASRIN resin preloaded with an N-Fmoc-protected amino acid (0.165 mmol, 0.3 g of resin, loading 0.55 mmol/g) was treated with a solution of 30% piperidine in NMP for 45 min in a standard peptide flask. The resin was rinsed with NMP or DMF, DCM, MeOH, and Et$_2$O and dried under reduced pressure. The resin was suspended in a mixture of DCM (1.5 mL) and trimethyl orthoformate (1.5 mL), and the aldehyde (2.3 mmol) was added. After 3 h of agitation, the resin was rinsed with DCM, MeOH, and Et$_2$O and dried under reduced pressure. The resin was transferred to a glass vial, suspended in DCM (3 mL), and cooled to 0°C. To the suspension was added triethylamine (3.3 mmol) followed by a slow addition of the acid chloride (2.5 mmol). The reaction mixture was left at 0°C for 5 min and agitated overnight at room temperature. The resin was filtered, rinsed (DMF, DCM, MeOH, Et$_2$O), and dried under reduced pressure. The product was cleaved from the support by treating the resin with a solution of 3% (v/v) TFA/DCM for 45 min. For products derived from amino acids requiring acid-labile side-chain protection, the crude material was subjected to a second TFA treatment (50% TFA/DCM) to remove these protecting groups. The solution was filtered, and after removal of the solvent the crude product was purified by preparative HPLC. From reactions where mixtures of two cis-β-lactams were formed, no attempt was made to separate the individual diastereomers or enantiomers.

REFERENCES FOR SECTION 4.7.33

1. Ruhland, B.; Bhandari, A.; Gordon, E. M.; Gallop, M. A. *J. Am. Chem. Soc.* **1996**, *118*, 253–254.
2. Look, G. C.; Murphy, M. M.; Campbell, D. A.; Gallop, M. A. *Tetrahedron Lett.* **1995**, *36*, 2937–2940.
3. Fitch, W. L.; Detre, G.; Holmes, C. P.; Shoolery, J. N.; Keifer, P. A. *J. Org. Chem.* **1994**, *59*, 7955–7956.

4.7.34   Dihydrobenzopyrans Utilizing the Split Synthesis Method and the Haloaromatic Tag-Encoding Strategy[1]

$X = O, S_2(CH_2)_2, OH$

Points of Interest

1. SAR of tag-encoded libraries were targeted toward carbonic anhydrase as a model. Compounds with nanomolar dissociation constants were identified.

2. Dihydrobenzopyrans were prepared from the condensation of ketones with hydroxyacetophenones attached to PEG–PS via a photolabile linker.

3. Support-bound ketones were converted to alcohols or dithioacetals.

4. The authors report that in general the dihydrobenzopyrans were isolated with purities of >80%; however, the exact number of compounds and the structures of the compounds were not provided. Few details of the library synthesis or analytical evaluation were provided.

Representative Examples

Three support-bound hydroxyacetophenones and seven ketones (four of which contained a protected amine functionality for further derivatization) were condensed. Both the amine and ketone functional groups were modified on support.

Literature Procedure

In solution, three dihydroxyacetophenones were coupled using Mitsunobu conditions to the *o*-nitrobenzyl (photolabile) linker, which was then coupled to the beads (TentaGel S-NH$_2$) under standard conditions (DIC/HOBt in DMF). The intermediates were cyclized on the resin with a set of seven ketones, four of which contained a protected amine functionality. Modification of the amine functionality (when present) with 31 different headgroups under standard conditions produced a library of ketones containing 381 members. The ketones were then reduced with NaBH$_4$, converted into the corresponding dithiolanes with ethane-1,2-dithiol, or left unaltered to produce a library of 1143 distinct compounds.

REFERENCES FOR SECTION 4.7.34

1. Burbaum, J. J.; Ohlmeyer, M. H. J.; Reader, J. C.; Henderson, I.; Dillard, L. W.; Li, G.; Randle, T. L.; Sigal, N. H.; Chelsky, D.; Baldwin, J. J. *Proc. Natl. Acad. Sci. U.S.A.* **1995**, *92*, 6027–6031.

4.7.35   Bicyclo[2.2.2]octane Derivatives via Tandem Michael Addition Reactions[1]

Points of Interest

1. Two carbon–carbon bonds were formed via tandem Michael reaction between cyclohexenone enolates and support-bound acrylates.

2. The intermolecular tandem Michael reaction (first reported in solution by Lee)[2] was adapted to Wang resin and followed by reductive amination and three modes of cleavage from support.

3. *Endo/exo* selectivities for the cyclization were usually greater than 9:1 *endo*. Reductive amination required ultrasound conditions and sodium sulfate; surprisingly only the *endo* products from cyclization reacted under these conditions. The authors speculated that perhaps the absence of any *exo* product was caused by a retro-Michael reaction under the acidic conditions.

Representative Examples

Tandem cyclizations were performed between cross-conjugated dienolate bases derived from 3-methyl-2-cyclohexenone or 3-ethoxy-2-cyclohexenone and aliphatic, aromatic, or heteroaromatic support-bound acrylic acid derivatives. Reductive amination of the support-bound ketones incorporated primary and secondary amines or aniline. Products were cleaved from support by three mechanisms to provide acids, amides, or alcohols.

Literature Procedure

Commercially available Wang resin was coupled with a series of acrylic acids (5 equiv) using DIC (3.5 equiv) and DMAP (catalyst) in DCM at room temperature. Reactions were monitored by IR and gel-phase ^{13}C NMR in CD_2Cl_2. The derivatized resins were treated with the cross-conjugate dienolate bases derived from 3-methyl-2-cyclohexenone and 3-ethoxy-2-cyclohexenone (5 equiv of LDA and 5 equiv of cyclohexenone) at −78°C and warmed to room temperature to provide polymer-bound bicyclo[2.2.2]octanones. Cleavage with DIBAL in toluene at −78°C provided a range of bicyclo[2.2.2]octanones in good yield (average 72.5%) and high *endo/exo* selectivities[3] (although DIBAL reduction of the support-bound ketone was less selective). Bicyclo[2.2.2]octanones were further manipulated by reductive amination. Reductive amination required ultrasound conditions and sodium sulfate; surprisingly only the *endo* products reacted under these conditions. The exact conditions employed the amine (15 equiv), NaBH(OAc)$_3$ (5 equiv), Na$_2$SO$_4$ (10 equiv), and AcOH/DCM (1%) with ultrasound. Cleavage by DIBAL reduction provided alcohols (average 63% yield), cleavage under acidic conditions with TFA/DCM (1:1, for 2 min) provided carboxylic acids as their TFA ammonium salts (average 72% yield), and cleavage with BnHNAlMe$_2$ (5 equiv) in DCM afforded secondary amides (average 26% yield).

REFERENCES FOR SECTION 4.7.35

1. Ley, S. V.; Mynett, D. M.; Koot, W. J. *Synlett* **1995**, 1017–1020.
2. Lee, R. A. *Tetrahedron Lett.* **1973**, *35*, 3333.
3. Bateson, J. H.; Smith, C. F.; Wilkinson, J. B. *J. Chem. Soc., Perkin Trans. 1* **1991**, 651–653.

4.7.36   Solid-Phase Synthesis of Tropane Derivatives via a Palladium-Mediated Three-Component Coupling Strategy [1]

Points of Interest

1. Tropane derivatives display diverse pharmacological properties and include muscarinic cholinergic antagonists, antiemetics, and anticognition disorder agents.[2]

2. The first step in the three-component coupling produces a stable support-bound palladium–arene complex. Because the palladium intermediate is isolated, split synthesis strategies may be employed for the generation of tropane libraries.

3. The palladium intermediates can react with diverse nucleophiles as shown in the reaction scheme.

4. Removal of the [[(trimethylsilyl)ethyl]oxy]carbonyl (Teoc) group can be monitored by IR analysis by following the disappearance of the carbamate carbonyl stretch. The resulting secondary amine was further derivatized by reductive amination or acylation with HATU.[3]

Representative Examples

Electron-rich and electron-poor aromatic substituents as well as alkynes and a hydride were incorporated into tropane derivatives by a three-component coupling. Acylation or reductive amination of the secondary amine afforded a range of tropane derivatives (average 64% yield).

Literature Procedure

To a slurry of dihydropyranyl-derivatized resin (7.70 g, 0.48 mequiv/g, 3.70 mmol)[4] and 4.00 g (14.9 mmol) of the Teoc-protected tropane scaffold in DCM (24 mL) was added 0.64 g (3.70 mmol) of p-toluenesulfonic acid. The reaction slurry was gently stirred at room temperature for 24 h. The resin was filtered and washed with DCM (2×), DCM/TEA (99:1, 3×), and DCM (3×) and dried *in vacuo*. The filtrate was washed with 1 N NaOH to remove p-toluenesulfonic acid, and the organic layers were dried and concentrated. Purification of the residue resulted in the recovery of 3.0 g of the starting secondary alcohol for reuse.

The loading of the resin was determined as follows. To a solution of pyridinium p-toluenesulfonate (0.12 g, 0.47 mmol) in butanol/dichloroethane (1:1) was added a portion of the derivatized resin (0.50 g, 0.24 mmol). The resulting slurry was heated at 60°C overnight. The resin was filtered and washed with DCM (5×), and the combined organic layers were washed with water (2×) and saturated. NaHCO$_3$ (2×). The organic layers were then dried and concentrated. The crude product was purified on silica gel eluting with ethyl acetate/hexanes (1:1) to give 40 mg of the starting alcohol. The calculated loading level was 0.30 mequiv/g.

The general procedure for the solid-phase synthesis of tropanes follows. In a 25-mL flask fitted with a magnetic stir bar was suspended 0.40 g (0.12 mmol) of derivatized resin in degassed THF (9 mL) or anisole (9 mL) followed by the addition of an aryl bromide or aryl iodide (0.36 mmol) and Pd(PPh$_3$)$_4$ (0.42 g, 0.36 mmol). The reaction apparatus was purged with nitrogen for 5 min. The flask was then fitted with a reflux condenser and the reaction mixture was heated at reflux for 48 h under nitrogen to form the support-bound palladium complex. The yellow/orange solution was removed from the resin by filtration cannula, the resin was rinsed with degassed THF (4×), and then degassed THF (9 mL) or anisole (9 mL) was added. To the resulting suspension were added an arylboronic acid (0.60 mmol), 2 N Na$_2$CO$_3$ (6 mL), and PPh$_3$ (94 mg, 0.36 mmol), generating a biphasic solution. The suspension was heated under nitrogen at 66°C for 54 h. The yellow solution was removed from the resin by filtration cannula and the resin was rinsed with DMF/H$_2$O (3×), DMF (3×), THF (3×), and DCM (3×). The resin was then dried *in vacuo* to a constant weight. To the resin suspended in 9 mL of distilled THF was added TBAF (256 mg, 0.60 mmol), and the reaction mixture was heated under nitrogen at 50°C for 20 h. The solution was removed from the resin by filtration cannula and the resin was rinsed with THF (3×), MeOH, and DCM (3×). The resin was dried *in vacuo* to a constant weight. Complete removal of the Teoc group could be established by IR (KBr) analysis by monitoring for complete disappearance of the carbamate carbonyl stretch.

Reductive amination of the secondary amine on support was performed as follows. The dried resin was added to a solution of aldehyde (0.60 mmol) in 1% HOAc in DMF (9 mL). The mixture was gently stirred for 1 h, then NaBH(OAc)$_3$ was added, and the resulting suspension was gently stirred for 35 h, at which point MeOH was added to the resin to quench the excess reductant and dissolve the borate salts. The solution was removed from the resin via filtration cannula and the resin was rinsed with DMF (3×), DMF/water (3×), THF (3×), and DCM (3×). The fully derivatized support-bound tropane was cleaved by stirring in 5 mL of TFA/water (95 : 5) for 30 min, filtered, rinsed with DCM, and concentrated to yield a crude yellow residue. The residue was dissolved in DCM (15 mL), and the organic layers were washed with 1 N NaOH (10 mL), dried, concentrated, and purified by silica gel chromatography.

REFERENCES FOR SECTION 4.7.36

1. Koh, J. S.; Ellman, J. A. *J. Org. Chem.* **1996,** *61,* 4494–4495.
2. Forder, G.; Dharanipragada, R. *Nat. Prod. Rep.* **1994,** *11,* 443–450.
3. Carpino, L. A. *J. Am. Chem. Soc.* **1993,** *115,* 4397–4398.
4. Thompson, L. A.; Ellman, J. A. *Tetrahedron Lett.* **1994,** *35,* 9333–9336.

4.7.37   Solid-Phase Synthesis of Azabicyclo[4.3.0]nonen-8-one Amino Acid Derivatives via Intramolecular Pauson-Khand Cyclization[1]

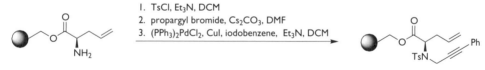

Note: Pauson-Khand cyclization
of propargyl glycine on
Wang resin was also performed.

Points of Interest

1. Either regioisomeric cyclopentenone derivative could be obtained depending on the allyl- or propargylglycine derivative used as the starting material.

2. Carbon–carbon bonds were formed on support via palladium- and cobalt-mediated reactions. A seven-step sequence, followed by chromatography afforded a mixture of diastereomers, in an overall yield of 48% based on initial loading.

Representative Examples

Bicyclic tosyl amino acid derivatives were produced in all cases.

Literature Summary

Unnatural amino acids were prepared in solution and loaded onto Wang resin as the mixed anhydride.[2] To the support-bound Pauson–Khand precursor (0.53 g, 0.34 mmol, 0.64 mmol/g) in DCM (10 mL) was added $Co_2(CO)_8$ (0.17 g, 0.51 mmol). The suspension was shaken under nitrogen for 2 h with periodic venting. The resin was filtered and washed with DCM (3×). The resin was suspended in DCM (10 mL), and N-methylmorpholine N-oxide (0.13 g, 1.11 mmol) was added. The mixture was shaken under nitrogen with periodic venting for 1 h, and another portion of NMO (0.13 g) was added. After an additional hour of shaking, the resin was filtered and washed with DCM (3×), AcOH/DCM (1:3 ratio, 3×), and DCM (3×). The resin was then shaken with TFA/DCM (1:1) for 1 h, filtered, and washed with DCM (3×). The combined filtrates were concentrated, reconcentrated with DCM (2×), and dried *in vacuo* to give 0.11 g of a bicyclic tosyl amino acid. Esterification with diazomethane provided the corresponding methyl ester.

REFERENCES FOR SECTION 4.7.37

1. Bolton, G. L. *Tetrahedron Lett.* **1996**, *37*, 3433–3436.
2. Sieber, P. *Tetrahedron Lett.* **1987**, *28*, 6148–6150.

4.7.38   Pictet–Spengler Reaction on an Oxime Resin: Synthesis of 1,2,3,4-Tetrahydro-β-carboline Libraries[1]

Points of Interest

1. β-Carbolines from a matrix synthesis were prepared in 71–95% yield by HPLC and LC–ES/MS.

2. The authors mentioned that other functionalized tryptophan analogs can be used in the Pictet–Spengler reaction and that unacylated β-carbolines can be cleaved with a wide variety of amines, although details were not provided.

Representative Examples

Aliphatic and aromatic aldehydes were condensed with tryptophan on oxime resin. β-Carbolines were acylated with acid chlorides, sulfonyl chlorides, or isocyanates.

Literature Summary

Oxime resin (6.0 g, 0.38 mmol/g) suspended in 50 ml of DCM was shaken for 15 min. Boc-L-Tryptophan (1.1 g, 3.5 mmol) was added followed by DIC (548 μL, 3.5 mmol). The resin was shaken for 16 h and washed with DCM (3×), 2-propanol (3×), and DCM (3×). Boc-L-Tryptophan-derivatized oxime resin (0.7 g, 0.38 mmol/g) was washed with DCM (3×). To remove the protecting group, 25% TFA in DCM (3 mL) was added and the resin shaken for 1 min. The resin was filtered to remove the TFA solution (N_2). An additional 3 mL of 25% TFA in DCM was added and the resin shaken for 6 h. The resin was again filtered to remove the second TFA solution. Fresh TFA (25% solution in DCM, 3 mL) was added followed by the aldehyde (6 mmol, 15 equiv, 2 M). The reaction mixture was shaken for 6 h. The resin was then filtered and washed with DCM (3×), DMF (3×), and DCM (3×). Cleavage of a resin aliquot with 0.5 mL of saturated ammonia in ethanol for 2 min followed by HPLC analysis indicated formation of the desired tetrahydro-β-carboline product. To the tetrahydro-β-carboline on solid support (175 mg, 0.067 mmol) in DCM (0.5 mL) was added pyridine (0.20 mL of a 1 M solution in DCM) followed by the acid chloride or sulfonyl chloride (0.20 mL of a 1 M solution in DCM). For the isocyanate, the pyridine was omitted. The resin was shaken for 2 h, filtered, and washed with DCM, DMF, and DCM. The resin was then treated for 2 h with 2 mL of a mixture (1:1, v/v) of DCM and saturated ammonia in MeOH. The resin was filtered, and the filtrate was collected and evaporated to afford the 1,2,3,4-tetrahydro-β-carboline as a solid.

4.7.39 Pictet–Spengler Reaction on a Nitrophenyl Carbonate Resin: Solid-Phase Synthesis of 1,2,3,4-Tetrahydro-β-carbolines[2]

or

Points of Interest

1. Support-bound 4-nitrophenyl chloroformate[3] (see p 62) was treated with excess diamine to minimize resin cross-linking.

2. The support-bound Pictet–Spengler reaction was performed with molecular sieves in contrast to the other studies[1,4,5] where a TFA catalyst was employed for cyclization.

3. Efforts to generate additional diversity by derivatizing the secondary nitrogen at position 2 of the tetrahydro-β-carboline were unsuccessful. For example, the acid chloride of Fmoc-glycine did not acylate the hindered secondary amino group.

Representative Examples

Nine 1,2,3,4-tetrahydro-β-carbolines were prepared with aromatic and aliphatic aldehydes (average 68% yield by HPLC).

Literature Summary

Nitrophenyl carbonate resin[3] (0.08 mmol) and 500 mg of piperazine in 1.8 mL of DCM were shaken for 4 h and washed with DMF and DCM. The secondary amino groups on the resin were acylated with Fmoc-Trp hydroxybenzotriazole ester or reductively alkylated by Fmoc-tryptophanal (detailed procedures for these steps were not provided) followed by Fmoc deprotection. In a typical procedure 200 mg of 2,4-dichlorobenzaldehyde (1.14 mmol) and 1 mL of DCM were added to 100 mg of derivatized resin in a closed reaction vessel containing 3-Å molecular sieves and the resulting mixture was shaken 2 h at 50°C. If the ninhydrin test showed the presence of a primary amino group, then the procedure was repeated. The 1,2,3,4-tetrahydro-β-carbolines were cleaved from the resin with TFA/EDT/H$_2$O (1 mL : 250 μL : 50 μL).

4.7.40 Pictet–Spengler Reaction on Merrifield Resin[4]

Points of Interest

1. The Pictet–Spengler reaction was initially investigated under neutral conditions in aprotic solvents such as benzene or toluene. The reaction was very

slow at room temperature. Running the reaction at 80°C overnight effected cyclization to give the desired product with modest purity (80%) after acidic cleavage. This approach was abandoned due to the inconvenience of heating and the insolubility of many aldehydes in toluene. Cyclization with 10 equiv of aldehyde in 1–50% TFA in DCM afforded the desired product in >90% purity after cleavage. The 1,2,3,4-tetrahydro-β-carbolines were cleaved from Merrifield resin with volatile primary amines.

2. To avoid *tert*-butylalkylation during Boc deprotection, excess indole or thioanisole was added to the cleavage cocktail (TFA/DCM).

3. Diastereomeric ratios were not determined after cleavage.

Representative Examples

Products were cleaved from resin with small, volatile amines to afford amides usually in >80% yields. Electron-rich and electron-poor aldehydes as well as aliphatic aldehydes were incorporated. Halides and boronic acids were incorporated to provide sites in the molecule for additional diversity. With the exception of 4-nitro substitution, substitution of the benzene ring has little effect on the product purity or yield.

Literature Summary

Commercially available Boc-Trp-Merrifield resin was treated with TFA/DCM (1:1) and excess indole to remove the protecting group. To a suspension of the derivatized resin (100 mg, 0.6 mmol/g) in 2 mL of 10% TFA/DCM with indole (2 mg, optional) was added benzaldehyde (60 μL). After overnight shaking, the resin was washed with DCM (3×), 5% DIEA/DCM (2×), 20% MeOH/DCM (3×), DCM (3×), and THF (4×). The product was cleaved from the resin by overnight shaking with a mixture of THF (1 mL) and ethylamine (70% in water, 1 mL). The resin was filtered and washed with MeOH/THF (3×). The filtrate and washings were concentrated to afford the product (19 mg) in 98% crude yield (94% purity by HPLC).

4.7.41   Pictet–Spengler Reaction on Wang Resin [5]

Points of Interest

1. The critical Pictet–Spengler reaction occurred under mild conditions with 1% TFA in DCM on Wang resin. The Pictet–Spengler reactions were monitored by the Kaiser ninhydrin test (see p 214) and were complete in 2–4 h for aldehydes and 48–72 h for ketones.

2. The 1,2,3,4-tetrahydro-β-carbolines were removed from Wang resin with 95% TFA/H_2O. HPLC was used to analyze the crude samples and revealed the presence of both diastereomers in a majority of the aldehyde-derived products. Purities of the crude compounds were typically in excess of 85% based on HPLC analysis; however, a number of methyl ketones exhibited purities of only around 50%. In these cases the sole contaminant was identified as unreacted tryptophan.

Representative Examples

Aromatic and aliphatic aldehydes as well as aliphatic and heteroaromatic ketones were incorporated into β-carbolines.

Literature Summary

A 1-g portion of Fmoc-L-tryptophan-Wang resin (available from a number of commercial sources) with a typical loading range of 0.5–0.7 mmol/g was treated with 20% piperidine in DMF (2×)to remove the Fmoc protecting group. The resin was then washed several times with DMF followed by DCM followed by the addition of 4 equiv of an aldehyde or ketone in a solution of 1% TFA in DCM. The reaction vessel was then agitated at room temperature for 2–4 h with aldehydes and for 48–72 h in the presence of ketones. The reaction could be monitored by the Kaiser ninhydrin test (see p 214). Cleavage from the support was accomplished by suspending and stirring the resin in neat TFA (5% H_2O) for 2 h at room temperature.

REFERENCES FOR SECIONS 4.7.38 to 4.7.41

1. Mohan, R.; Chou, Y.-L.; Morrissey, M. M. *Tetrahedron Lett.* **1996,** *37,* 3963–3966.
2. Kljuste, K.; Unden, A. *Tetrahedron Lett.* **1996,** *37,* 9211–9214.
3. Dressman, B. A.; Spangle, L. A.; Kaldor, S. W. *Tetrahedron Lett.* **1996,** *37,* 937–940.
4. Yang, L.; Guo, L. *Tetrahedron Lett.* **1996,** *37,* 5041–5044.
5. Mayer, J. P.; Bankaitis-Davis, D.; Zhang, J.; Beaton, G.; Bjergarde, K.; Anderson, C. M.; Goodman, B. A.; Herrera, C. J. *Tetrahedron Lett.* **1996,** *37,* 5633–5636.

4.7.42 Fischer Indole Synthesis on Solid Support[1]

Points of Interest

1. Reaction temperature, reaction time, and resin type were all important variables for the clean production of indole products on support. Optimal cyclization conditions were found in acetic acid with either TFA or $ZnCl_2$.

2. At higher temperatures required for reactions employing electron-deficient hydrazines (e.g., 2,4-dichlorophenylhydrazine hydrochloride), the yields began to decline and impurities related to the PEG–PS support became problematic. When polystyrene resin (Advanced ChemTech (aminomethyl)polystyrene resin loaded with HMB handle using PyBOP) was used, the desired 5,7-dichloroindole was prepared in excellent purity from 2,4-dichlorophenylhydrazine hydrochloride and the support-bound phenyl ketone at 70°C for 18 h.

3. A small library was also prepared by dendrimer-supported combinatorial chemistry employing the Fischer indole synthesis. Reactions are perfomed in solution and dendrimeric intermediates are separated by size-selective methods such as size exclusion chromatography (SEC) or ultrafiltration.[2]

Representative Examples

A variety of phenylhydrazine hydrochlorides were utilized in this reaction. Monosubstitution with alkyl and halo groups was well tolerated. 2,4-Dichlorophenylhydrazine also afforded the desired product in high chemical purity. Even 2,5-disubstituted hydrazines which produced sterically hindered 4,7-disubstituted indoles work moderately well, yielding final products in 74% purity.

Literature Summary

A double coupling was performed on 50 mg of PS-HMB resin using 4 equiv of Fmoc-Phe and EDIC in the presence of catalytic DMAP in 2 mL of THF/DCM (1:1) for 2 h each time. The resin was rinsed with THF/DCM (4×) and DMF (4×). Piperidine–DMF (1:1) was added to the resin and after 25 min the resin was rinsed with DMF (8×). Another double coupling was performed with 4 equiv of 4-benzoylbutyric acid, PyBOP, and HOBt in the presence of 10 equiv of DIEA in 2 mL of DMF for 1 h each time. The resin was then rinsed with DMF (4×), DCM (4×), and acetic acid (4×). To the derivatized resin was added 2 mL of phenylhydrazine hydrochloride (0.5 M) and zinc chloride (0.5 M) in glacial acetic acid and the reaction mixture was heated to 70°C. After 18–20 h at 70°C, the resin was cooled and washed with glacial acetic acid (4×), THF/DCM (4×), DMF (4×), DCM (4×), and MeOH (4×). The indole was cleaved from support by methanolysis with 2 mL of MeOH/triethylamine (9:1) at 50°C for 20 h. The sample was allowed to cool and then filtered, concentrated to dryness, and lyophilized from acetonitirile/water to yield 11.6 mg (98% crude yield) of the desired indole.

4.7.43   The Heck Reaction in the Solid-Phase Synthesis of Indole Analogs[3]

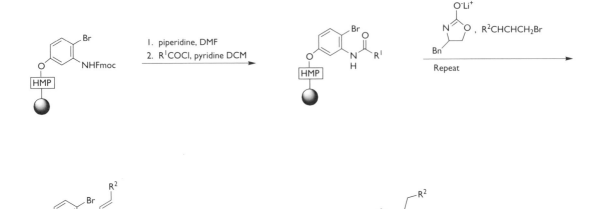

Points of Interest

1. In most cases, the Heck reaction proceeded via a *5-exo-trig* transition state to afford the exocyclic double bond that then underwent *exo* to *endo* double-bond migration to provide the desired indole derivatives.

2. Indole precursors were alkylated under reaction conditions previously employed for the alkylation of 1,4-benzodiazepines. The alkylation reactions were routinely repeated to ensure complete alkylation. In addition, the anthranilate scaffold was prepared in solution and loaded onto resin under reaction conditions previously employed for loading 2-aminobenzophenones onto resin (see p 148).[4]

3. Indoles have also been prepared by an intramolecular Wittig reaction under anhydrous conditions (see p 70).

Representative Examples

Support-bound 4-bromo-3-aminophenyl ether was acylated with aliphatic and aromatic acid chlorides followed by alkylation of the corresponding anilide with substituted allyl bromides.

Literature Summary

Support-bound 4-bromo-3-Fmoc-aminophenyl ether was suspended in 20% piperidine in DMF and vortexed for 30 min to remove the protecting group. The resin was washed with DMF and DCM and treated with the acid chloride (5 equiv) and pyridine (7 equiv) in DCM for 2 h. The resin was washed with DMF, *i*-PrOH, and DCM and then slurried in DMF, and a solution of lithium benzyloxazolidinone (15 equiv) in THF was added. The reaction was vortexed at room temperature for 1 h followed by the addition of the allylic halide (30 equiv). After 6 h of vortexing, the alkylation procedure was repeated. The resin was rinsed with DMF, *i*-PrOH, and DCM. Then Pd(PPh$_3$)$_4$ (0.5 equiv), Ph$_3$P (2 equiv), Et$_3$N (13 equiv), and DMA were added to the derivatized resin. The reaction tube was evacuated, sealed under N_2, and heated to 85–90°C for 5 h. This procedure was repeated. The resin was then rinsed with DMF, a solution of $Et_2NCS_2Na \cdot 3H_2O$ in DMF (0.2 M) to remove palladium, DMF, *i*-PrOH, and DCM. Cleavage with 95% TFA in H_2O for 3 h provided the indole analogs. An analytical sample was obtained by flash chromatography.

REFERENCES FOR SECTIONS 4.7.42 AND 4.7.43

1. Hutchins, S. M.; Chapman, K. T. *Tetrahedron Lett.* **1996,** *37,* 4869–4872.
2. Kim, R. M.; Manna, M.; Hutchins, S. M.; Griffin, P. R.; Yates, N. A.; Bernick, A. M.; Chapman, K. T. *Proc. Natl. Acad. Sci. U.S.A.* **1996,** *93,* 10012–10017.
3. Yun, W.; Mohan, R. *Tetrahedron Lett.* **1996,** *37,* 7189–7192.
4. Bunin, B. A.; Ellman, J. A. *J. Am. Chem. Soc.* **1992,** *114,* 10997–10998.

4.7.44 Solid-Phase Synthesis of Olomoucine (Purine) Derivatives[1]

Solid-phase synthesis of 2-(acylamino)-6-aminopurines

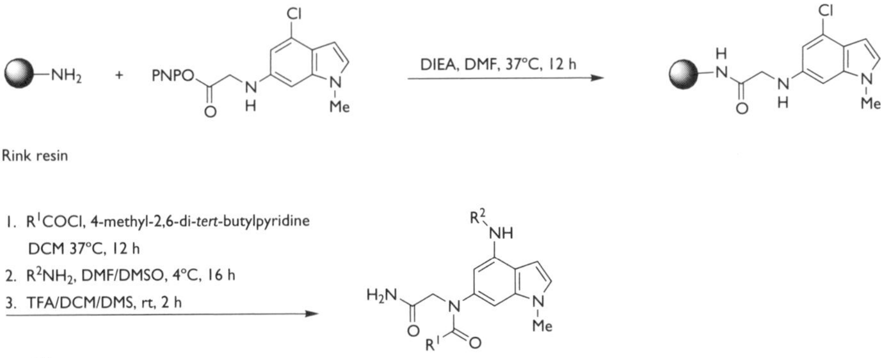

Rink resin
DIEA, DMF, 37°C, 12 h
1. R1COCl, 4-methyl-2,6-di-tert-butylpyridine DCM 37°C, 12 h
2. R2NH2, DMF/DMSO, 4°C, 16 h
3. TFA/DCM/DMS, rt, 2 h

or

Solid-phase synthesis of 2,6-diaminopurines

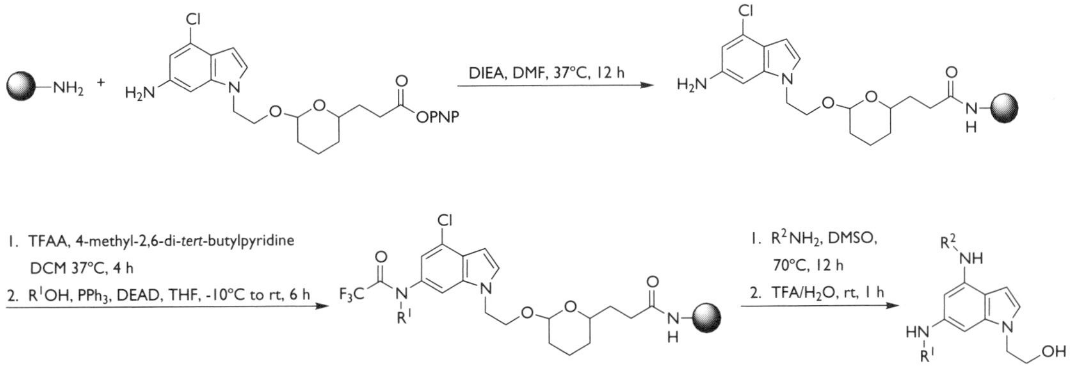

DIEA, DMF, 37°C, 12 h
1. TFAA, 4-methyl-2,6-di-tert-butylpyridine DCM 37°C, 4 h
2. R1OH, PPh3, DEAD, THF, -10°C to rt, 6 h
1. R2NH2, DMSO, 70°C, 12 h
2. TFA/H2O, rt, 1 h

Points of Interest

1. A small library of 36 2-(acylamino)-6-aminopurines was synthesized on Geysen's pin apparatus[2] in 30–75% yield by HPLC (289 nm). A library of 348

2-(acylamino)-6-aminopurines was then synthesized and evaluated for CDKZ inhibitors. An inhibitor with an IC_{50} value of 600 nm was found.

2. The C-2 amino group was alkylated under Mitsunobu conditions between the trifluoranilide and alcohols. Subsequent amination of the purine was accompanied by aminolysis of the trifluoroacetamide to provide the fully derivatized 2,6-diaminopurines. A small library of 16 alkylated aminopurines was prepared on polyethylene pins using primary and benzylic alcohols.

3. The PNP ester and THP derivatized olomoucine derivatives were prepared respectively in six and four steps in solution prior to solid-phase synthesis.

4. The (hydroxyethyl)-2,6-diaminopurines were linked to resin via a modified THP linker (see p 55).[3]

Representative Examples

Sterically hindered acid chlorides as well as both primary and secondary amines were incorporated into 2-(acylamino)-6-aminopurines. Primary and benzylic alcohols were incorporated under Mitsunobu conditions into 2,6-diaminopurines.

Literature Procedures

For the Solid-Phase Synthesis of 2-(Acylamino)-6-aminopurines. The protecting group was removed from Rink amide resin (1.23 g, 0.638 mmol) under standard conditions (20% piperidine in DMF), and the resin was rinsed with DMF (8×) and DCM (6×) and then dried *in vacuo*. Dry deprotected Rink amide resin (0.838 mmol) was added to a 25-mL silanized round-bottom flask. Distilled DCM (12 mL) was added to the reaction flask, and the resulting slurry was allowed to stand for 15 min to allow for solvation of the resin. DIEA (450 μL, 2.59 mmol) and the *p*-nitrophenyl active ester of the 6-chloropurine derivative (1.00 g, 2.50 mmol) were added, and the flask was sealed with a plastic cap. The flask was shaken at room temperature for 17 h, at which point a ninhydrin test (see p 214)[4] on ca. 10 mg of the solid support demonstrated that coupling was complete. The solution was removed by filtration, the derivatized resin was washed with DCM (6×), DMF (3×), and DCM (3×), and the red-orange resin was dried *in vacuo*.

Dry support-bound chloropurine (0.11 mmol) was placed in a 3-mL silanized vial. Anhydrous DMF (1 mL) containing 2,6-di-*tert*-butyl-4-methylpyridine (133 mg, 0.65 mmol) was added by syringe followed by the appropriate acid chloride (0.5 mmol). The vial was sealed with a plastic cap, and the reaction mixture was shaken at 37°C for 62 h. The solvent and reagents were removed from the vial by filtration cannula and the support-bound intermediate was washed with DMF (5×) and DCM (3×). After drying *in vacuo*, the resin was treated with a 4°C 1:1 DMF–DMSO mixture (900 μL). To this slurry was added ca. 0.23 mmol of the appropriate amine. The vial was sealed with a plastic cap, and the reaction mixture was shaken for 20 h at 4°C. The solution was removed by filtration cannula, and the support-bound aminopurine was washed with DMF (6×) and DCM (6×) and dried *in vacuo* to remove any remaining volatile material. The purine analog was cleaved from support with DCM/TFA/DMS (80:15:5) for 1 h. The resin and cleavage solution were transferred to a Pasteur pipette containing glass wool and the filtrate was collected in a tared round-bottom flask. The resin was rinsed with several small portions of the cleavage cocktail followed by DCM. The solution was concentrated down followed by azeotropic distillation with acetonitrile (3×). Selected purine derivatives were purified by flash chromatography (eluting with DCM/MeOH) and characterized.

REFERENCES FOR SECTION 4.7.44

1. Norman, T. C.; Gray, N. S.; Koh, J. T.; Schultz, P. G. *J. Am. Chem. Soc.* **1996**, *118*, 7430–7431.
2. Geysen, H. M.; Rodda, S. J.; Mason, T. J.; Tribbick, G.; Schoofs, P. G. *J. Immunol. Methods* **1987**, *102*, 259–274.
3. Thompson, L. A.; Ellman, J. A. *Tetrahedron Lett.* **1994**, *35*, 9333–9336.
4. Sarin, V. K.; Kent, S. B. H.; Tam, J. P.; Merrifield, R. B. *Anal. Biochem.* **1981**, *117*, 147–157.

4.7.45 Synthesis of Aromatic 1,2-Diazines by Inverse Electron Demand Diels–Alder Reaction of Polymer-Supported 1,2,4,5-Tetrazines[1]

Points of Interest

1. A methyl sulfide-substituted polymer-supported tetraazadiene complex reacted with various electron-rich olefins in dioxane at room or elevated temperature to give the resin-bound cycloaddition products. The methyl sulfide was oxidized on solid support to the corresponding sulfone. The azadiene complex

bearing the sulfone group was more reactive than the sulfide in the Diels–Alder reactions, allowing more efficient conversion with less reactive dienophiles.

2. The regiochemical course of the Diels–Alder reaction could be inverted by the conversion of SMe to SO_2Me, since the trend of the electron-withdrawing groups is $-SO_2Me > -NBoc > -SMe$.

Representative Examples

Inverse Diels–Alder reactions were performed between support-bound aza-diene complexes and electron-rich dienophiles including pyrrolidino enamines, enols, and terminal alkynes. The overall yields ranged from 22 to 82% based on the initial loading level of carboxylic acid.

Literature Summary

The oxidation of an aryl methyl sulfide to the corresponding sulfone was performed with 2.5 equiv of *m*CPBA in DCM for 6 h. The reaction between the support-bound azadiene (50 mg of resin, 0.077 mmol) and 10 equiv of dieno-phile in 4.0 mL of dioxane was performed for 24 h (room temperature or re-flux). The resin was washed with dioxane (3×) and DCM (5×) to afford the support-bound Diels–Alder product, which was then stirred slowly in excess TFA/DCM (1:4) for 1 h to remove the Boc group. The resulting resin was washed with DCM (3×) and MeOH/THF (2×) and then the product was cleaved from resin by treatment with MeOH/THF and solid K_2CO_3 for 12 h. This reaction mixture was filtered and the filtrate was extracted with EtOAc. The organic phase was concentrated to yield the 1,2-diazine derivative.

REFERENCES FOR SECTION 4.7.45

1. Panek, J. S.; Zhu, B. *Tetrahedron Lett.* **1996**, *45*, 8151–8154.

4.7.46 Library Generation through Successive Substitution of Trichlorotriazine[1]

Points of Interest

1. The decreasing reactivity of tri-, di-, and monochlorotriazine was utilized for the solid-phase construction of a combinatorial library.

2. Although both the first and second chlorine atoms can be substituted at room temperature, the kinetics of both reactions are sufficiently different so that with a large excess of trichlorotriazine in solution the support-bound amino acids reacted selectively to afford monosubstituted chlorotriazines (i.e., cross-linking was not observed).

3. The library was assayed on support or selectively removed from resin prederivatized with methionine by cyanogen bromide-mediated cleavage (see p 233).[2]

Representative Examples

A range of natural and unnatural support-bound amino acids were reacted with excess cyanuric chloride followed by aromatic substitution with amines, anilines, and hydrazines. Mixtures of products were identified, but yields were not reported.

Literature Procedure

PEG–PS support-bound amino acids were washed with DCM (2×) and cooled to 2°C for 30 min. A solution of cyanuric chloride (5.8 g, 31.3 mmol) in 100 mL of DCM was added to the resin, followed by the addition of DIEA (5.5 mL, 31.3 mmol) in 20 mL of DCM over a period of ca. 15 min. The suspension was agitated with nitrogen at room temperature until the ninhydrin test was negative (ca. 1 h). The resin was washed with DCM (3×). For the second derivatization, the resin was divided into 30 equal portions and distributed in Wheaton glass vials (0.6 g, 0.25 mmol/vial). A solution of the amine (0.85 mmol) in DCM (5 mL) was added to the resin and the suspension was shaken for 2 h at room temperature. Vials 1–15 and 16–30 were mixed together to give group A and group B. Resin (A and B separately) was washed with DCM (3×) and dioxane (1×). For the third derivatization, the resin (group A) was divided into 20 equal portions and distributed in V-shaped 4-mL Wheaton glass vials (0.45 g, 0.18 mmol/vial). A solution of an amine or hydrazine (5.2 mmol) in 0.6 mL of dry dioxane was added to each resin portion and the suspension was shaken for 2 h at 90°C. The resins were then transferred to syringes equipped with polypropylene frits. Each syringe was washed with DCM (5×). The same procedure was performed with group B.

REFERENCES FOR SECTION 4.7.46

1. Stanková, M.; Lebl, M. *Mol. Diversity* **1996**, *2*, 75–80.
2. Youngquist, R. S.; Fuentes, G. R.; Lacey, M. P.; Keough, T. *J. Am. Chem. Soc.* **1995**, *117*, 3900–3906.

4.7.47 Other Examples of the Solid-Phase Synthesis of Heterocycles

1. The combinatorial derivatization of the indole alkaloids yohimbine and rauwolscine on solid support has been reported by Jacobs and coworkers.[1]

2. The solid-phase synthesis of three different substituted 1,2,3-triazoles (from three variable building blocks) has been described.[2]

3. Teicoplanin aglycon has been used as a molecular scaffold for the solid-phase synthesis of combinatorial libraries.[3]

4. The selective solid-phase synthesis of substituted 3-aminothiophenes and 2-methylene-2,3-dihydrothiazoles has been reported by Zaragoza.[4]

REFERENCES FOR SECTION 4.7.47

1. Atuegbu, A.; Maclean, D.; Nguyen, C.; Gordon, E. M.; Jacobs, J. W. *Bioorg. Med. Chem.* **1996**, *4*, 1097–1106.
2. Zaragoza, F.; Petersen, S. V. *Tetrahedron* **1996**, *52*, 10823–10826.
3. Seneci, P.; Sizemore, C.; Islam, K.; Kocis, P. *Tetrahedron Lett.* **1996**, *37*, 6319–6322.
4. Zaragoza, F. *Tetrahedron Lett.* **1996**, *37*, 6213–6216.

5
ANALYTICAL METHODS FOR SOLID-PHASE SYNTHESIS

Because chemical reactions on solid support can be very general and high yielding, they are often used for the construction of chemical libraries. Solid-phase synthesis is particularly useful for the preparation of libraries involving multiple reactions and variables. However, if support-bound reactions are low yielding, reaction optimization can be more challenging on support than in solution. A number of reliable analytical and purification techniques, such as chromatography and ^{1}H NMR, are only available when the organic chemist is working in solution. Organic chemists sometimes feel that they have lost their eyes and ears when they perform chemical reactions on solid support. However, the solid support does not have to be a black box. The most rigorous method to evaluate a reaction on solid support is to cleave the products off of support, obtain mass balance yields, and fully characterize the product, but this is not always efficient. In certain cases cleavage from support is not even possible, let alone time efficient, because intermediates in a multistep synthesis may be unstable to cleavage conditions. Fortunately, there are alternative methods for evaluating reactions on support. This chapter will focus on the numerous standard techniques that are available for monitoring reactions on support and the novel, highly sensitive techniques being developed for the identification of small quantities of material on resin.

Quantitative and qualitative techniques for monitoring reactions on support have been developed, such as the ninhydrin or bromophenol blue tests for support-bound amines. The manipulation of other functional groups on support creates a need, and thus provides research opportunities, to develop new and selective tests for these functional groups. Another straightforward method to monitor reactions on support is to quantitate the release of protecting groups from support. The dimethoxytrityl and fluorenylmethoxycarbonyl protecting groups can be monitored spectrophotometrically, and the same approach can be applied to any auxiliary molecule that can be selectively removed from support and quantitated.

Spectroscopic techniques can be used directly to identify different functional groups on support. The primary caveat is that the polymeric matrix may complicate spectral analysis. Standard techniques for infrared and ^{13}C NMR spectroscopy have been used to identify molecules on solid support. The resin can be pressed into a KBr pellet for the identification of functional groups with distinctive vibrations. Single-bead FT-IR microspectroscopy, although recently developed, is an efficient technique to monitor the progress of organic reactions on solid supports. The quality of ^{13}C NMR can be improved with a magic-angle-spinning (MAS) probe, presaturation, and/or ^{13}C-labeled building blocks. Often, diagnostic ^{13}C NMR spectra may be obtained with a standard probe. MAS and nanoprobes have been used to obtain meaningful 1D and 2D ^{1}H NMR spectra. If the functional groups of interest contain elements not present in the polymeric matrix (i.e., N, P, F), then the products can often be detected by NMR or elemental analysis. There are many techniques available for monitoring reactions on solid support, and if the reaction products are not amenable to any of the available techniques, one can cleave the molecule (or a more stable derivative) off of support for analysis in solution. The specific cases where none of these existing methods are viable options provide research opportunities to develop new techniques.

5.1 COLORMETRIC ASSAYS ON SOLID SUPPORT

5.1.1 Ninhydrin Color Test for Detection of Free Terminal Amino Groups in the Solid-Phase Synthesis of Peptides[1]

Points of Interest

1. The ninhydrin test is a facile, standard technique for detection of a free amine on support.

2. In addition to being a standard qualitative method to evaluate an amide coupling reaction on solid support, the ninhydrin solution is also a useful TLC stain for monitoring reactions in solution involving amines.

Literature Procedure

Stock solutions were prepared of (1) 500 mg of ninhydrin in 10 mL of ethanol, (2) 80 g of phenol in 20 mL of ethanol, and (3) 2 mL of a 0.001 M solution of KCN diluted to 100 mL with pyridine. A small sample of the resin (10–20 mg) was placed in a test tube and 2–3 drops of each stock solution was added, and the mixture was heated at 100°C for 5 min with occasional swirling. Resin not subjected to the reaction (i.e., still containing the support-bound amine) was compared as a control. For complete reaction, the beads remain yellow; for incomplete reaction, the beads turn blue. For nearly complete reaction, the beads will be very light blue. Even 5 μmol/g of free terminal amine group can still be detected by this method.

5.1.2 Quantitative Monitoring of Solid-Phase Peptide Synthesis by the Ninhydrin Reaction[2]

Points of Interest

1. For routine monitoring, an effective extinction coefficient ϵ' of 1.5×10^4 M^{-1} cm^{-1} may be used to estimate the level of free amine based on Ruhemann's purple complex. For a more quantitative result, the exact ϵ' for the particular peptide must be known. Amine loading is quantitated with the equation $[\mu mol/g = (A_{570} \times V_{mL}/\epsilon'_{570} \times wt) \times 10^6]$, where ϵ' is the observed molar absorbance produced by the peptide and wt is the weight in milligrams. The extinction coefficients for the ninhydrin reaction with various terminal amino acids were tabulated.

2. The rate and extent of color development were very sensitive to heating time, temperature, and individual peptide differences. Deviations from 100°C and 10 min will cause irreproducible results.

3. The precision of the assay was ±4% throughout the range 0.005–1 μmol.

6. Quantitation of the amine group on the resin can also be established by picric acid titration.[3]

5. Imino acid proline does not release a chromophore into the solution, but the resin was yellow-orange.

6. Rinsing the support with 0.5 M Et_4NCl in methylene chloride affords colorless beads with all the anionic ninhydrin chromophore in solution where it can be readily quantitated. The rinses are not necessary for resins without residual chloromethyl groups.

7. This procedure can also be used to quantitate the reactive halide sites on a polymeric support, although the standard Volhard titration is also straightforward.

Recommended Stock Solutions

(a) Mix 40 g of phenol with 10 mL of absolute ethanol. Warm until dissolved. Stir with 4 g of Amberlite mixed-bed resin MB-3 for 45 min and then filter. Also prepare a solution from 65 mg of KCN in 100 mL of water. Dilute 2 mL of the KCN solution to 100 mL with pyridine (freshly distilled from ninhydrin). Stir with 4 g of Amberlite mixed-bed resin MB-3 and then filter. Mix the KCN and phenol solutions. (b) Dissolve 2.5 g of ninhydrin in 50 mL of absolute ethanol. Stopper and store in the dark under nitrogen.

Samples

Samples of synthetic peptide-resins after either the deprotection or the coupling reaction are washed twice with 5% diisopropylethylamine in DCM, washed three times with DCM, and dried *in vacuo* at room temperature.

Literature Procedure

(a) For Low Levels (0–0.1 μmol) of Amine.   Weigh a 2- to 5-mg sample of dry resin into a test tube. Add 100 μL of reagent a and 25 μL of reagent b (for routine monitoring use 6 and 2 drops, respectively). To another tube add only the reagents. Mix well and place both tubes in a heating block preadjusted to 100°C. After 10 min place the tubes in cold water. Add 1 mL of 60% ethanol in water and mix thoroughly. Filter through a Pasteur pipette containing a tight plug of glass wool. Rinse twice with 0.20 mL of 0.5 M Et$_4$NCl in methylene chloride. Make the solution to 2.00 mL with 60% ethanol. Measure the absorbance of the sample filtrate against the reagent blank at 570 nm.

(b) For High Levels (0.1–2 μmol) of Amine. Treat resin samples (2–5 mg) as before but dilute with 2 mL of 60% ethanol, wash twice with 0.50 mL of 0.5 M Et$_4$NCl in methylene chloride, and dilute to 25.0 mL with 60% ethanol. For larger samples of resin, the volume of reagents **a** plus **b** should be increased by 20 μL/mg.

REFERENCES FOR SECTIONS 5.1.1 AND 5.1.2

1. Kaiser, E.; Colescott, R. L.; Bossinger, C. D.; Cook, P. I. *Anal. Biochem.* **1970,** *34,* 595–598.
2. Sarin, V. K.; Kent, S. B. H.; Tam, J. P.; Merrifield, R. B. *Anal. Biochem.* **1981,** *117,* 147–157.
3. Gisin, B. F. *Anal. Chim. Acta* **1972,** *58,* 248.

5.1.3 Bromophenol Blue Test for Noninvasive, Continuous Detection of Free Amines[1]

Points of Interest

1. An indicator that would be deprotonated by free amino groups on resin and whose ionized form is deeply colored was developed. Among the several indicators tested, bromophenol blue (3′,3″,5′,5″-tetrabromophenolsulfonphthalein) displayed the best properties: the color changes dramatically from yellow (λ_{max} 429 nm) to dark blue (λ_{max} 600 nm) and the sensitivity is high (extinction coefficient ϵ_{600} 91,800 M^{-1} cm^{-1}). Bromocresol green and bromocresol purple also display satisfactory properties.

2. Acylation reactions can be monitored continuously even in a parallel format (a solution of bromophenol blue is simply added to the acylating agent or to the last washing solvent before coupling) or discontinuously (a sample of the resin is withdrawn from the reactor and treated with bromophenol blue). The test is not destructive and the excess bromophenol blue can be washed out by a solution of a base.

3. The number of free amine groups were determined quantitatively by the bromophenol blue test and provided results that were consistent with quantitative ninhydrin and picric acid tests.

4. The bromophenol blue test is highly sensitive; consequently triethylamine (TEA) is not compatible with the monitoring method due to significant background coloring. On the other hand, diisopropylethylamine (DIEA) does not cause a significant side reaction.

Literature Procedures

Qualitative Continuous Monitoring in Solid-Phase Peptide Synthesis Performed Batchwise The syntheses were carried out in an ordinary shaker reactor. After the addition of the protected amino acid, hydroxybenzotriazole, and

DCC in DMF (Note: DMF must be free of dimethylamine) or DCM, 3 drops of a 1% solution of bromophenol blue in dimethylacetamide were added. The suspension turned dark blue. After the suspension in the reaction vessel turned greenish-yellow, the next step of the synthesis was carried out.

Quantitative Monitoring A sample of washed and dried resin (2–5 mg) was loaded into a small vessel with sintered-glass bottom and washed with dimethylacetamide and DCM. Then it was treated with a solution of bromophenol blue in dimethylacetamide (5% solution, 0.2 mL, 2 x 30 s) and a saturated solution of bromophenol blue in DCM (0.2 mL, 30 s). The resin first turned dark blue and then slowly turned orange (excess of indicator). The resin was washed carefully with 5% ethanol in DCM until the indicator excess was removed. Elution of the coloring was carried out with 5% DIEA in DCM or DMF (2 mL, or until the beads became colorless), the solution was adjusted to an appropriate volume (25–100 mL) with ethanol, and absorbance at 600 nm was read. The substitution was calculated according to the equation

$$S = AV/91,800 \text{wt}$$

where S is substitution of the resin (in mmol/g), A is the absorbance of the solution, V is volume (in mL), and wt is the weight of the resin sample (in g).

REFERENCES FOR SECTION 5.1.3

1. (a) Krchñák, V.; Vágner, J.; Lebl, M. *Int. J. Pept. Protein Res.* 1988, *32*, 415–416. (b) Krchñák, V.; Vágner, J.; Safár, P.; Lebl, M. *Collect. Czech. Chem. Commun.* 1988, *53*, 2542–2548.

5.1.4 Quantitation of Amines with the Picric Acid Test.[1,2]

Points of Interest

1. Picric acid is both an explosive reagent and a strong acid. The picric acid test is therefore incompatible with acid-labile linkers.

Literature Procedure:

A known quantity of resin (~50 mg) was transferred to a peptide flask and rinsed with ca. 5 mL each of the following solvents: DCM (2x), 15% Et_3N in DCM (3x), DCM, MeOH (2x), DCM (2x), a saturated solution of picric acid (*Caution!* Picric acid is explosive and should be handled carefully) in DCM (2 x 30 s), DCM (until absorbance at 340 nm was less than 0.05), and MeOH (2x). A beaker was then used to collect the following washes (the sequence was repeated twice): 15% Et_3N in DCM (30 s), DCM, and MeOH. The spectrophotometer was zeroed with a blank containing 1 mL of DCM diluted to 10 mL with ethanol. The combined washes were diluted to 50 mL with DCM, and 1.00 mL of the resulting bright yellow solution was diluted to 10 mL with absolute ethanol. The absorbance was measured at 358 nm and the substitution level was given by

$$\text{substitution (mmol/g)} = A_{358} \text{ x } 34.48/\text{wt}$$

where A_{358} is the absorbance at 358 nm, 34.48 is a constant that includes the extinction coefficient for picric acid and a dilution factor, and wt is the weight of the resin used in milligrams. Two runs and two dilutions are generally performed.

REFERENCES FOR SECTION 5.1.4

1. Oded, A.; Houghten, R. A. *Pept. Res.* 1990, *3*, 42.
2. Bachem California Catalog Procedure Number QC-160.

5.1.5   NPIT for Quantitatively Monitoring Reactions of Amines in Combinatorial Synthesis.[1]

Points of Interest

1. NPIT (nitrophenyl isothiocyanate-O-trityl) was developed to efficiently monitor reactions involving less reactive amines during combinatorial synthesis of library mixtures. NPIT was able to detect 1% of sterically hindered primary and secondary amines as well as anilines.

2. The NPIT reagent consists of an activated isothiocyanate that selectively reacts with amines, coupled to a trityl ether reporting group carrying a latent trityl cation which is released through mild acid hydrolysis.

3. The NPIT reagent was prepared in five steps in solution.

Literature Summary

The resin-bound amine was coupled for 10 min with 0.1 M NPIT in DMF. The resin was washed with DMF (2x) and DCM (3x). The DMT (dimethoxy-trityl) group was removed with 2% TFA in DCM (2 × 3 min) for quantitation and the resin was washed with DCM (5x). The combined cleavage and washed fractions were diluted to 250 mL with 0.2% TFA in CH_3CN and the absorption was measured at 498 nm (dimethoxytrityl cation $\epsilon_{498} = 76,000$).

REFERENCES FOR SECTION 5.1.5

1. Chu, S. S.; Reich, S. H. *Bioorg. Med. Chem. Lett.* 1995, *5*, 1053–1058.

5.1.6   Other Examples of Colorimetric Assays on Solid Support

1. The Ellman test[1] has been used to detect the presence of free thiols on solid support.[2]

REFERENCES FOR SECTION 5.1.6

1. Ellman, G. L. *Arch. Biochem. Biophys.* 1959, *82*, 70–77.
2. Virgilio, A. A.; Ellman, J. A. *J. Am. Chem. Soc.* 1994, *116*, 11580–11581.

5.2 IDENTIFICATION OF COMPOUNDS RELEASED FROM RESIN

5.2.1 DMT Quantitation of Free Amines or Alcohols[1]

Points of Interest

1. DMT quantitation of amines can be integrated into automated peptide synthesizers.

2. The color of the released trityl cation measured spectrophotometrically at 498 nm (dimethoxytrityl cation $\epsilon_{498} = 76,000$ M^{-1}cm^{-1}) determines the loading of amino functionalities on the starting resin.

3. DMT quantitation is commonly used to quantitate support-bound alcohols; the procedure is routinely used in solid-phase oligonucleotide synthesis (see p 87).[2]

4. The tritylation of amino groups was complete in 30 s and detritylation with 2% DCA was complete within 2 min.

Literature Procedure

After prewashing with DCM (3x), the resin was tritylated with a stock solution of 0.25 M DMTCl, 0.25 M (n-Bu)$_4$NClO$_4$, and 3% collidine in DCM for 5 min. After extensive washing with DCM (7x), the resin was treated with 2% DCA (dichloroacetic acid) in DCM (7x) for 10 min. The acid washings were collected and the absorbance of the trityl cation was measured at 498 nm.

REFERENCES FOR SECTION 5.2.1

1. Reddy, M. P.; Voelker, P. J. *Int. J. Pept. Protein Res.* 1988, *31*, 345–348.
2. *Applied Biosystems Model 391 DNA Synthesizer Manual*; Section 6: Chemistry for Automated DNA Synthesis.

5.2.2 Fmoc Quantitation.[1,2]

Points of Interest

1. Fmoc quantitation is commonly used to determine loading levels on solid support. Upon Fmoc cleavage, the resulting piperidine-dibenzylfulvene adduct can be quantitated by its absorbance at 301 nm.

Literature procedure

To a round-bottom flask with stir bar was added 0.100 g of (aminomethyl)polystyrene resin. In a separate flask, Fmoc-Ala-OH (311 mg, 1.00 mmol), HOBt (153 mg, 1.00 mmol), N-methyl-2-pyrrolidinone (4 mL), and DIC (156 μL) were mixed. The amino acid solution was then added to the resin and stirred for 4 h, at which point the ninhydrin test (see p 214) showed no free amine groups present on the resin. The resin was transferred to a peptide flask and rinsed with DCM (10x). A brief description of the Millipore procedure for Fmoc quantitation follows. A known quantity of resin (~5 mg) was weighed into a 10-mL volumetric flask. To the flask were added 0.4 mL of piperidine and 0.4 mL of DCM. The mixture was allowed to cleave for 30 min, at which point MeOH (1.6 mL) and DCM (7.6 mL) were added to bring the total volume to 10 mL. The spectrophotometer was zeroed with a blank solution containing 0.4 mL of piperidine, 1.6 mL of MeOH, and DCM to make 10 mL. The absorbance was measured at 301 nm and the loading level was given by

$$loading\ (mmol/g) = A_{301} \times 10\ mL/7800 \times wt$$

where A_{301} is the absorbance at 301 nm, 7800 is the extinction coefficient of the piperidine–fluorenone adduct, and wt is the weight of resin used in milligrams. Fmoc quantitation is generally performed in duplicate or triplicate.

REFERENCES FOR SECTION 5.2.2

1. *Novabiochem Catalog and Peptide Synthesis Handbook*, 1997/1998, method 16.
2. MilliGen Technical Note 3.10.

5.2.3   Other Examples of the Identification of Compounds Released from Resin

1. The Dorman test is a nondestructive method for the detection of amines employing a pyridine hydrochloride titration (see p 75).[1]
2. The TNBS test[2] has also been used to detect support-bound amines.[3]
3. The Volhard titration has been employed to quantitate support-bound chlorides.[4]

REFERENCES FOR SECTION 5.2.3

1. Dorman, L. C. *Tetrahedron Lett.* 1969, *28*, 2319–2321.
2. Hancock, W. S.; Battersby, J. E. *Anal. Biochem.* 1976, *71*, 261.
3. *Novabiochem Catalog and Peptide Synthesis Handbook*, 1997/1998, methods 14 and 15.
4. Chen, C.; Ahlberg Randall, L. A.; Miller, R. B.; Jones, A. D.; Kurth, M. J. *J. Am. Chem. Soc.* 1994, *116*, 2661–2662.

5.3   NMR TECHNIQUES FOR SOLID-PHASE SYNTHESIS

5.3.1   Application of Gel-Phase ^{13}C NMR To Monitor Solid-Phase Peptide Synthesis[1]

Points of Interest

1. Optimal conditions to acquire standard gel-phase ^{13}C NMR spectra of copoly(styrene–1% divinylbenzene) were determined. The technique was used to characterize polystyrene-based starting supports and to determine the degree of functionality and purity.
2. Stage by stage ^{13}C NMR characterization of a tripeptide was described; the spectra were used to monitor coupling and deprotection reactions.

Literature Summary

Polymers were dried to constant weight under vacuum at 40°C or at room temperature. Then 300 mg was weighed into a 10-mm-o.d. NMR tube and the solvent was pipetted in. Complete swelling of the polymers was usually achieved by allowing the resin to equilibrate with the solvent for 30 min after agitation with a vortex shaker. However, sometimes ultrasonic agitation was necessary to obtain homogeneous samples. ^{13}C NMR spectra were obtained at 50.31 MHz on a Varian XL-200 spectrometer equipped with quadrature phase detection and broad-band decoupling. The 90° pulse width was 12.5 μs. The ^{2}H resonance of the solvent was used for the field-frequency lock and sample spinning rates were between 15 and 20 Hz. Standard conditions were as follows: 16 or 32 data points. 0.7-s acquisition time, zero delay, 12-μs pulse width, 10,000 transients, 1-Hz line broadening. Under these conditions the polymer signals were not saturated; however, Boc and other substrate-related signals can show appreciable saturation.

5.3.2 ^{13}C NMR Relaxation Times of a Tripeptide Methyl Ester and Its Polymer-Bound Analogs.[2]

Points of Interest

1. The carbon-13 line widths and relaxation times T_1 of the peptide esters (methyl, insoluble cross-linked polystyrene (PS), soluble poly(oxyethylene) (POE), and the graft copolymer poly(oxyethylene)–polystyrene–divinylbenzene (POE–PS)) were compared.

2. T_1 relaxation times decrease in the sequence methyl ester > POE ester $\geq$ POE–PS ester > PS ester. The line widths in the ^{13}C NMR spectra of the tripeptide ester followed the same trend.

3. The relaxation times were correlated with the mobility of the peptides.

Literature Procedure and Results

Relaxation data were obtained by using the inversion–recovery 180–τ–90° pulse sequence. T_1 relaxation times decrease in the sequence methyl ester > POE ester $\geq$ POE–PS ester > PS ester. The line widths in the ^{13}C NMR spectra of the tripeptide ester followed the same trend.

REFERENCES FOR SECTIONS 5.3.1 AND 5.3.2

1. Giralt, E.; Rizo, J.; Pedroso, E. *Tetrahedron* 1984, *40*, 4141–4152.
2. Bayer, E.; Albert, K.; Willisch, H.; Rapp, W.; Hemmasi, B. *Macromolecules* 1990, *23*, 1937–1940.

5.3.3   Reaction Monitoring by Gel-Phase ^{31}P NMR[1]

Points of Interest

1. Phosphonoacetamide esters bound to solid-supported peptides or peptide synthesis supports are active Horner–Wadsworth–Emmons (HWE) reagents with a variety of aldehydes (see p 102).

2. Gel phase ^{31}P NMR was demonstrated to be a highly sensitive technique to monitor reactions on solid support.

3. Gel phase ^{31}P NMR showed the resonance of the starting resin-bound diethylphosphonoacetamide as a narrow multiplet at δ 22. After the reaction was complete, the δ 22 peaks had disappeared and a new broad resonance for diethyl phosphate appeared near δ 0.

Literature Summary

Shifts were determined relative to H_3PO_4. A sample of ca. 50 mg of resin in acetonitrile in a regular 5-mm NMR tube was used. Addition of ca. 25% deuterated solvent was adequate for a lock signal. The spectra were acquired at 122 MHz and a recycle time of 1.3 s. Spectra with adequate signal-to-noise ratios required a few hundred acquisitions, but less than 10 min of actual spectrometer time.

REFERENCES FOR SECTION 5.3.3

1. Johnson, C. R.; Zhang, B. *Tetrahedron Lett.* 1995, 36, 9253–9256.

5.3.4 ^{19}F NMR Monitoring of an S$_N$Ar Reaction on Solid Support[1]

Points of Interest

1. The displacement of fluoride from support-bound 4-fluoro-3-nitrobenzamide by ethyl 3-aminopropionate was examined to establish that ^{19}F NMR[2] can be used to monitor solid-phase reaction kinetics. Several spectroscopic methods were used to follow the solid-phase S$_N$Ar reaction; however, ^{19}F NMR spectroscopy on solvent-swollen resin was the most efficient technique.

Literature Summary

For typical gel phase NMR spectra, it took 1–1.5 h to acquire data with a sufficient signal-to-noise ratio. MAS NMR data could be obtained with a shorter experiment (4 min for ^{19}F).

REFERENCES FOR SECTION 5.3.4

1. Shapiro, M. J.; Kumaravel, G.; Petter, R. C.; Beveridge, R. *Tetrahedron Lett.* 1996, 37, 4671–4674.
2. Manatt, S. L.; Amsden, C. F.; Bettison, C. A.; Frazer, W. T.; Gudman, J. T.; Lenk, B. E.; Lubetich, J. F.; McNelly, E. A.; Smith, S. C.; Templeton, D. J.; Pinnell, R. P. *Tetrahedron Lett.* 1980, 21, 1397–1403.

5.3.5 High-Resolution ^{1}H NMR in Solid-Phase Organic Synthesis[1]

Points of Interest

1. A Nano-NMR probe was designed to exploit magic-angle spinning (MAS) for very small (<40 μL) samples.

2. MAS had previously been applied to solid-state NMR to remove line broadening.

3. A significant improvement in the resolution (versus a conventional probe) allowed the identification of individual resonances from specific small molecules on TentaGel beads.

Representative Example

TentaGel beads (10 mg, containing about 3 μmol or 1.5 mg of the molecule of interest) were suspended in 30 μL of DMSO-d_6 and transferred to a Nano-NMR probe cell. Sixteen transients were sufficient to assign most of the peaks of the molecule of interest (broad peaks at 6–8 ppm are from the immobile polystyrene matrix and the large resonances between 3 and 4 ppm are due to water and the solvent-accessible poly(ethylene glycol).

REFERENCES FOR SECTION 5.3.5

1. Fitch, W. L.; Detre, G.; Holmes, C. P.; Shoolery, J. N.; Keifer, P. A. *J. Org. Chem.* 1994, *59*, 7955–7956.

5.3.6   MAS CH Correlation in Solvent-Swollen Resin[1]

Points of Interest

1. Magic-angle-spinning (MAS) ^{13}C–^{1}H-correlated NMR of organic molecules attached to solvent-swollen polymers was demonstrated using a standard solid-state probe.

2. A nonaromatic molecule was examined to avoid complications in the spectrum arising from the polymer.

3. Solid granulated resin in benzene-d_6 was used to swell the beads (as opposed to a slurry for gel-phase NMR); MAS ^{13}C NMR data for carbonyl signals were examined. Different linker lengths and subtle functional differences (i.e., between an amide and ester) were examined.

41. The MAS ^{13}C solid state NMR was used to semiquantitatively detect the 60:40 *exo/endo* ratio of resin-coupled norbornane-2-carboxylic acid based on the peak intensities. Average carbon resolution was 13 Hz for all the carbons in the norbornyl system. The observed resonances from the linker provide an internal reference for transformations on support.

5. The signals in the MAS proton NMR spectrum had a line width of 26 Hz; although broad for high-resolution studies, it allows structural information via CH-correlated data. A CH-correlated spectrum allowed the epimers of no-bornane-2-carboxylic acid to be assigned.

Literature Procedure and Results

The MAS ^{13}C solid-state NMR was used to semiquantitatively detect the 60:40 *exo/endo* ratio of resin-coupled norbornane-2-carboxylic acid based on the peak intensities. Average carbon resolution was 13 Hz for all the carbons in the norbornyl system. The observed resonances from the linker provide an internal reference for transformations on support. The signals in the MAS proton NMR spectrum had a line width of 26 Hz; although broad for high-resolution studies, it allows structural information via CH-correlated data. A CH-correlated spectrum allowed the epimers of nobornane-2-carboxylic acid to be assigned.

REFERENCES FOR SECTION 5.3.6

1. Anderson, R. C.; Jarema, M. A.; Shapiro, M. J.; Stokes, J. P.; Ziliox, M. *J. Org. Chem.* 1995, 60, 2650–2651.

5.3.7 Utilization of Magic-Angle-Spinning HMQC and TOCSY NMR Spectra in the Structure Determination of Wang-Bound Lysine[1]

Points of Interest

1. Total structural and NMR assignment of a compound on resin was reported. Previously the same authors distinguished diastereomers with the acquisition of magic-angle-spinning (MAS) data in a conventional solid-state probe using solvent-swollen resin.

2. HMQC and TOCSY data for Wang-bound Fmoc-lysine-Boc was obtained and the NMR spectrum completely assigned. Complete proton and carbon-13 resonances for the aliphatic resonances were assigned from 2-D NMR.

Literature Summary

Data were obtained on a Bruker DMX-400 wide-bore instrument using standard solution experiments and in a 7-mm rotor with MAS at 5 kHz. TOCSY data were collected with a 70-ms mixing time and 16 scans per increment with a data matrix of 1024×256. The HMQC was collected with 8 scans per increment and a 1024×256 data matrix.

REFERENCES FOR 5.3.7

1. Anderson, R. C.; Stokes, J. P.; Shapiro, M. J. *Tetrahedron Lett.* 1995, 36, (30), 5311–5314.

5.3.8 Use of Spin Echo Magic-Angle-Spinning ^{1}H NMR in Reaction Monitoring in Combinatorial Organic Synthesis[1]

Points of Interest

1. Magic-angle spectra on support were complicated by the presence of large peaks from the polystyrene matrix. These resonances are much broader, presumably because of short T_2.

2. The use of a spin echo sequence can distinguish between narrow and broad lines. The authors extended the MAS ^{1}H NMR technique by demonstrating the combination of MAS along with a spin echo sequence to follow a LAH reduction on support.

3. All the NMR spectra of compounds attached to the Wang resin were

collected using a Varian UnityPlus spectrometer operating at a ^{1}H frequency of 500 MHz using a Nano-NMR probe.[2]

4. Five to six milligrams of beads containing $5-6$ μmol of compounds was transferred into a Nano-NMR probe cell, and 40 μL of CD_2Cl_2 was then added. The spectra were collected using the spin echo sequence ($90°-\tau-180°-\tau$), with different τ values, along with magic-angle spinning at a speed of 2.0 kHz. (2D COSY and TOCSY spectra of resin-bound compounds were also described.)

Literature Procedure

All the NMR spectra of compounds attached to the Wang resin were collected using a Varian UnityPlus spectrometer operating at a ^{1}H frequency of 500 MHz using a Nano-NMR probe.[2] Five to six milligrams of beads containing $5-6$ μmol of compounds was transferred into a Nano-NMR probe cell, and 40 μL of CD_2Cl_2 was then added. The spectra were collected using the spin echo sequence ($90°-\tau-180°-\tau$), with different τ values, along with magic-angle spinning at a speed of 2.0 kHz. (2D COSY and TOCSY spectra of resin-bound compounds were also described.)

REFERENCES FOR SECTION 5.3.8

1. Garigipati, R. S.; Adams, B.; Adams, J. L.; Sarkar, S. K. *J. Org. Chem.* 1996, *61*, 2911–2914.
2. Barbara, T. M. *J. Magn. Reson. A* 1994, *109*, 265

5.3.9 Influence of Resin Structure, Tether, Length, and Solvent upon the High-Resolution ^{1}H NMR Spectra of Solid-Phase Synthesis Resins[1]

Points of Interest

1. NMR line widths obtained with a Nanoprobe can vary significantly from sample to sample. Nine different SPS resins and seven different solvents were evaluated to find various combinations that generate ^{1}H NMR spectra having narrow line widths.

2. Because sensitivity is not a limiting factor with a Nanoprobe, line width serves as a measure of spectral quality. Resin structure is the dominant factor, with the solvent playing a secondary role. TentaGel resins, which have the longest tethers and the most mobile moieties, provided the narrowest line widths. Wang, PA-500, and Rink resins produced spectra of intermediate quality. Conventional polystyrene resins with no tethers and little regional mobility never generated high-quality spectra.

3. Presaturation can reduce background signals due to the resins. DMF, CD_2Cl_2, and DMSO (in decreasing order) typically generated the best NMR data.

REFERENCES FOR SECTION 5.3.9

1. Keifer, P. A. *J. Org. Chem.* 1996, *61*, 1558–1559.

5.3.10   An NMR Method to Identify Nondestructively Chemical Compound Bound to a Single Solid-Phase Synthesis Bead for Combinatorial Chemistry Applications[1]

Points of Interest

1. Using ^{13}C labeled building blocks, the compound of interest was selectively detected from a single bead using isotope-filtered ^{1}H NMR (HMQC) in a two-coil, high-resolution, magic-angle-spinning (MAS) probe. These techniques are particularly useful for split-and-mix methods for identification with "one compound on one bead" (0.1–1 nmol).

2. Previous studies have examined ^{13}C NMR with labeled compounds to monitor the progress of reactions on solid support.[2] Magic-angle spinning can narrow the ^{13}C NMR line widths of resin samples. A combination of MAS with high-resolution probe technology (a Nano-NMR probe) extends the analytical technique to ^{1}H NMR.

3. Spectra with line widths of 4–5 Hz are obtained from 5 mg of beads. Total experiment time was 170 min for ^{13}C-filtered ^{1}H NMR spectra. 2D NMR results confirmed the 1D NMR spectra. The methodology requires ^{13}C-labeled compounds and a high-resolution probe.

REFERENCES FOR SECTION 5.3.10

1. Sarkar, S. K.; Garigipati, R. S.; Adams, J. L.; Keifer, P. A. *J. Am. Chem. Soc.* 1996, *118*, 2305–2306.
2. Look, G. C.; Holmes, C. P.; Chinn, J. P.; Gallop, M. A. *J. Org. Chem.* 1994, *59*, 7589–7590.

5.3.11   A Comparison of NMR Spectra Obtained for Solid-Phase Synthesis Resins Using Conventional High-Resolution, Magic-Angle-Spinning, and High-Resolution Magic-Angle-Spinning Probes[1]

Points of Interest:

1. Relatively narrow line widths were obtained with a Nano-NMR probe, which combines moderate-speed MAS (2 kHz) with high-resolution magnetic-susceptibility-matching probe technology.

2. Line widths of <6 Hz were obtained from ^{1}H NMR spectra and <3 Hz from ^{13}C NMR spectra on 2 mg of TentaGel S PHB Cys(Trt)Fmoc in CD_2Cl_2. How narrow the line widths need to be for routine structure assignment purposes must still be determined. A conventional CP/MAS probe previously had been used to generate ^{13}C NMR spectra having 13-Hz line widths and ^{1}H NMR spectra having 26-Hz line widths on 75 mg of derivatized swollen polystyrene resin.[2]

3. This study compared the spectral resolution and sensitivity using all available probe geometries (5- and 3-mm ^{1}H and ^{13}C high-resolution probes, a 5-mm broad-band CP/MAS probe, and ^{1}H and ^{13}C Nano-NMR probes).

Background

1. Solvent swelling is just one requirement for high-resolution NMR spectra on resin; other factors include the magnetic-susceptibility interface differ-

ences within the sample and the magnetic-susceptibility interfaces inherent in the construction of the probe. Because SPS resins are heterogeneous mixtures, the magnetic-susceptibility effects arising from different microscopic regions can affect the resonance frequencies of nearby nuclei and cause a distribution of chemical shifts (which will appear as an increase in the NMR line widths).

2. This magnetic-susceptibility broadening vanishes when the sample is spun about the magic angle (54° relative to the static magnetic field).

3. The Nano-NMR probe spins very small samples ($<40~\mu$L) at the magic angle to eliminate any line broadening caused by magnetic-susceptibility discontinuities either around (typically a problem for small samples) or within the sample.

Literature Procedure and Results

The ^{1}H NMR spectrum of the solvent-swollen resin in a conventional high-resolution liquid-style probe consists only of extremely broad lines (line widths from 100- to 300-Hz wide). At the magic angle in a standard CP/MAS probe, the resonances from the beads narrow up, but the residual solvent resonance is broader (about 65-Hz wide). With the ^{1}H nanoprobe the narrowest moiety resonances are 11 Hz, and the solvent line width is only 3.5 Hz. The results with resin parallel the minimum line widths attainable with solvents for each probe. The nanoprobe produces higher sensitivity because the sample cell ($40~\mu$L) constrains the entire sample to lie within the active region of the receiver coil. Similar trends are observed with the analogous ^{13}C probes; however, because carbon spectra have greater chemical-shift dispersion and the line widths are typically narrower, ^{13}C NMR spectra of SPS resins can be obtained on nonspinning samples. The ^{13}C nanoprobe has 50–55% of the sensitivity of the CP/MAS probe while using only 32% as much sample. Solvent effects were significant and temperature effects were minimal. Multidimensional NMR were also obtained on support.

REFERENCES FOR SECTION 5.3.11

1. Keifer, P. A.; Baltusis, L.; Rice, D. M.; Tymiak, A. A.; Shoolery, J. N. *J. Magn. Reson. A* 1996, *119*, 65–75.
2. Anderson, R. C.; Jarema, M. A.; Shapiro, M. J.; Stokes, J. P.; Ziliox, M. *J. Org. Chem.* 1995, *60*, 2650–2651.

5.3.12   Other Applications of NMR Spectroscopy to Solid-Phase Synthesis

1. Rapid ^{13}C NMR spectroscopy has been used with ^{13}C-enriched building blocks to enhance the NMR spectral signals.[1]

2. MAS NMR has been used to monitor the progress of a cyclization reaction on solid support.[2]

3. Diffusion-encoded spectroscopy (DECODES) has been applied to mixture analysis in combinatorial chemistry. DECODES combines PFG (pulsed field gradient) technology with TOCSY methodologies.[3]

REFERENCES FOR SECTION 5.3.12

1. Look, G. C.; Holmes, C. P.; Chinn, J. P.; Gallop, M. A. *J. Org. Chem.* 1994, *59*, 7589–7590.
2. Wehler, T.; Westman, J. *Tetrahedron Lett.* 1996, *37*, 4771–4774.
3. Lin, M.; Shapiro, M. J. *J. Org. Chem.* 1996, *61*, 7617–7619.

5.4   IR TECHNIQUES FOR SOLID-PHASE SYNTHESIS

5.4.1   Examples of Solid-Phase Infrared Spectroscopy with Standard KBr Pellets

Solid-Phase Synthesis of Oligosaccharides.[1]

Points of Interest

1. IR spectroscopy was used for characterization of intermediates in the preparation of a resin with 10% allyl alcohol side chains.

2. Oxidation of chloromethylated resin with DMSO and $NaHCO_3$ provided an aldehyde resin that showed an intense carbonyl absorption at $1690 \, cm.^{-1}$

3. Oxime and Wittig products were formed from the aldehyde intermediate on support.

4. A support-bound phosophine ylide was also prepared in an alternative synthesis of allyl alcohol linker.

5. An allyl alcohol linker has been employed for palladium-catalyzed cleavage (see p 34).

Unidirectional Dieckmann Cyclizations on a Solid-Phase and in Solution[2]

Points of Interest

1. Infrared difference spectra of functionalized resin relative to untreated polystyrene provided the best analytical information. When the reference beam was attenuated by a control sample containing an appropriate quantity of untreated resin, the difference spectrum of each of the functionalized resins was substantially simplified.

2. Chlorides, bromides, methyl ethers, acetates, and alcohols all had significant infrared spectral peaks on resin that corresponded with their cymyl analogs in solution.

3. A rigorous rinsing protocol to remove salts and solvents (which can interfere with the spectra) from resin is provided.

"Analogous" Organic Synthesis of Small-Compound Libraries.[3]

Points of Interest

1. All functional group changes were monitored by FT-IR (KBr pressed windows of ground polystyrene beads) as follows: appearance of O–H stretch at 3572 cm^{-1}, C=O stretches at 1724 and/or 1712 cm^{-1}, conjugated C=O stretch at 1674 cm^{-1}.

Other examples of IR Spectroscopy on KBr Pellet of Resin

1. An early example of IR spectroscopy on ~10-mg beads was described by Mazur and Jayalekshmy.[4]

2. Hauske and Dorff have reported that much better IR spectra were obtained with Wang resin than with TentaGel matrices (see p 58).[5]

5.4.2   Infrared Spectrum of a Single Resin Bead for Real-Time Monitoring of Solid-Phase Reactions.[6,7]

Reaction I

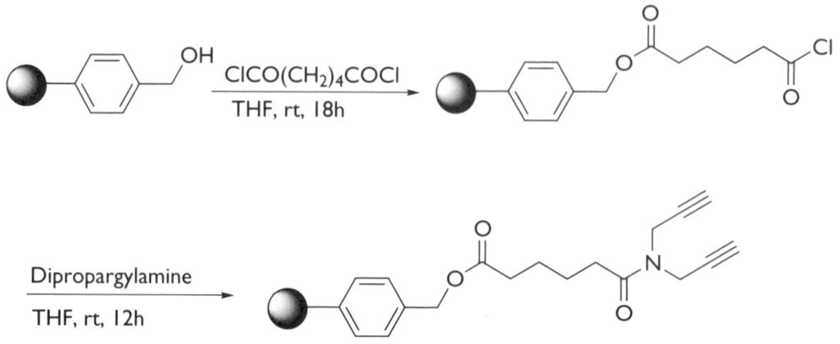

Points of Interest

1. Single-bead FT-IR microspectroscopy can measure solid-phase reactions during the course of the reaction. The specific instruments necessary for single-bead FT-IR microspectroscopy are given in the literature procedure. For larger scale reactions, FT-IR spectra may be obtained from a standard KBr pellet with 10 mg of beads.

Reaction II

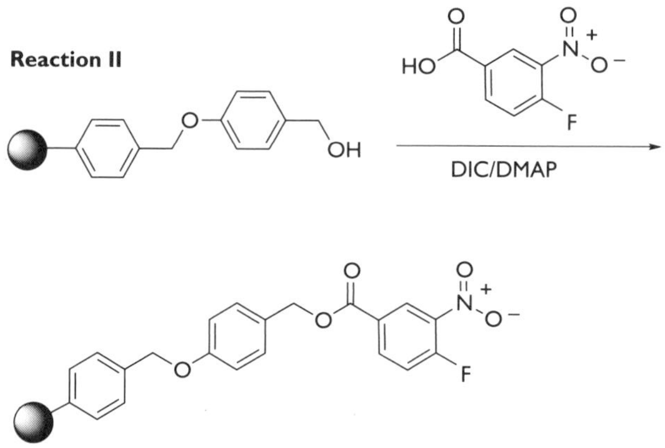

Literature Results

With FT-IR microspectroscopy a high signal-to-noise ratio was obtained from products on a single-bead in <5 min. Two different esterification reactions and an aminolysis were monitored. A >95% conversion was estimated by observing the disappearance of an OH stretch vibration and the emergence of the carbonyl. Distinctive alkyne stretches were also detected during aminolysis of a support-bound acid chloride with dipropargylamine. Four different linker/resin combinations were examined and single-bead spectra were reproducibly obtained in all cases. Both the percent conversion and rate of an esterification with DIC/DMAP were monitored on support. This particular reaction reached steady state within 20 min.

Literature Summary

A drop of resin suspension was removed from the reaction and rinsed with THF (3×) or DMF (3×) and then with MeOH. The resin sample was dried under vacuum for 30 min. Spectra were collected on a BIO-Rad FTS-40 spectrophotometer, using an SPC-3200 data station. The optical bench is coupled with a UMA-300 IR microscope. The microscope is equipped with a liquid nitrogen cooled mercury–cadmium–telluride (MCT) detector and a 36X Cassegrain objective. The total visual magnification in the view mode is 360×, which facilitated locating a single bead. Data collection could be performed in both the transmission and reflectance modes.

After collection of the background spectrum, a few resin beads were then put on a well-polished NaCl window. The view mode and x–y platform of the microscope were used to focus the incident radiation on a single resin bead. The diameter of the individual bead was measured under the microscope. Data were collected with 4-wavenumber resolution as the average of 64 scans. Single-bead FT-IR spectra were reproducible for 30 randomly chosen beads to demonstrate that a single bead is a representative sample size.

REFERENCES FOR SECTIONS 5.4.1 AND 5.4.2

1. Frechet, J. M.; Shuerch, C. *J. Am. Chem. Soc.* 1971, *93*, 492–496.
2. Crowley, J. I.; Rapoport, H. *J. Org. Chem.* 1980, *45*, 3215–3227.
3. Chem, C.; Ahlber Randall, L. A.; Miller, R. B.; Jones, A. D.; Kurth, M. J. *J. Am. Chem. Soc.* 1994, *116*, 2661–2662.
4. Mazur, S.; Jayalekshmy, P. *J. Am. Chem. Soc.* 1979, *101*, 677–683.
5. Hauske, J. R.; Dorff, P. *Tetrahedron. Lett.* 1995, *36*, 1589–1592.
6. Yan, B.; Kumaravel, G.; Anjaria, H.; Wu, A.; Petter, R. C.; Jewell, C. F.; Wareing, J. R. *J. Org. Chem.* 1995, *60*, 5736–5738.
7. Yan, B.; Kumaravel, G. *Tetrahedron* 1996, *52*, 843–848.

5.4.3 Progression of Organic Reactions on Solid Supports Monitored by Single-Bead FT-IR Microspectroscopy.[1]

Points of Interest

1. The time courses of five solid-phase reactions were obtained by single-bead FT-IR. The experimental results challenge many of the postulates regarding reaction kinetics on solid support. Specifically, diffusion into resin was not rate limiting for the reactions that were studied.

2. Solid-phase reaction rates were faster than expected. For example, the S_N2 reaction of potassium acetate with Merrifield resin was actually faster on support than the analogous reaction in solution. The authors hypothesized that the reactant may be concentrated within the polymer bead, and this high local concentration may play a role in accelerating the reaction.

3. A reaction on TentaGel resin (PEG–PS) was shown not to be faster than the same reaction on Wang resin (polystyrene–1% divinylbenzene), suggesting that diffusion of the substrate into the bead is not rate limiting.

4. For the reactions studied, the rates were similar on the surface and in the interior of the bead.

5. The authors provide a number of references to previous reports dealing with reaction time courses of solid-phase synthesis including the UV–vis spectrophotometric method,[2] IR on a KBr pellet of ~10-mg beads,[3] titration using 100–200 mg of beads,[4] and chromatographic analysis of the reaction.[5]

6. By monitoring the steady state for reactions on solid support, optimization time can be reduced, libraries can be evaluated, and prolonged reaction times (i.e., after the reaction is essentially complete) may be avoided.

7. The catalytic oxidation of support-bound alcohols to aldehydes and ketones with NMO/TPAP has also been monitored by single-bead FT-IR.[6]

Literature Procedure

After the beads were rinsed with DMF (3×), MeOH (3×) and dried under vacuum (15 min), all spectra were collected on a Bio-Rad FTS-40 spectrophotometer, using an SPC-3200 data station as previously described. The IR spectra of the bead surface were collected on a Magna-IR system 550 coupled with a Nic-Plan microscope with a germanium attenuated total reflection (ATR) objective (SpectraTech Inc., Shelton, CT).

REFERENCES FOR SECTION 5.4.3

1. Yan, B.; Fell, J. B.; Kumaravel, G. *J. Org. Chem.* 1996, *61*, 7467–7472.
2. (a) Merrifield, B. *Br. Polym. J.* 1984, *16*, 173–178. (b) Gisin, B. F. *Anal. Chim. Acta* **1972**, *58*, 248–249.
3. Mazur, S.; Jayalekshmy, P. *J. Am. Chem. Soc.* 1979, *101*, 677–683.
4. Lu, G.-S.; Mojsov, S.; Tam, J. P.; Merrifield, R. B. *J. Org. Chem.* 1981, *46*, 3433–3436.
5. Blanton, J. R.; Salley, J. M. *J. Org. Chem.* 1991, *56*, 490–491.
6. Yan, B.; Sun, Q.; Wareing, J. R.; Jewell, C. F. *J. Org. Chem.* 1996, *61*, 8765–8770.

5.4.4 Other Applications of IR Spectroscopy to Solid-Phase Synthesis.

1. Photoacoustic FT-IR (PA-FTIR) spectroscopy has been reported as a nondestructive method for sensitive analysis of solid-phase organic chemistry.[51] Because only absorbed radiation is measured, PA-FTIR spectroscopy eludes the effects of light scattering and reflection. In contrast to preparing KBr pellets, the resin sample is directly placed into a photoacoustic cell and purged with an inert gas (e.g., He).

2. Internal reflection FT-IR spectroscopy of injection-molded high-density polystyrene ("Mega-Crowns") has been described.[52]

REFERENCES FOR SECTION 5.4.4

1. Gosselin, F.; Di Renzo, M.; Ellis, T. H.; Lubell, W. D. *J. Org. Chem.* 1996, *61*, 7980–7981.
2. Gremlick, H.-U.; Berets, S. L. *Appl. Spectrosc.* 1996, *50*, 532.

5.5 MASS SPECTROMETRY TECHNIQUES FOR SOLID-PHASE SYNTHESIS

5.5.1 The Application of Solid-Phase *In Situ* Mass Spectrometry for Reaction Analysis with a Range of Traditional Peptide Linkers[1]

Points of Interest

1. Reactions were directly monitored from single beads using MALDI-TOF mass spectrometry on a range of linkers. Upon cleavage from resin, products with a variety of different termini (amines, carboxylates, and homoserine lactones) were produced.

2. Peptides were detected by MALDI-TOF MS after cleavage with TFA vapor from HMPB and Wang linkers. Treatment of the resin beads with CNBr/TFA vapor converted peptides containing methionine into C-terminal homoserine lactones.

3. The chlorotrityl linker[2] was found to be suitable as an amine-linking functionality. After an acylation and Wittig reactions, the support-bound trityl amine was cleaved from support and detected by MALDI-TOF MS.

4. Cyanogen bromide-mediated cleavage has also been employed for the rapid sequencing of peptide libraries by mass spectrometry.[55]

Literature Procedures

For Loading onto Resins. Peptides were prepared via standard Fmoc-amino acid synthesis on a variety of linkers. For amino terminal detection, 2-chlorotrityl chloride resin (0.71 mmol/g, 0.1 g) was shaken overnight with 1,4-diaminobutane (putrescine, 4 equiv) in DCM (4 mL) and then washed with DCM (4 × 4 mL), DMF (4 × 4 mL), EtOH (4 × 4 mL) and DCM (4 × 4 mL). The derivatized resin was shaken with 2 equiv each of 4-carboxybenzaldehyde, HOBt, and DIPEA in 4 mL of DMF/DCM (1:2). A modified Wittig reaction on the support-bound aldehyde employing 10 equiv of $Ph_3PCMe-CH_2CO_2Et$ in DCM overnight was monitored by MALDI-TOF MS.

For Cleavage and MALDI-TOF MS Detection A single bead or beads (ca. 10–100) were placed into a well of the sample plate and then placed onto a metal rack in a glass dish that contained 10 mL of slowly stirred TFA or TFA-CNBr (20 mg). The glass dish was sealed with a plate of glass and the sample plate was left in the TFA vapor for 5–30 min or overnight with CNBr. Matrix solution and internal standards were added to each well [2 μL of a saturated 2,5-dihydroxybenzoic acid solution (100 mg/mL) in water/acetonitrile/TFA (70/30/0.1) plus 1 μL of a solution of Bradykinin (10 μg/mL in water/TFA (100:0.1))]. These were left to cocrystallize around the bead(s) for 15–30 min at room temperature before the plate was loaded into a GSG fOCUS Benchtop II linear-laser desorption time-of-flight mass spectrometer. The sample was irradiated with a nitrogen laser (337 nm, 3-ns pulse width, 20 Hz) with an acceleration voltage of 20 kV. The spectra were generated from the sum of 10–50 scans with a laser power just above the threshold of the least ionizable component.

REFERENCES FOR SECTION 5.5.1

1. Egner, B. J.; Cardno, M.; Bradley, M. *J. Chem. Soc., Chem. Commun.* **1995**, 2163–2164.
2. Barlos, K.; Gatos, D.; Papaphotiu, G.; Schafer, W.; Wenqing, Y. *Tetrahedron Lett.* 1989, *30*, 3947.
3. Youngquist, R. S.; Fuentes, G. R.; Lacey, M. P.; Keough, T. *J. Am. Chem. Soc.* 1995, *117*, 3900–3906.

5.5.2   Direct Monitoring by MALDI-TOF MS. A Tool for Combinatorial Chemistry[1]

Points of Interest

Matrix-assisted laser desorption/ionization time-of-flight mass spectrometry (MALDI-TOF MS) allows the detection of reaction products from both single beads and a small number of beads (10–100). The method is useful for monitoring solid-phase reactions performed with the Rink amide linker (or another linker that can be readily cleaved with TFA vapors). The method was used for reaction optimization where failing or side reactions were dominating the reaction process. The authors compared different literature procedures for the palladium-catalyzed allyl ester deprotection and peptide cyclization on resin.

Literature MS Procedure

A single bead (removed using microtweezers and a microscope) or beads (ca. 10–100 removed by pipetting from a suspension in CH_2Cl_2) were placed into a well of the sample plate. The stainless steel sample tray was placed onto a metal rack in a glass dish that contained 10 mL of slowly stirred TFA. The glass dish was sealed with a plate of glass, and the sample plate was left in the TFA vapor for typically 30 min at room temperature. Cleavage of the peptide causes a red coloration of the bead(s) due to cation formation, the intensity of which could be used to determine if a longer TFA treatment was necessary. The sample plate was removed from the glass dish, and the matrix solution and internal standards were added to each well (2 μL of a saturated solution of 2,5-dihydroxybenzoic acid (100 mg/mL) in water/acetonitrile/TFA (70/30/0.1) plus 1 μL of a solution of gramicidin S (10 μL/mL in water/TFA (100/0.1)). These were left to cocrystallize around the bead(s) for 15–20 min at room temperature before the plate was loaded into a GSG fOCUS Benchtop II linear-laser desorption time-of-flight mass spectrometer and the sample irradiated with a 337-nm laser with an acceleration voltage of 20 kV.

Literature Results

The authors compared different literature procedures for the palladium-catalyzed deprotection of an (Asp) allyl ester by analysis of the products by MS. The first method proved to be very unreliable, with only very low levels of allyl deprotection being observed by MS analysis.[2] A second literature procedure proved to be a much more reliable method of allyl deprotection to provide the free acid + sodium.[3]

Side reactions from attempted peptide cyclizations were also observed. Peptide cyclization with excess DIC/HOBt showed, by MS analysis, what is presumably the formation of N-acylurea. Peptide cyclization with PyBOP/DIEA initially gave the piperidyl amide. The latter side reaction was eliminated by washing the deblocked peptide resin with 0.4% concentrated HCl in DMF to remove resin-bound piperidine. Coupling with PyBOP/HOBt/DIEA gave the head-to-tail cyclic product to the exclusion of previously observed side products.

REFERENCES FOR SECTION 5.5.2

1. Egner, B. J.; Langley, J.; Bradley, M. *J. Org. Chem.* 1995, *60*, 2652–2653.
2. Lloyd-Williams, P.; Jou, G.; Albericio, F.; Giralt, E. *Tetrahedron Lett.* 1991, *32*, 4207.
3. Kates, S. A.; Daniels, S. B.; Albericio, F. *Anal. Biochem.* 1993, *212*, 303.

5.6   OTHER ANALYTICAL METHODS FOR COMBINATORIAL SYNTHESIS

5.6.1   Quantitation of Combinatorial Libraries of Small Organic Molecules by Normal-Phase HPLC with Evaporative Light-Scattering Detection.[1]

Points of Interest

1.   Light scattering was reported as a detection method independent of the sample chromophore.

2.   Evaporative light-scattering detection provided less compound to compound variation than UV detection within similar structural classes. The quantitation errors, based on a single external standard, for a series of steroids, hydantoins, and Boc- and Fmoc-protected amino acids by normal-phase HPLC with evaporative light-scattering detection averaged ca. ±10%.

Representative Literature Procedure for Light Scattering Detection

Isocratic normal-phase HPLC of the steroids was performed on a 250 x 4.6 mm Alltima CN column (Alltech, Deerfield, IL), eluting with 70% hexane/ 20% ethyl acetate/10% 2-propanol (v/v/v) at a flow rate of 1 mL/min. A 10-μL injection volume was employed, and evaporative light-scattering detection (ELSD) was carried out at 75°C and a nitrogen gas flow rate of 2.00 standard liters per minute (SLPM). The chromatographic separations were performed on a Perkin-Elmer (Norwalk, CT) Turbo-LC-Plus liquid chromatograph consisting of a Model ISS-200 autosampler, Model LC235C photodiode-array detector, and Model 200 quaternary pump. A Varex (Burtonsville, MD) Model MK-III evaporative light-scattering detector was connected to the outlet of the UV detector.

REFERENCES FOR SECTION 5.6.1

1. Kibbey, C. E. *Mol. Diversity* **1995**, *1*, 247–258.

6

PREPARATION OF SOLUTION LIBRARIES AND COMBINED APPROACHES AT THE SOLUTION/SOLID-PHASE INTERFACE

Although there are many differences between the solution and solid-phase strategies for generating libraries, in both cases the synthetic challenge is to develop reaction conditions that are general and high yielding. Perhaps the most obvious difference between the two methods is the linker to solid support. The linker can be considered as a protecting group employed throughout a combinatorial solid-phase synthesis. Another distinction between the two methods is that for solid-phase synthesis, a large excess of reagents can be used to drive reactions to completion, and the unreacted material may be simply filtered away. For solution synthesis, an unreacted excess of reagents cannot always be simply removed. There are exceptions: excess reagents can be used in high-throughput solution synthesis when the excess reagent is volatile or can be removed from the product (i.e., by extraction). Advances in high-throughput purification techniques could allow a large excess of reagents to be more generally employed in the synthesis of solution libraries. Finally, variable solubility properties often complicate the isolation of the products in a solution library.

Only a limited number of examples of solution libraries have been published so far. However, the ease of preparation and early success ensure that work will continue in this growing field. Support-bound reagents, soluble polymers, and resin capture strategies have also been developed that combine aspects of both solution and solid-phase strategies. In the following, some selected examples are provided to illustrate these different strategies.

Multicomponent coupling reactions, such as the Ugi reaction, have been used for generating solution libraries. The Ugi condensation involves the reaction of four different building blocks in one pot. Upon reaction completion, the product of the Ugi condensation, often in high yield, precipitates out of solution.

The Ugi condensation has also been performed with one of the four building blocks on support to facilitate isolation of the desired product.

The Ugi four component coupling (4CC) has been used to generate libraries in solution

Other high-yielding reactions have been employed to generate diversity in solution. For example, the optimization of a known lead structure with activity against the herpes simplex virus was performed in solution with a parallel synthesis and extractive workup. High-throughput condensations, substitutions, and palladium-catalyzed reactions were all utilized in the parallel solution synthesis for lead optimization.

Three related libraries for optimization of an in-house aminothiazole derivative with activity against herpes simplex virus.

Since long before the development of combinatorial synthesis, support-bound reagents have been known and used. In some cases, support-bound reagents can be used in ways that would be impossible for traditional reagents in solution. For example, mutually incompatible ("antagonistic") reagents, such as support-bound borohydride and peroxide, can be used in a single pot. Although reducing and oxidizing agents would be incompatible in solution, on solid support only a few sites on the surface of the beads can come in contact with each other.

Antagonistic support-based reagents for oxidation and reduction.

Parlow has demonstrated that three antagonistic reagents can be used in one pot. Even when support-bound reagents are used for a single step (as opposed to the antagonistic support-bound reagents), they are still attractive for combinatorial synthesis because they can be used in excess to drive reactions to completion and simply filtered away.

In contrast to employing an excess of support-bound reagent, resin capture strategies have been developed to selectively remove the desired product from a mixture prepared in solution. For example, support-bound aryl iodides can selectively capture borylalkenes from a mixture of reagents in solution. As a side point, the Suzuki cross-coupling reaction has been particularly reliable for combinatorial synthesis in a number of different contexts.

A resin capture strategy for the isolation of a borylalkene from a mixture in solution.

This chapter describes solution libraries as well as support-bound reagents, soluble polymers, and resin capture strategies. Often the nature of the target molecules will dictate the appropriate strategy for library generation.

6.1   SOLUTION LIBRARIES

6.1.1   Synthesis and Screening Procedures for Solution Libraries on Tetrasubstituted Templates[1].

Rigid cores

Points of Interest

1. A pooled library was prepared from a rigid-core molecule possessing multiple reactive functional groups and a mixture of small reactive organic molecules. A trypsin inhibitor was identified from the mixture in solution.

2. *tert*-Butyl amino esters were converted to the free acids to achieve water solubility required for evaluation. Hydrazides and hydroxyl amides prepared from hydrazinolysis or hydroxyl aminolysis of methyl esters were too insoluble in water for biological evaluation.

1. Mixture of 19 amino *tert*-butyl esters
Extraction with acid and base to remove unreacted starting material

2. Cleavage with reagent K to afford free acids with solubility in water/DMSO (9:1)

3. The diversity of the mixtures was qualitatively assessed by HPLC. Smaller representative libraries were prepared to determine if the expected statistical number of compounds were present. Of the theoretically expected 55 compounds, 42 were observed by FAB-MS.

4. An inhibitor of trypsin-catalyzed cleavage of an amide bond was obtained from the larger libraries. Smaller mixtures were examined to identify a competitive inhibitor for trypsin ($K_i = 9 \times 10^{-6}$ M).

5. The solution-phase generation of tetraurea libraries on similar scaffolds has also been described.[2]

Representative Examples

L-Valine, L-asparagine, L-aspartic acid, L-cysteine, L-histidine, L-lysine, L-methionine, L-proline, L-serine, L-tryptophan, and L-tyrosine amino acids were incorporated with the scaffolds.

Literature Procedure

The tetrachloride (100 mg, 0.217 mmol), a mixture of amines (0.88 mmol total), and 5 mL of DCM under argon were treated with 1 mL of triethylamine. The reaction mixture was stirred under argon for 3 h, diluted with DCM (50 mL), and washed with 1 M citric acid solution (2 × 75 mL) and saturated sodium hydrogen carbonate solution (1 × 100 mL). The organic phase was dried over MgSO$_4$ and concentrated to afford a tan oil that formed a tan foam *in vacuo*. The protecting groups were removed by stirring with reagent K at room temperature for 4 h (reagent K: TFA, water, phenol, thioanisole, and ethanedithiol (82.5:5:5:2.5). The solution was concentrated *in vacuo* and the libraries were precipitated by adding cold diethyl ether and *n*-hexane. The white precipitate was filtered off, washed five times with cold diethyl ether, and dried *in vacuo*.

REFERENCES FOR SECTION 6.1.1

1. Carell, T.; Wintner, E. A.; Bshir-Hashemi, A.; Rebek, J. *Angew. Chem., Int. Ed. Engl.* **1994**, *33*, 2059–2060.
2. Shipps, G. W.; Spitz, U. P.; Rebek, J. *Bioorg. Med. Chem.* **1996**, *4*, 655–657.

6.1.2 A Novel Solution-Phase Strategy for the Synthesis of Triamide Libraries Containing Small Organic Molecules[1]

84-86%

65-99%

16-100%

Points of Interest

1. High purity was obtained (>90% pure) after extractions regardless of overall yields (16–100%).

2. Thiols and alcohol could be used instead of the first amine or aniline; however, separation of the neutral byproducts required extraction of the carboxylic acid product into 10% aqueous NaOH, reacidification, and extraction for isolation.

3. The method did not employ orthogonal protecting groups.

4. In general, for high-throughput amide bond formation the side products are removed by either solution- or solid-phase extraction. Both extraction methods can be automated.

Representative Examples

R^1 = primary and secondary amines, ortho-substituted anilines, amino alcohols, and aminophenols. R^2 = anilines, tertiary amines, and amino esters. R^3 = aliphatic acids, electron-rich and electron-poor aromatic acids and hetero-aromatic acids, and protected amino acids.

Literature Procedure

A solution of N-(($tert$-butyloxy)carbonyl)iminodiacetic acid (0.349 g, 1.50 mmol) in DMF (15 mL) was treated with EDCI (0.294 g, 1.54 mmol) at 25°C. The mixture was stirred at 25°C for 1 h before the amine (R^1NH_2, 1 equiv) was added, and the solution was stirred for 20 h. The reaction mixture was poured into 10% aqueous HCl (60 mL) and extracted with EtOAc (100 mL). The organic phase was washed with 10% HCl (40 mL) and saturated aqueous NaCl (2 × 50 mL), dried (Na$_2$SO$_4$), filtered, and concentrated *in vacuo* to yield the pure N-(($tert$-butyloxy)carbonyl)iminodiacetic acid monoamides. Each of the N-(($tert$-butyloxy)carbonyl)iminodiacetic acid monoamides was dissolved in anhydrous DMF (20 mL/mmol) and was divided into three equal portions in three separate vials. Each solution was treated with one of three amines (R^2NH_2, 1 equiv), diisopropylethylamine (2 equiv), and PyBOP (1 equiv). The solution was stirred at 25°C for 20 h. The mixture was poured into 10% aqueous HCl and extracted with EtOAc. The organic phase was washed with 10% aqueous HCl, saturated aqueous NaCl, 5% aqueous NaHCO$_3$, and saturated aqueous NaCl. The organic layer was dried (Na$_2$SO$_4$), filtered, and concentrated to yield the diamides (65–99%). Each of the N'-((*tert*-butyloxy)carbonyl)-N,N-disubstituted iminodiacetic acid diamides was dissolved in 4 N HCl–dioxane (32 mL/mmol), and the mixture was stirred at 25°C for 45 min. The solvent was removed *in vacuo*, and the residue was dissolved in anhydrous DMF (28 mL/mmol), divided into three equal portions, and placed in three separate vials. The solution was treated with one of three carboxylic acids (R^3CO_2H, 1 equiv) followed by diisopropylethylamine (3 equiv) and PyBOP (1 equiv). The solution was stirred for 20 h. The mixture was poured into 10% aqueous HCl and extracted with EtOAc. The organic phase was washed with 10% aqueous HCl, saturated aqueous NaCl, 5% aqueous $NaHCO_3$, and saturated aqueous NaCl. The organic layer was dried (Na_2SO_4), filtered, and concentrated *in vacuo* to yield the final products (16–100%).

Similar methods (shown in the following) were described on another scaffold in an earlier communication; regardless of overall yield the triamides were >85% pure.[2]

REFERENCES FOR SECTION 6.1.2

1. Chang, S.; Comer, D. D.; Williams, J. P.; Myers, P. L.; Boger, D. L. *J. Am. Chem. Soc.* **1996**, *118*, 2567–2573.
2. Boger, D. L.; Tarby, C. M.; Myers, P. L.; Caporale, L. H. *J. Am. Chem. Soc.* **1996**, *118*, 2109–2110.

6.1.3　Solution Synthesis and Biological Evaluation of a Library Containing Potentially 1600 Amides/Esters[1]

Points of Interest

1. A library was constructed by carrying out all the possible reactions between a set of 40 acid chlorides and a set of 40 nucleophiles (amines and alcohols). The reactions are very simple and produce molecules with relatively low molecular weight as leads.

2. The 1600 synthesized compounds were presented for biological screening in two sets of sample mixtures. In the first set, each pure acid chloride (A) was reacted with a stoichiometric amount of an equimolar mixture of nucleophiles (N_{1-40}), and in the other set, each pure nucleophile (N) was reacted with the acid chlorides (A_{1-40}) in an analogous fashion. A positive result in a biological assay for any given sample identifies one-half of an active dimer; thus by evaluating the positives from both sets of sample mixtures, one can identify the precise structure of active components. The authors mention the size (1600) and structural (dimers) limitations of this type of library. Because many of the components within each mixture are quite similar, the observed biological activity of a sample is often likely to be a sum of several moderate activities rather than due to a singly active entity.

3. Active components have been detected from larger mixtures in a similar manner using indexed and orthogonal libraries.[2,3]

4. Each nucleophile contained a tertiary amine moiety to neutralize the HCl liberated in the reaction (Boger and coworkers have used triethylamine to serve the same function; see p 241).[4] The expected products were observed in 75% of the cases by EI-MS.

Representative Examples

The nucleophiles used included 4-substituted piperazines and other diamines or anilines containing a tertiary amine, as well as primary and secondary alcohols. Acid chlorides (RCOCl) were employed where R = saturated or unsaturated (simple or branched, cyclic), aryl, substituted aryl, heteroaryl, aryl ether, and ester functional groups.

Literature Summary

Stock solutions of each reactant (A or N) were prepared in DCM (10 mL, 0.5 M). Half of each stock solution was then removed and mixed with an equal volume of each of the solutions of the other 39 components of the same type. This produced 200 mL of A_{1-40} and N_{1-40} (0.0125 M in each reactant). Coupling was achieved by mixing the remaining 5 mL of each pure reactant solution (A or N) with 5 mL of the mixture from the other reactant (N_{1-40} or A_{1-40}). The mixtures were allowed to react for 48 h at room temperature. Any last small traces of unreacted acid chlorides were destroyed with methanol (10 mL), and the solvent was then evaporated in air over an additional 24–48 h to provide residual gums, which were dried in a vacuum desiccator and screened directly. Model couplings on 4- and 10-component mixtures were carried out and monitored by NMR and HPLC to optimize the reaction conditions.

REFERENCES FOR SECTION 6.1.3

1. Smith, P. W.; Lai, J. Y. Q.; Whittington, A. R.; Cox, B.; Houston, J. G.; Stylli, C. H.; Banks, M. N.; Tiller, P. R. *Bioorg. Med. Chem. Lett.* **1994,** *4,* 2821–2824.
2. Pirrung, M. C.; Chen, J. *J. Am. Chem. Soc.* **1995,** *117,* 1240–1245.
3. Deprez, B.; *et al. J. Am. Chem. Soc.* **1995,** *117,* 5405–5406.
4. Chang, S.; Comer, D. D.; Williams, J. P.; Myers, P. L.; Boger, D. L. *J. Am. Chem. Soc.* **1996,** *118,* 2567–2573.

6.1.4 Automated Parallel Solution-Phase Synthesis and Purification of Amides[1]

1. Amidopiperidines

Points of Interest

1. The products were purified from starting materials and byproducts via automated robotic filtration in good yields and purities.

2. A mixture of 2:1 DCM/DMF as reaction solvent provided reasonable reaction rates and homogeneous reaction product mixtures. In a typical run, ca. 5% of the reactions did not provide homogeneous solutions after reaction completion, either due to limited solubility of the starting acid or due to limited solubility of the resulting product. These reactions were carried out manually in cases in which the starting acid had limited solubility. In cases in which the product precipitated, filtration and trituration were done to purify the product.

Representative Examples

Diamines were coupled with acids containing free alcohols, ethers, diols, ureas, amides, and phenols.

Literature Procedure

Diamines were reacted with 4 equiv of acid and 1.5 equiv of DIC and HOBt in 2:1 DCM/DMF for 24 h (average yield 75%, average HPLC purity 90%) and purified with silica support functionalized with sulfonic acid (SCX, 0.6 mequiv/g, available from Varian Sample Preparation Products). Side products were removed with MeOH and 0.1 M ammonia in MeOH. Products were eluted with 4–7 mL of 1–2 N ammonia in MeOH. A total of 225 amides were prepared in this manner and up to 0.4 mmol of product was obtained.

2. Neutral analogs

Points of Interest

1. Neutral amides were formed with *p*-nitrophenol esters and excess amines.

2. An anion-exchange resin (SAX, 0.7 mequiv/g, bonded silica support functionalized with propyltrimethylammonium chloride; the chloride resin was prewashed with methanolic KOH to increase basicity) was used to remove the *p*-nitrophenol, followed by cation exchange (SCX) to remove the excess amine.

3. Reaction conditions were general for amines, but not anilines. Eleven separate robotic washes of the columns were used in the procedure. A total of 150 analogs were prepared in this manner (as in the foregoing; average yields 75%, average purity 90%).

Representative Examples

Nitrophenyl esters were aliphatic (except Boc-Phe); sterically hindered amines as well as amino alcohols and amino esters were employed (*tert*-butyl-amine required heating, aniline did not react).

REFERENCES FOR SECTION 6.1.4

1. Lawrence, R. M.; Fryszman, O. M.; Biller, S. A.; Poss, M. A. *Am. Biotechnol. Lab.* Oct. 1996, 12–17.

6.1.5   Solution and Solid-Phase Synthesis of 5-Alkoxyhydantoins [1]

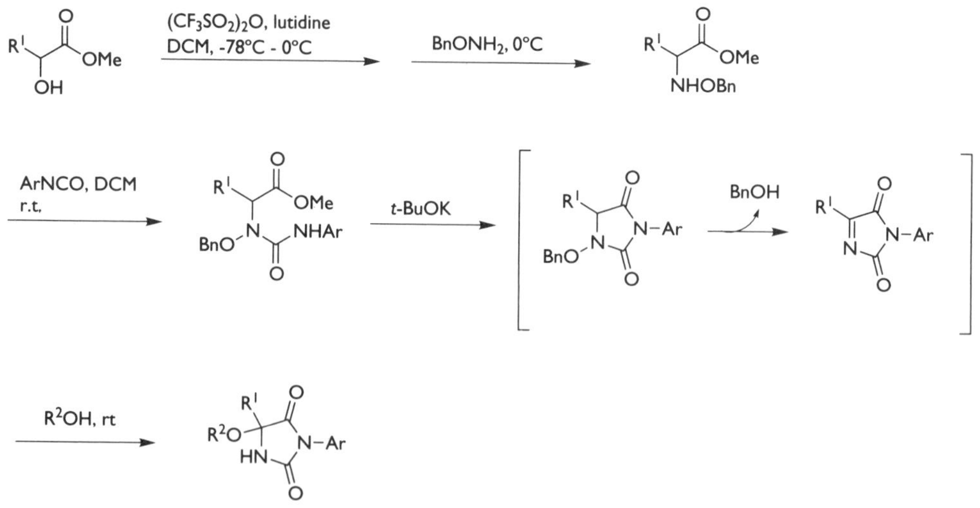

Points of Interest

1. N-Benzyloxy α-amino esters were prepared in one pot and used as precursors for the preparation of 5-alkoxyhydantoins. A library of 50 discrete 5-alkoxyhydantoins with three functional group variations was prepared in solution and six analogs were prepared on Merrifield resin (see p 160).

2. Each of the 50 5-alkoxyhydantoins prepared in solution was fully characterized by ^{1}H, ^{13}C, and mass spectrometric techniques. The 5-alkoxyhydantoins prepared on support also exhibited satisfactory ^{1}H and ^{13}C NMR spectra. HPLC analysis of crude products cleaved from the resin showed purities ranging from 89 to 96%.

Representative Examples

5-Alkoxyhydantoins were prepared on support (average yield 56%) and in solution (average yield 80%).

Literature Summary

α-Hydroxy esters were transformed into the corresponding N-benzyloxy amino ester by treatment with trifluoromethanesulfonic acid anhydride in the presence of lutidine in DCM, followed by the addition of O-benzylhydroxylamine. The resulting α-N-benzyloxy amino acid esters were condensed with different aryl isocyanates in DCM to provide urea derivatives. Treatment with potassium *tert*-butoxide in a variety of alcohols afforded the expected 5-alkoxyhydantoins (average yield 80%).

REFERENCES FOR SECTION 6.1.5

1. Hanessian, S.; Yang, R. Y. *Tetrahedron Lett.* **1996**, *37*, 5835–5838.

6.1.6 Liquid-Phase Combinatorial Synthesis on Poly(ethylene glycol)Monomethyl Ether (MeO-PEG) [1]

Points of Interest

1. MeO-PEG has a strong propensity to crystallize; thus as long as the polymer remains unaltered during the construction of the library, purification by crystallization can be accomplished at each stage of the combinatorial process.

2. MeO-PEG is soluble in a variety of aqueous and organic solvents. All manipulations, including split synthesis, may be carried out under homogeneous conditions.

3. MeO-PEG is precipitated from DCM by the addition of ether and then filtered away from the excess reagents. However, the purification technique is obviously limited to reagents that are soluble after the addition of ether. Combinatorial applications of soluble polymers generally require specific conditions under which excess reagents or desired products display solubility characteristics distinct from those of the polymer for purification.

4. Dendrimer-supported combinatorial chemistry has been reported. Reactions are performed in solution and dendrimeric intermediates are separated by size-selective methods such as size exclusion chromatography (SEC) or ultrafiltration.[2]

Representative Examples

A peptide library was synthesized and evaluated against a monoclonal antibody elicited against β-endorphin. An arylsulfonamide library consisting of six members was constructed.

Literature Procedure for the Synthesis of Six Arylsulfonamides

O-(MeO-PEG) N-[4-(chlorosulfonyl)phenyl]carbamate was prepared as follows. 4-(Chlorosulfonyl)phenyl isocyanate (0.653 g, 3 mmol) was added to MeO-PEG (5 g, 1 mmol) in DCM and 2 drops of dibutyltin laurate were added. After 5 h of stirring at room temperature, diethyl ether was slowly added to the vigorously stirred reaction mixture. The precipitate was collected on a glass filter and thoroughly washed with diethyl ether. The precipitate was dried under vacuum to yield the desired product quantitatively.

The N-[4-((alkylamino)sulfonyl)phenyl]carbamate of MeO-PEG was prepared by continuously bubbling ammonia gas through O-(MeO-PEG) N-[4-(chlorosulfonyl)phenyl]carbamate (0.5 g, 96 μmol) in DCM (5 mL) containing pyridine (20 equiv) for 24 h at room temperature (method A), by stirring O-(MeO-PEG) N-[4-(chlorosulfonyl)phenyl]carbamate (0.5 g, 96 μmol) with an excess amine (15 equiv) in DCM (5 mL) containing pyridine (20 equiv) for 24 h at room temperature (method B), or by heating the reaction mixture in pyridine solvents at 65°C for an hour (method C). The MeO-PEG polymer was precipitated from the homogeneous solution by the addition of diethyl ether, washed with ethanol, and dried under vacuum to give the desired product.

O-(MeO-PEG) N-[4-((alkylamino)sulfonyl)phenyl]carbamate (0.45 g) was dissolved in 0.5 M NaOH (10 mL) and heated at 90°C for 30 min. The reaction mixture was cooled to 4°C and neutralized to pH 6–8 with concentrated HCl. The reaction mixture was extracted with ethyl acetate (3×) and backwashed with brine, and the organic layer was dried with MgSO₄. The removal of solvent gave pure product by NMR.

REFERENCES FOR SECTION 6.1.6

1. Han, H.; Wolfe, M. M.; Brenner, S.; Janda, K. D. *Proc. Natl. Acad. Sci. U.S.A.* **1995**, *92*, 6419–6423.
2. Kim, R. M.; Manna, M.; Hutchins, S. M.; Griffin, P. R.; Yates, N. A.; Bernick, A. M.; Chapman, K. T. *Proc. Natl. Acad. Sci. U.S.A.* **1996**, *93*, 10012–10017.

6.1.7　A Fluorous Tin Hydride Reagent for Liquid-Phase Combinatorial Synthesis[1]

$$(C_6F_{13}CH_2CH_2)_3 SnH$$

Points of Interest

1. Tris(2-(perfluorohexyl)ethyl)tin hydride [$(C_6F_{13}CH_2CH_2)_3 SnH$] was introduced as a prototypical example of a fluorous reagent. This reagent behaves like normal tin hydride reagent in ionic and radical reductions, yet it can be separated from organic products by liquid–liquid extraction.

2. Benzotrifluoride (BTF, $C_6H_5CF_3$, (trifluoromethyl)toluene) was introduced as a partially fluorinated solvent to provide a homogeneous reaction medium in lieu of organic/fluorous solvent mixtures.

3. The reduction of alkyl halides and other functional groups to alkanes with catalytic fluorous tin reagent and stoichiometric $NaCNBH_3$ was reported.

4. Three halides were cross-coupled with three electron-poor alkenes (used in excess). Products were simply "purified" by three-phase liquid–liquid extraction (conducted in the original reaction vial) and evaporation. The crude products contained no identifiable starting materials or side products by capillary GC.

6.1.8　Stille Couplings with Fluorous Tin Reactants for Liquid-Phase Combinatorial Synthesis[2]

$$(C_6F_{13}CH_2CH_2)_3 - Sn \cdot Ar^1 \quad + \quad R^2X \quad \xrightarrow[\substack{LiCl \\ DMF/THF (1:1) \\ 80°C,\ 22\ h}]{PdCl_2(PPh_3)_2} \quad Ar^1-R^2$$

Points of Interest

1. The fluorous strategy retains the attractive features of organotin reagents with less of its liabilities (toxicity, separation, disposal).

2. An improved procedure was developed to prepare over 50 g of a fluorous tin reactant.

3. The Stille coupling with fluorous tin reactants was successfully performed on a half-gram scale. The ease of separation and reuse of the tin reactants are attractive features for preparative organic synthesis.

4. If desired, a fluorous reagent may be used in excess and separated by extraction for liquid-phase combinatorial synthesis.

Representative Examples

Electron-rich, electron-poor, and benzylic halide and triflates were coupled with three different fluorous tin reactants in good yield. The crude products from the pyridyl tin reactant were not very clean.

Literature Procedure

A mixture of 1.2 equiv of tin reagent, 1 equiv of halide or triflate, 2 mol % of $PdCl_2(PPh_3)_2$, and 3 equiv of LiCl in 1:1 DMF/THF (1 mL) was heated at

80°C. Reactions were conducted in individual vessels in groups of five (one tin reagent with all five partners). After 22 h, each mixture was evaporated to remove some of the solvent and then partitioned in a three-phase extraction between water (top), DCM (middle), and FC-72 (bottom). Evaporation of the FC-72 phase provided the tin chloride ($(C_6F_{13}CH_2CH_2)_3SnCl$, 80–90%), which was routinely recycled. Evaporation of the organic phase provided a crude organic product that was further purified by preparative TLC to provide the major cross-coupled biaryl or diaryl methane along with small amounts of the symmetrical biaryl derived from the tin reactant (a common byproduct of Stille couplings).

REFERENCES FOR SECTIONS 6.1.7 AND 6.1.8

1. Curran, D. P.; Hadida, S. *J. Am. Chem. Soc.* **1996**, *118*, 2531–2532.
2. Curran, D. P.; Hoshino, M. *J. Org. Chem.* **1996**, *61*, 6480–6481.

6.1.9   Solution-Phase Combinatorial Synthesis of Polyazacyclophane Scaffolds and Tertiary Amine Libraries[1]

Points of Interest

1. Monoprotected polyazacyclophanes were synthesized with an orthogonal Boc and 2-nitrobenzenesulfonyl protecting groups to allow for further derivatization.

2. Libraries of 100 compounds were prepared in pools of 10. Mixtures were purified from reagents by chromatography.

Representative Examples

Free amino sites on three different polyazacyclophanes were selectively reacted with benzyl bromides in solution.

Literature Summary

A mixture of the triprotected triamine (18.6 mmol), anhydrous Cs_2CO_3 (4 equiv), and 2,6-bis(bromomethyl)pyridine (1 equiv) in 500 mL of anhydrous DMF was stirred for 24 h. After workup and flash chromatography, the triprotected polyazacyclophane was obtained (80% yield). A stirred mixture of the triprotected polyazacyclophane (6 mmol) and anhydrous K_2CO_3 (8 equiv) in 80 mL of anhydrous DMF was treated with thiophenol (2.4 equiv) for 2 h. After aqueous workup and flash chromatography, the desired mono-t-Boc-protected polyazacyclophane scaffold was obtained in 94% yield. A solution containing equal amounts of 10 electrophiles (benzyl bromide derivatives, 1.44 mmol each for a total of 14.4 mmol, 2.4 equiv) in CH_3CN was added to a stirred mixture of the monoprotected polyazacyclophane (6.0 mmol) and anhydrous K_2CO_3 in CH_3CN. The reaction mixture was stirred overnight and the crude product was purified by flash chromatography on a silica gel column to afford the library mixture in 94% yield. The two reactive sites were combinatorialized with 10 functionalities, resulting in 100 compounds. Deprotection of the library with TFA was followed by chromatographic purification in 93% yield. The remaining reactive site was treated sequentially with each of 10 electrophiles (1.3 equiv) under the same conditions. This afforded libraries in yields of 60–98% after preparative thin-layer chromatographic purification.

REFERENCES FOR SECTION 6.1.9

1. An, H.; Cook, P. D. *Tetrahedron Lett.* **1996**, *37*, 7233–7236.

6.1.10   Parallel-Compound Solution-Phase Synthesis for Lead Optimization [1]

Three related libraries for optimization of an in-house aminothiazole derivative with activity against herpes simplex virus:

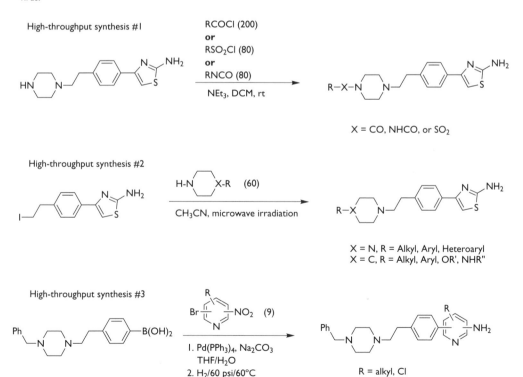

Points of Interest

1. Parallel synthesis was used to rapidly prepare three compound libraries containing 400 analogs of the lead compound. Activity at the micromolar level was readily measured even on the crude library compounds; the presence of byproducts such as triethylamine hydrochloride did not significantly affect the biological assay. Exact IC_{50} values for the most active compounds in the library were confirmed by resynthesis and evaluation of purified and fully characterized compounds. A compound 18 times more potent than the lead compound in the herpes simplex virus (HSV) helicase plaque reduction assay was identified from the parallel synthesis.

2. Microwave irradiation was used for the simultaneous heating of substitution reactions.

3. The Suzuki reaction was followed by catalytic hydrogenation without the addition of any further catalyst.

Literature Summary

General Procedure for the Parallel Synthesis of 90 Compounds. The substrate (1 mmol) was dissolved in 10 mL of a suitable solvent (0.1 M solution) and an aliquot (100 μL, 10 μmol) was added to 90 reaction vessels. Each reactant (0.1 mmol) was dissolved in the chosen solvent (1 mL, 0.1 M solution) and an aliquot (110 μL, 11 μmol) of each reactant was added to the 90 reaction vessels, a different reactant to each vessel. Any additional reagents necessary for the reaction were also added to the reaction vessels and then subjected to the chosen reaction conditions. The crude products were analyzed by TLC and MS to provide an indication of the average extent of the reaction and number of products formed.

For the HSV-1 High-Throughput Synthesis #1. The starting substrate was acylated (with acid chlorides, isocyanates, or sulfonyl chlorides) in the presence of triethylamine in DCM for 18 h. The majority of products showed a single new product on TLC and the expected molecular ion by MS.

For the HSV-1 High-Throughput Synthesis #2. The starting substrate was treated with 4-substituted piperazines (27 compounds) and 4-substituted piperidines (21 compounds) in acetonitrile. The reaction mixtures were heated in sealed polypropylene vials under microwave irradiation for 4 h. The majority showed a single new product on TLC and the expected molecular ion by MS.

For the HSV-1 High-Throughput Synthesis #3. The small library was prepared by reaction of [4-(2-(4-benzylpiperazin-1-yl)ethyl)phenyl]boronic acid (10 μmol for each reaction) with substituted bromonitrobenzenes (9 compounds, 10 μmol of each). To a solution of the boronic acid (0.1 M, 100 μL, 10 μmol) in THF was added a solution of a halonitrobenzene (0.1 M, 100 μL, 10 μmol) in THF and an aqueous solution of sodium bicarbonate and $Pd(PPh_3)_4$ (5 mol %). The mixture stood at room temperature for 4 h, and then in one pot, with the same palladium catalyst, each reaction mixture was hydrogenated with hydrogen gas at 60 psi and 60°C for 8 h. The majority of products showed a single new product on TLC and the expected molecular ion by MS.

Four related libraries for optimization of a known NK$_2$ antagonist:

High-throughput synthesis #4

RCOCl (200)
or
RSO$_2$Cl (80)
or
RNCO (80)

NEt$_3$, DCM, rt

X = CO, NHCO, or SO$_2$
R = Alkyl, Aryl, Heteroaryl

High-throughput synthesis #5

RCOCl (200)
or
RSO$_2$Cl (80)
or
RNCO (80)

NEt$_3$, DCM, rt

X = CO, NHCO, or SO$_2$
R = Alkyl, Aryl, Heteroaryl

High-throughput synthesis #6

H-N⟷X-R   (100)

High-throughput synthesis #7

RCOCl (200)
or
RSO$_2$Cl (80)
or
RNCO (80)

X = CO, NHCO, or SO$_2$,
R = Alkyl, Aryl, Heteroaryl

Points of Interest

1. Extensive SAR against NK$_2$ receptors was derived from screening four libraries.

2. These seven high-throughput syntheses were used for combinatorial lead optimization.

Literature Summary

Acylations were performed as described in the preceding paragraphs. The mesylate was reacted with 4-substituted piperazines, 4-substituted piperidines, and secondary alicyclic amines (86 reactants in total) in acetonitrile as the solvent. The mixtures were heated in sealed vials. The products were analyzed by TLC and MS; the majority showed a single new product on TLC and the expected molecular ion by MS.

REFERENCES FOR SECTION 6.1.10

1. Selway, C. N.; Terrett, N. K. *Bioorg. Med. Chem.* **1996**, *4*, 645–654.

6.1.11   Robotic and Solution-Phase Derivatization of Dichlorotriazines[1]

R^1 = Me, Et

Points of Interest

1. Solution-phase robotic synthesis was performed on a modified version of the commercially available HP 7686 PrepStation. Reaction conditions were programmed for the synthetic sequence on a PC computer terminal using a Windows-based program called Bench Supervisor.

2. Corticotropin-releasing factor$_1$ (CRF) receptor binding data were obtained on known quantities of over 350 triazine analogs. The most potent CRF receptor antagonist had a K_i value of 57 nM.

3. A range of cyclopropylamines were prepared by reduction of cyclopropylamides that were in turn prepared from the reaction of cyclopropanecarbonyl chloride with primary amines.

Representative Examples

2-Methyl- and 2-ethyl-4,6-dichlorotriazines were used as scaffolds. Both anilines and amines were added to the dichlorotriazines. Approximately two-thirds of the desired products were isolated in 70–95+% purity as determined by gas chromatography (GC–MS), and only these compounds were used in SAR determinations.

Literature Procedures

Modifications of the Original PrepStation.   The PrepStation was modified by heating one of the sample trays with hot water from a heating bath circulator. This allows as many as 25 reactions to be heated to temperatures of up to 80°C at any one time. In addition, the reaction heating block supplied on the original PrepStation had a heating limitation of temperatures which could not exceed 90°C. This was modified to allow temperatures of up to 125°C. This upgrade is currently supplied with the new machines.

Interfacing the PrepStation with a GC–MS. The PrepStation was interfaced to a HP 5890 GC with a 30-m capillary column (packed with cross-linked 5% phenylmethylsilicone) and a HP 5972 EI-MS detector. The interface programs used were HP Bench Supervisor and HP Chem Station with sample delivery via the HP 7673A Sample Tray and HP 18593B Autoinjector in the standard setups. Data listing the sample identity, reaction conditions, GC chromatogram, and percent purity for each sample is automatically reported from the HP Chem Station software.

Reaction Conditions. 2-Methyl- and 2-ethyl-4,6-dichlorotriazines[2] were made as 0.5 M stock solutions in THF. The anilines were made as 1.5 M stock solutions in THF containing 3 M DIPEA. All the commercially available amines were used either as neat solutions or as 1 M THF solutions in the case of solids. The appropriate dichlorotriazine (0.04 mL) was added in series by the robot to each 1.8-mL vial (septum sealed) followed by addition of 0.026 mL of the aniline. The reactions were allowed to stand for 10 min in parallel. The resulting solutions were treated with 10–20 equiv of amines in series at 65°C in parallel for 1 h. The robot dispensed EtOAc (0.4 mL) and 1 N HCl (0.5 mL) to each of the reaction vials and transferred the organic layer to an empty vial. The products were analyzed in series by GC–MS for purity and identity and reactions were evaporated to dryness with a stream of nitrogen in parallel to provide oils or solids. Most compounds were >70% pure and only those were used to determine SAR. A program disk is available for this synthetic sequence and will be supplied on request from the authors.

REFERENCES FOR SECTION 6.1.11

1. Whitten, J. P.; Xie, Y. F.; Erickson, P. E.; Webb, T. R.; De Souza, E. B.; Grigoriadis, D. E.; McCarthy, J. R. *J. Med. Chem.* **1996**, *39*, 4354–4357.
2. Hirt, R.; Nidecker, H.; Berchtold, R. *Helv. Chim. Acta* **1950**, *179*, 1365–1369.

6.1.12 Multiple-Component Condensation Strategies in Solution-Phase Combinatorial Synthesis[1,2]

Points of Interest

1. The Ugi four-component coupling (4CC, shown in the reaction scheme) was initially developed over 30 years ago. Recently, a number of researchers have focused on solution-phase and solid-phase synthesis of the Ugi reaction to provide libraries of small molecules.[3]

2. A range of multiple-component condensations (MCC) are discussed and contrasted with the other methods for combinatorial synthesis (see ref 2). A short list of MCC with potential combinatorial applications includes the Strecker amino acid synthesis, the Hantsch pyrrole and pyridine, the Biginelli dihydropyrimidine, the Mannich reaction, the Passerini three-component coupling, the Bucherer/Bergs hydantoin synthesis, the Asinger condensations, and the Pauson–Khand reaction, in addition to the Ugi four-component coupling.

3. Azinomycin analogs were prepared via a Passerini MCC.[2]

4. A limited number of isocyanides are commercially available. Isocyanides can be prepared from amines, but they are nontrivial to work with due to their reactivity, toxicity, and odor. Convertible isocyanides have been developed for efficient postcondensation modification (see pp 166 and 257).[4,5]

5. The Ugi 4CC has been applied to the combinatorial synthesis of aminoglycoside antibiotic mimetics linked to poly(ethylene glycol).[6]

Literature Procedure

General Procedure for Ugi Four-Component Condensation. The carboxylic acid (1.25 equiv), amine (1.25 equiv), and aldehyde (1.0 equiv) were dissolved in methanol to an approximate concentration of 1 M in each. This solution was allowed to stand for 10 min and then was added in one portion to a flask containing the isocyanide (1.0 equiv). The resulting solution was allowed to stir at room temperature for 12 h. When the reaction was complete by TLC (1–5% methanol in DCM), the solvent was removed *in vacuo,* and the residue was purified (if desired) by flash column chromatography on silica gel, eluting with a 0–5% MeOH in DCM gradient. With a convertible isocyanide the Ugi reaction can be followed by methanolysis or resin capture with Wang resin.[4]

REFERENCES FOR SECTION 6.1.12

1. Ugi, I.; Domling, A.; Horl, W. *Endeavour* **1994,** *18,* 115–122.
2. Armstrong, R. W.; Combs, A. P.; Tempest, P. A.; Brown, D.; Keating, T. A. *Acc. Chem. Res.* **1996,** *29,* 123–131.
3. (a) See ref 1. (b) Keating, T. A.; Armstrong, R. W. *J. Am. Chem. Soc.* **1995,** *117,* 7842–7843. (c) Weber, L.; Walbaum, S.; Broger, C.; Gubernator, K. *Angew. Chem., Int. Ed. Engl.* **1995,** *34,* 2280–2282.
4. Keating, T. A.; Armstrong, R. W. *J. Am. Chem. Soc.* **1996,** *118,* 2574–2583.
5. Mjalli, A. M. M.; Sarshar, S.; Baiga, T. J. *Tetrahedron Lett.* **1996,** *37,* 2943–2946.
6. Park, W. K. C.; Auer, M.; Jaksche, H.; Wong, C. H. *J. Am. Chem. Soc.* **1996,** *118,* 10150–10155.

6.1.13 Postcondensation Modifications of Ugi Four-Component Condensation Products [1]

Points of Interest

1. There are many commercially available acids, amines, and aldehydes, but few isocyanides. The convertible isocyanide 1-isocyanocyclohexene compensates for the lack of commercially available isocyanides.

2. Acetyl chloride provides an anhydrous form of HCl for the formation of esters of the Ugi four-component coupling (4CC) products.

3. The pyrroles are believed to be formed from the cycloaddition of the intermediate oxazolinium-5-one (münchnone) with a dipolarophile. Imidazoles were not observed from cycloaddition of münchnones with electron-deficient nitriles.

4. Anthranilic acids were used without protecting groups in the construction of 1,4-benzodiazepine-2,5-diones (also see p 150).[2,3]

5. Resin capture of the Ugi 4CC products with a support-bound alcohol afforded products with >95% purity after cleavage from support.

6. Phenyl isocyanate (PhNC) and 2-pyridyl isocyanate (2-PyrNC) have been used for the conversion of Ugi 4CC products on support into pyrroles (see p 166).[4]

Representative Examples

For the esters, methanol, benzyl alcohol, *tert*-butyl alcohol, ethanethiol, and water could be used as nucleophiles. For the pyrroles, bis-substituted electron-poor acetylenes worked best. Anthranilic acids are commercially available and could be used to increase the diversity accessible in the synthesis of 1,4-benzodiazepine-2,5-diones.

Literature Procedures

For the Ugi 4CC. The carboxylic acid (1.25 equiv), amine (1.25 equiv), and aldehyde (1.0 equiv) were dissolved in methanol to an approximate concentration of 1 M in each. This solution was allowed to stand for 10 min and then was added in one portion to a flask containing the isocyanide (1.0 equiv). The resulting solution was allowed to stir at room temperature for 12 h. When reaction was complete by TLC (1–5% methanol in DCM), the methanol volume was doubled and acetyl chloride (10 equiv) was added in one portion. The flask was equipped with a reflux condenser and heated to 55°C for 3 h. When TLC showed complete conversion to the methyl ester, the reaction was cooled to room temperature, the solvent was removed *in vacuo*, and the residue was taken up in DCM and filtered. Products were purified by flash column chromatography on silica gel, eluting with a 0–5% methanol in DCM gradient.

For Acidic Hydrolysis of Cyclohexenamide Products. The starting cyclohexenamide (0.1 mmol) was dissolved in 1 mL of a stock solution of 0.5 mL of concentrated HCl in 9.5 mL of THF. This reaction mixture was stirred for 12–18 h, neutralized with solid sodium bicarbonate, and then filtered. The solvent was removed *in vacuo*, and the residue was dissolved in pH 10 water and washed with DCM. The aqueous layer was then acidified and reextracted with DCM. The organics were combined and dried over sodium sulfate, and the solvent was evaporated.

For Pyrrole Synthesis. The cyclohexenamide (0.05 mmol) was azeotropically dried with toluene and then dissolved in 1 mL of toluene. The acetylene (0.25 mmol, 5 equiv) was then added followed by HCl (3 equiv as a 1.0 M solution in anhydrous ether). The flask was then capped and heated to 100°C for 4 h. Alternatively, THF can be used as the solvent at 55°C. The solvent was then evaporated after cooling, the residue was taken up in DCM and filtered, and the soluble portion was purified by preparative TLC (0.5-mm thickness, 20 cm × 20 cm, 1–5% MeOH in DCM as eluent).

For Polymer Capture and Cleavage Using Wang Resin. The cyclohexenamide (0.11 mmol) was azeotropically dried with toluene, dissolved in THF, and then added to a flask containing Wang resin (0.027 mmol, 0.25 equiv, or 0.073 mmol, 0.67 equiv). HCl (0.55 mmol as a 1 M solution in anhydrous ether) was added and the flask was capped and then heated to 55°C for 5 h. After cooling, the resin was filtered off and rinsed with DCM (3×), MeOH (3×), and then with DCM (3×) again. The carboxylic acid product was cleaved from the resin by incubating with 20% trifluoroacetic acid in DCM solution for 20 min, at which point the pink to dark purple resin was washed three times with DCM. The solvent was evaporated, and the product was characterized.

For a Representative 1,4-Benzodiazepine-2,5-dione. Isobutyraldehyde (9 μL, 0.1 mmol) and *p*-methoxybenzylamine (16 μL, 0.125 mmol) were combined in 500 μL of methanol with 4-Å molecular sieves. The solution was stirred at 23°C for 1 h, and then 1-isocyanocyclohexene (100 μL of a 1 M solution in hexanes, 0.1 mmol) was added followed by anthranilic acid (14 mg, 0.1 mmol). After 18 h of stirring at 23°C, the solution was filtered through

Celite, the solvent was removed *in vacuo,* and the residue was purified via flash column chromatography (silica, hexanes to 1 : 1 hexanes/ethyl acetate gradient) to yield 32 mg (73%) of a clear glass.

REFERENCES FOR SECTION 6.1.13

1. Keating, T. A.; Armstrong, R. W. *J. Am. Chem. Soc.* **1996**, *118,* 2574–2583.
2. Boojamra, C. G.; Burow, K. M.; Ellman, J. A. *J. Org. Chem.* **1995**, *60,* 5742–5743.
3. Keating, T. A.; Armstrong, R. W. *J. Org. Chem.* **1996**, *61,* 8935–8939.
4. Sarshar, S.; Mjalli, A. M. M.; Siev, D. *Tetrahedron Lett.* **1996**, *37,* 835–838.

6.2   RESIN CAPTURE

6.2.1   Solid-Phase, Parallel Synthesis by the Ugi Multicomponent Condensation[1]

Points of Interest

1. A 96-member library was generated from 12 acids, 8 aldehydes, 1 isocyanide, and the support-bound Rink amine.[2]

2. To overcome limited commercial availability, isocyanides were also prepared *in situ* from α-lithiated benzyl isocyanide (1 equiv) and a slight excess (1.1 equiv) of either an alkylating agent or an aldehyde followed by an anhydride.[3]

3. The Ugi 4CC was conducted with the amino group tethered to solid support, since a variety of amino-functionalized matrices are readily available. Zhang and coworkers have previously tethered the isonitrile to solid support, because the number of commercially available or readily accessible isonitrile inputs are limited.[4] A support-bound isocyanide has been prepared via dehydration of the corresponding formylamine. The support-bound isocyanide was then used for the solid-phase synthesis of tetrasubstituted imidazoles (see p 165).

4. C-Glycoside peptide ligands were prepared employing an Ugi 4CC with the amino group tethered to support. A 96-compound library directed toward Sialyl Lewis X was constructed employing resin capture methodology.[5]

Representative Examples

In parallel with solution reactions, the product yields were most sensitive to the structure of the aldehyde inputs. Reactions with aliphatic aldehydes and electron-rich aromatic aldehydes afforded products in significantly higher yield than the corresponding electron-poor aromatic aldehydes. For the carboxylic acid inputs, the reactions involving phenolic derivatives gave generally low yields, partially due to precipitation over 24 h. Yields for four selected products carried out on larger scale (0.12 mmol) ranged from 68 to 80% (>90% purity).

Literature Procedure

Ninety-six individual reactions were performed in a single 96-well microtiter plate to produce one product per well. For the reaction, 0.014 mmol of Rink amine resin per well in MeOH/DCM (1:2) served as the amine input, and methyl isocyanoacetate (10 equiv) as the single isocyanide input. Twelve carboxylic acids (10 equiv) in columns 1–12 and eight aldehydes (10 equiv) in rows A–H were used for the construction of the library as follows. Rink Fmoc-amide resin (2.88 g, 0.46 mmol/g) was deblocked with piperidine and partitioned equally into a 96-well polyethylene microtiter plate. The eight aldehydes (1 M in DCM, 138 μL, 10 equiv) and twelve carboxylic acids (1 M in MeOH, 138 μL, 10 equiv) were each added to the appropriate wells. After 30 min methyl isocyanoacetate (1 M in DCM, 138 μL, 10 equiv) was added to all the wells. The plate was capped for 24 h and then allowed to evaporate to dryness. The resin was transferred to a 96-well filtration apparatus (details of the apparatus were not described) and washed liberally with MeOH and DCM. The products were removed from the resin, washed into a second 96-well plate with TFA in DCM (2 × 600 μL), and then rinsed with DCM (200 μL) and MeOH (200 μL). The solvents were then removed in a reduced-pressure oven.

REFERENCES FOR SECTION 6.2.1

1. Tempest, P. A.; Brown, S. D.; Armstrong, R. W. *Angew. Chem., Int. Ed. Engl.* **1996**, *35*, 640–642.
2. Rink, H. T. *Tetrahedron Lett.* **1994**, *18*, 115–123.
3. Böll, W. A.; Gerhart, F.; Nürrenbach, A.; Schöllkopf, U. *Angew. Chem.* **1970**, *82*, 482–483. *Angew. Chem., Int. Ed. Engl.* **1970**, *9*, 458–459.
4. Zhang, C.; Moran, E. J.; Woiwode, T. F.; Short, K. M.; Mjalli, A. M. M. *Tetrahedron Lett.* **1996**, *37*, 751–754.
5. Sutherlin, D. P.; Stark, T. M.; Hughes, R.; Armstrong, R. W. *J. Org. Chem.* **1996**, *61*, 8350–8354.

6.2.2 Synthesis of Tetrasubstituted Ethylenes on Solid Support via Resin Capture[1]

Points of Interest

1. The bis(boryl)alkenes are more reactive than the monoaddition product. The first coupling is fast in comparison to the second coupling; thus, most of the bis(boryl)alkene is consumed in the reaction. The mixture prepared in solution was combined with a Rink resin-bound aryl iodide to initiate a second Suzuki reaction without any further addition of palladium catalyst.

2. This methodology provided sterically hindered tetraphenylethylenes in high yield (75–95%) after cleavage with TFA.

3. When unsymmetrical bis(boryl)alkenes were used, mixtures of isomers (e.g., 2.3:1) were produced.

4. After resin capture, an alcohol was further modified under solid-phase Mitsunobu conditions (see p 106),[2] without any protecting group manipulations.

5. The Pt-catalyzed reaction was also used to prepare a support-bound arylboronic ester as shown in the following reaction scheme.

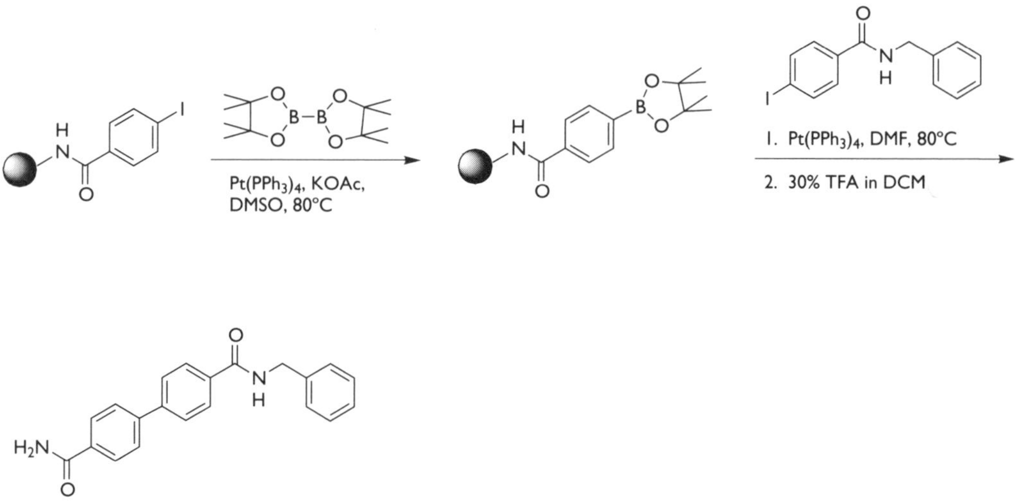

Representative Examples

Tetrasubstituted ethylene benzamides were prepared in high yield (75–95% prior to chromatography), including tamoxifen benzamide in a 2.5 to 1 ratio of regioisomers (95% yield). Both ethyl- and phenyl-substituted symmetrical alkynes were used as starting materials. After resin capture, further synthetic transformations (such as the Mitsunobu condensation) were performed.

Literature Procedures

For the Preparation of Tetrasubstituted Ethylenes. Bis(boryl)alkenes were synthesized according to Suzuki and Miyaura's procedure.[3] A small test tube was charged with bis(boryl)alkene (10 equiv), organohalide (15 equiv), $PdCl_2(PPh_3)_4$ (0.3 equiv), 3 M KOH (20 equiv), and enough DME to bring the concentration of bis(boryl)alkene to 0.5 M. The test tube was covered with a septum and flushed with N_2. The reaction mixture was heated overnight in a sand bath under N_2. Another test tube was charged with 100 equiv of KOH and 1 equiv of the support-bound aryl iodide and flushed with N_2. The DME/KOH solution was syringed into the tube containing the polymer and heated overnight. The polymer was filtered out of the solution and washed successively with H_2O, MeOH, EtOAc, and DCM. The products were cleaved from the polymer with 30% TFA in DCM.

For the Preparation of Biaryls. A flask was charged with polymer-bound aryl iodide (0.28 mmol), diborating reagent (1.4 mmol, 5 equiv), KOAc (4.2 mmol, 15 equiv), and $Pt(PPh_3)_4$ (0.04 mmol, 0.15 equiv) in a N_2-filled glovebag. DMSO (5 mL) was added and the reaction was heated to 80°C for 2 h under a N_2 atmosphere. The resin was filtered off and washed with MeOH, DCM, and EtOAc. The products were cleaved from the polymer with 30% TFA in DCM.

REFERENCES FOR SECTION 6.2.2

1. Brown, S. D.; Armstrong, R. W. *J. Am. Chem. Soc.* **1996,** *118,* 6331–6332.
2. Richter, L. S.; Gadek, T. R. *Tetrahedron Lett.* **1994,** *35,* 4705–4706.
3. (a) Ishiyama, T.; Matsuda, N.; Miyaura, N.; Suzuki, A. *J. Am. Chem. Soc.* **1993,** *115,* 11018–11019. (b) Ishiyama, T.; Matsuda, N.; Murata, M.; Ozawa, F.; Suzuki, A.; Miyaura, N. *Organometallics* **1996,** *15,* 713–720.

6.3 SUPPORT-BOUND REAGENTS

6.3.1 Support-Bound Activated Nitrophenyl Esters [1]

X = F, Cl, or OMe

Points of Interest

1. Polymer-bound 4-hydroxy-3-nitrobenzophenone was employed because it displayed a reactivity in between that of the *o*-nitrophenyl active ester, which takes hours for reaction completion, and that of the hydroxybenzyltriazole (HOBt) active ester, which is too water sensitive.

active
ester

relative
acylating   1   40   8000
reactivity

2. The polymeric acyl-transfer reagent combined fast coupling with high yields (95% yield). The polymeric reagent can be reused.

3. Carpino and coworkers have reported that the corresponding support-bound nitrophenyl sulfone is 5 times more active than the ketone.[2]

4. Friedel–Crafts acylation of polystyrene resin has also been performed with a FeCl₃ catalyst.[3]

Representative Examples

Boc-Phe, Boc-Gly, Boc-Trp, Boc-Tyr(OBzl), Boc-Met, Boc-Asp, and Boc-Lys(CBz) were all loaded onto resin in high yield (95% loading), but Boc-Gln was lower (30% loading). Boc-Tyr(OBzl)-Gly-Gly-Phe-Leu-OBz was synthesized in 92% crude yield.

Literature Procedures

Preparation of the Polymer. To a mixture of 50 g of macroreticular polystyrene (XE-305, Rohm & Haas) and 100 g of 4-chloro-3-nitrobenzoyl chloride was added a solution of 25 g of aluminum trichloride in 300 mL of dry nitrobenzene. The mixture was stirred mechanically at 60°C for 5 h and poured into a mixture of 150 mL of DMF, 100 mL of concentrated HCl, and 150 g of ice. The beads slowly turned white. They were washed with 300-mL portions of DMF/water (3:1) until the washings were colorless, then with warm (60°C) DMF, and finally with six portions of 300 mL of DCM/MeOH (2:1). The dried polymer weighed 82 g (1.88 mmol/g). Hydrolysis was carried out with a mixture of 130 mL of 40% benzyltrimethylammonium hydroxide in water, 130 mL of water, and 260 mL of dioxane for 8 h at 90°C. The polymer was filtered, and the process was repeated. The beads were then washed with four portions of warm (60°C) dioxane. Acetic acid (30 mL) was added with stirring for 15 min. The polymer was washed with dioxane until the washings were neutral, followed by six portions of 300 mL of DCM/MeOH (2:1). Anal. Calcd: Cl, <0.1; N, 2.38 (1.7 mmol/g). The amount of available hydroxide groups was determined by esterifying with a threefold excess of benzoyl chloride and pyridine in dry chloroform at 0–10°C for 30 min, washing with chloroform, and reacting with excess benzylamine. The polymer was washed with chloroform, and excess benzylamine was extracted with hydrochloric acid. The organic phase afforded pure N-benzylbenzamide, and from its weight the loading on the polymer was determined to be 1.7–1.8 mmol/g, assuming quantitative reactions.

Polymeric Active Esters. Esters of simple acids were prepared from the acid chloride and pyridine as described. Active esters of Boc-protected amino acids were prepared by the symmetric anhydride method as follows: 4 mmol of the Boc-protected amino acid was dissolved in 6 mL of DCM (THF was added in cases of poor solubility). The solution was cooled to −10 to 0°C and 2 mmol of DCC was added. After 30 min at 0°C, the mixture was filtered directly into a vessel containing 1 g of the polymer-bound 4-hydroxy-3-nitrobenzophenone. Pyridine (0.5 mL) was added, and the mixture was shaken for 1 h at room temperature. The polymer was washed with six to eight 10-mL portions of chloroform. Active esters of Boc-glycine, Boc-phenylalanine, and Boc-O-benzyltyrosine were thus prepared. Determination of the loading by reacting the polymers with an excess of benzylamine and weighing the resulting amide showed that 70–80% of the available OH groups underwent esterification.

REFERENCES FOR SECTION 6.3.1

1. Cohen, B. J.; Karoly-Hafeli, H.; Patchornik, A. *J. Org. Chem.* **1984**, *49*, 922–924.
2. Carpino, L. A.; Cohen, B. J.; Lin, Y.; Stephens, K. E.; Triolo, S. A. *J. Org. Chem.* **1990**, *55*, 251–259.
3. Zikos, C. C.; Ferderigos, N. G. *Tetrahedron Lett.* **1995**, *36*, 3741–3744.

6.3.2   Mediator Methodology: A Two-Polymeric System [1,2]

Points of Interest

1. Imidazole acts as a shuttle between two functionalized polymers. The method is based on transferring a support-bound electrophilic donor (such as a polymeric *o*-nitrophenyl ester) to an insoluble nucleophilic acceptor (such as a polymer-bound amino ester) with the aid of a soluble mediator molecule (the "shadchan" method).[3]

2. The reactions were monitored by UV absorbance of the intermediates in a flow cell connected between the polymer reservoirs.

3. A "bank" of activated resins were stored at −10°C for several months without decomposition.

4. Carpino and coworkers have used a slightly different procedure called Inverse Merrifield Synthesis for the preparation of peptides in solution with support-bound reagents.[4]

Representative Examples

Both active nitrophenyl esters and DMAP salts were employed in the mediator methodology (note: support-bound DMAP is now commercially available from Aldrich). Esters, thioesters, a sulfonamide, and a dioligonucleotide were prepared with support-bound DMAP.

Literature Procedures

Polymeric 4-(Dialkylamino)pyridine. This compound was prepared by a modification of the method of Shinkai.[5] Macroporous (chloromethyl)polystyrene (5 mequiv/g) was prepared (from Rohm & Haas XE 305 resin) by the method of Rossey.[6] The dry polymer (19 g) was suspended in 30 mL of DMF saturated with methylamine gas at 0°C.[7] The vessel was sealed and agitated for 24 h. The polymer was washed successively with dioxane, ethanol, 2 N NaOH/ *i*-PrOH (1:1), water (until eluate neutral), ethanol, and ether. After drying *in vacuo*, the polymer (3.7 g/3.8 mequiv of amino groups/1 g of dry weight) was suspended in a mixture of water (1.5 mL), ethanol (0.5 mL), triethylamine (7 mL), and 4-chloropyridine hydrochloride (4.7 g) in a glass pressure vessel, and the vessel was sealed and heated for 4 days at 140°C. The polymer was washed as before, and unreacted amino groups were blocked by acetylation (acetic anhydride in DCM, then hydroxide base wash to regenerate the support-bound DMAP). The washed DMAP polymer was dried at 150°C *in vacuo* until constant weight. Incorporation of pyridine groups was determined by potentiometric chloride titration of the hydrochloride salt bound to the polymer: 2.53 mequiv/g compared to 3.15 mequiv/g prior to acetylation.

Polymeric 1-Acyl-4-(dialkylamino)pyridinium Chlorides. In a typical experiment, the anhydrous DMAP polymer was swelled in DCM (freshly distilled from P_2O_5 under argon) and treated with excess benzoyl chloride at 0°C. The polymer was filtered and washed with DCM under anhydrous conditions until the washings contained negligible amounts of benzoyl chloride. The polymer was dried under vacuum at room temperature and was stable at –10°C for several months. After treatment with a primary amine in DCM, a pure amide was recovered by filtration and acid/base wash. The amount of amide corresponded to 0.8 mequiv/g of acyl substitution on the polymer.

Preparation of Polymeric Active Ester. A solution containing 2.45 mequiv of the Boc-protected amino acid and 2.5 mequiv of Et_3N in 3 mL of DCM (THF was added in case of low solubility) was shaken for 1 h at 25°C with 1.0 g of polymer-bound 4-hydroxy-3-nitrobenzophenone[8] containing 2.23 mequiv of available OH groups. The mixture was cooled in a dry ice/isopropyl alcohol bath, and 2.9 mequiv of DCC was added. After 1 h, the temperature was set to –10 to 0°C for another 10 h. The polymer was then washed with DCM (6–8×) until no Boc-amino acid could be detected. Active esters of Boc-glycine, Boc-phenylalanine, Boc-O-benzyltyrosine, and Boc-O-(3,6-dichlorobenzyl)tyrosine were thus prepared. Determination of the loading[4] showed that 85–98% of the available OH groups underwent esterification.

Peptide Synthesis in the Shadchan System.   A single cycle of the shadchan system was performed as follows. The polymeric nitrophenyl active ester of Boc-Phe-OH (0.8 g, 0.2 mequiv) was placed in one column and the polymer H$_2$N-Leu-OCH$_2$-Ⓡ (200 mg, 0.1 mequiv) was placed in a separate column (both columns at 0°C). The circulating solvent in the system was DCM (total volume 10–15 mL) containing imidazole (5 mg, 0.073 mequiv) and diisopropyl-ethylamine (8.7 μL, 0.05 mequiv). Circulation was started with a pump rate of 4 mL/min (60–30% of the imidazole was acylated at pump rates of 0.4–4 mL/min and less than 5% at pump rates greater than 10 mL/min). Monitoring of the reaction, using UV detection, was possible due to the low concentration of the acylimidazole in the system and was carried out at 310 nm. By this method, one flow cell of the detector was connected to the entrance (substrate cell) and another flow cell to the exit (reference cell) of column II. Because imidazole was added in a concentrated solution at the entrance of column II, the circulating solvent was not homogeneous-maximum and minimum peaks were obtained at the first cycle. After 24 h, column II was washed, and 2 mg of polymer was transesterified with Et$_3$N/MeOH for 48 h. Less than 2% of unacylated Leu (starting material) was observed by TLC after cleavage.

REFERENCES FOR SECTION 6.3.2

1. Shai, Y.; Jacobson, K. A.; Patchornik, A. *J. Am. Chem. Soc.* **1985**, *107*, 4249–4252.
2. Patchornik, A. *CHEMTECH* **1987**, 58–63.
3. "Shadchan" is the Hebrew term for a matchmaker, go-between, or agent.
4. Carpino, L. A.; Cohen, B. J.; Lin, Y.; Stephens, K. E.; Triolo, S. A. *J. Org. Chem.* **1990**, *55*, 251–259.
5. Shinkai, S.; Tsuji, Y.; Hara, Y.; Manabe, O. *Bull. Chem. Soc. Jpn.* **1981**, *54*, 631.
6. Warshawsky, A.; Deshe, A.; Rossey, G.; Patchornik, A. *React. Polym.* **1984**, *2*, 301.
7. Standard gel-form Merrifield resin has also been used to form support-bound DMAP; TBAI was added during the chloride displacement: Bunin, B. A., unpublished results. Support-bound DMAP is now commercially available from Aldrich.
8. Cohen, B. J.; Karoly-Hafeli, H.; Patchornik, A. *J. Org. Chem.* **1984**, *49*, 922–924.

6.3.3 Polymer-Bound EDCI: A Convenient Reagent for Formation of an Amide Bond[1]

Points of Interest

1. Aqueous workup can be avoided by attaching EDCI to a polymer. EDCI was bound on chloromethylated polystyrene–divinylbenzene resin via alkylation of the nucleophilic dimethylamino group of EDCI. Because the byproducts of the reaction would remain on the polymer, the desired product can be isolated by filtration.

2. The reagent with the best coupling properties was prepared from Merrifield resin that was 2% divinylbenzene (200–400 mesh, 0.8 mequiv of Cl/g). The reagent derived from resin with higher mequiv of Cl/g resulted in reduced swelling in chloroform with a detrimental effect on the coupling reaction.

Representative Examples

A variety of primary and secondary amides as well as anilides were prepared in 72–100% yield.

Literature Procedures

For the Preparation of Polymer-Bound EDCI. To a stirred solution of EDCI (15.7 g, 101 mmol) in DMF (800 mL) was added chloromethylated polystyrene–2% divinylbenzene resin (105 g, 84 mequiv of Cl; 200–400 mesh, 0.8 mequiv of Cl/g).[2] After the mixture was allowed to stir overnight at 100°C, it was cooled and filtered. The polymer beads were washed (200 mL × 3) each with DMF, THF, and ether and dried (P$_2$O$_5$) under reduced pressure, giving 118 g of polymer-bound EDC.

General Procedure for Coupling. To a suspension of polymer-bound EDCI (650 mg, 0.5 mmol of EDCI) in chloroform (4 mL) were added a carboxylic acid (0.22 mmol) and an amine (0.2 mmol). After the reaction mixture was shaken overnight at room temperature, it was filtered. The resin was washed with chloroform (3 × 3 mL), and the combined filtrates were evaporated to dryness to yield amides in 72–100% yield.

REFERENCES FOR SECTION 6.3.3

1. Desai, M. C.; Stramiello, L. M. S. *Tetrahedron Lett.* **1993**, *34*, 7685–7688.
2. The commercially available EDC hydrochloride (Fluka Inc.) was free based with aqueous potassium carbonate or 10% ammonium hydroxide and the aqueous phase was extracted with DCM.

6.3.4　A Highly Reactive Acylsulfonamide Linker [1,2]

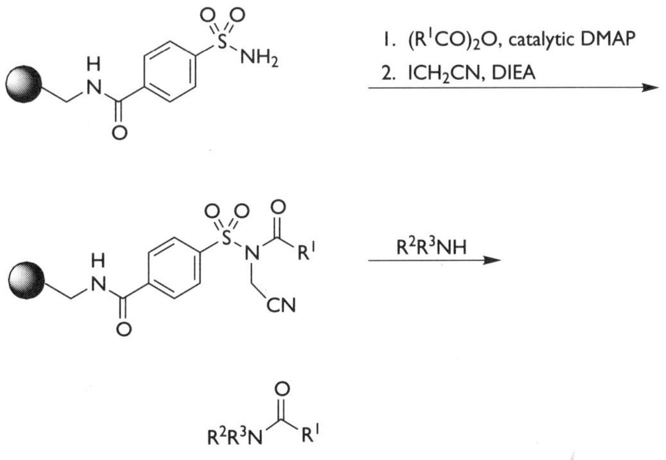

Points of Interest

1. The cyanomethyl linker is highly labile to nucleophilic displacement, with a $t_{1/2}$ of <5 min for displacement with 0.007 M benzylamine in DMSO. In comparison, the $t_{1/2}$ for the corresponding N-methyl derivative under the same conditions is 790 min.

2. Cyanomethylation provides a highly activated linker that can be readily cleaved with amines. This is particularly useful for combinatorial synthesis because additional diversity is incorporated during the cleavage step. Furthermore, because the linker is highly activated, equal quantities of amides can be obtained from limiting quantities of pools of amines in the cleavage cocktail.

Representative Examples

The highly activated linker reacts with a range of amines, including relatively poor nucleophiles such as aniline and *tert*-butyl amine, to provide the corresponding analytically pure amides in quantitative yield. The solid-phase synthesis of arylacetic acid derivatives on the sulfonamide "safety-catch" linker was described in an earlier study.[2]

Literature Procedure

For Loading onto the Linker. Sulfonamide-derivatized macroreticular resin was acylated with the symmetric anhydride of a carboxylic acid and catalytic DMAP as follows (this method has proven successful with a number of carboxylic acids and represents an improvement over the PFP ester method initially reported). To a 100-mL round-bottom flask were added 3-(3,4,5-trimethoxyphenyl)propanoic acid (8.6 g, 36 mmol), DCM (60 mL), and DIC (2.8 mL, 18 mmol). After 8 h of stirring, the solution was cooled with an ice bath and the urea precipitate was filtered. The filtrate was added to a 250-mL round-bottom flask fitted with an overhead stirrer and containing the sulfonamide-derivatized macroreticular resin (10 g, 3.9 mmol), DMAP (44 mg, 0.36 mmol), DIEA (2.1 mL, 12 mmol), and DCM (10 mL). The slurry was stirred 39 h and filtered, and the resin was washed with THF (200 mL), 5% TFA in THF (to protonate the support-bound acylsulfonamide), THF (200 mL), and MeOH (200 mL). The resin was dried with rotary evaporation and then under high vacuum for 24 h. The acylation of the sulfonamide could be monitored by the ninhydrin test.[3]

For Cleavage. A highly activated acylsulfonamide linker was activated as follows. To a 25-mL round-bottom flask were added the resin (500 mg, 0.17 mmol), DMSO (4 mL), DIEA (136 µL, 0.85 mmol), and bromoacetonitrile or iodoacetonitrile (4.0 mmol). After the mixture was stirred for 24 h, the resin was filtered and washed with DMSO (5 × 5 mL) and THF (3 × 5 mL). To the resin were added THF (3 mL) and an amine (3 mmol) or a pool of amines (3 mmol total), and the mixture was stirred for 12 h, after which the resin was filtered and washed with DCM (3 × 5 mL).[4] The filtrate and the washes were combined, washed with 1 M HCl (2 × 20 mL), dried with sodium sulfate, and concentrated with rotary evaporation. In order to remove particulates, the unpurified products were chromatographed.

REFERENCES FOR SECTION 6.3.4

1. Backes, B. J.; Virgilio, A. A.; Ellman, J. A. *J. Am. Chem. Soc.* **1996**, *118*, 3055–3056.
2. Backes, B. J.; Ellman, J. A. *J. Am. Chem. Soc.* **1994**, *116*, 11171–11172.
3. A positive test for unreacted amine gives an intense blue color and a positive test for unreacted sulfonamide gives a pale red color. Kaiser, E.; Colescot, R. L.; Bossinger, C. D.; Cook, P. I. *Anal. Biochem.* **1970**, *34*, 595–598.
4. After alkylation the linker is activated enough for expedient cleavage with a limiting quantity of amine or a pool of amines.

6.3.5   Use of Anion-Exchange Resins for the Solution-Phase Synthesis of Combinatorial Libraries Containing Aryl and Heteroaryl Ethers[1]

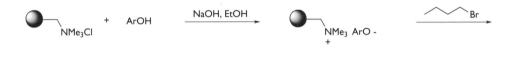

Points of Interest

1. In contrast to their soluble counterparts, polymer-bound reagents can be used in excess to drive the reaction to completion and simply separated by filtration. Anion-exchange resins, specifically, benefit from increased nucleophilicity due to the ionic nature of the species.

2. To minimize inconsistent product representation, the mixture of nucleophiles used to prepare the resins was selected by their relative reactivity. Mixture selection was important because an excess of polymeric reagent was used in ether formation. For example, a set of 10 electron-rich phenols were placed on a single batch of polymer, or a set of 10 phenols containing electron-withdrawing groups were attached on a batch of polymer, etc.

3. Ether formation was performed with excess resin relative to the electrophile. This allowed addition of polymer without precise measurement for each reaction, which becomes advantageous during automation. The excess of resin also ensures the total consumption of the electrophile.

Representative Examples

Mixtures of polymer-bound phenols or heteroaryl oxides were reacted with a limiting quantity of *n*-butyl bromide. All expected ethers were identified by GC–MS from the pools. Other electrophiles were explored with the same results as those cited by Gelbard[2] in 62–100% yields.

Literature Summary

For the Preparation of Amberlite IRA-900 (ArO-) Resins.   Amberlite IRA-900 (Cl-) (1000 mL) was packed in a column and flushed with 1 L of water, 3 L of 10% aqueous sodium hydroxide, water until effluent was neutral, 1 L of 95%

ethanol, and then 1 L of absolute ethanol. To load the phenols onto resin a solution of an equimolar mixture of 10 phenols (0.10 mol of each) or 10 hydroxy heterocycles (0.10 mol of each) in ethanol was circulated through the column for 5 h, and the resin was rinsed with ethanol, THF, and ether and dried *in vacuo.*

For the Preparation of Mixtures of Ethers.   Amberlite IRA-900 (ArO-) (0.40 g, excess, loading 1.3–3.0 mequiv/g) and *n*-butyl bromide (90.4 mg, 0.66 mmol) in THF were reacted for 6 h at 65°C. The solution was filtered and the polymer was rinsed with THF until no more product was eluted.

REFERENCES FOR SECTION 6.3.5

1. Parlow, J. J. *Tetrahedron Lett.* **1996,** *37,* 5257–5260.
2. Gelbard, G.; Colonna, S. *Synthesis* **1977,** *2,* 113–116.

6.3.6   Simultaneous Multistep Synthesis Using Polymeric Reagents [1]

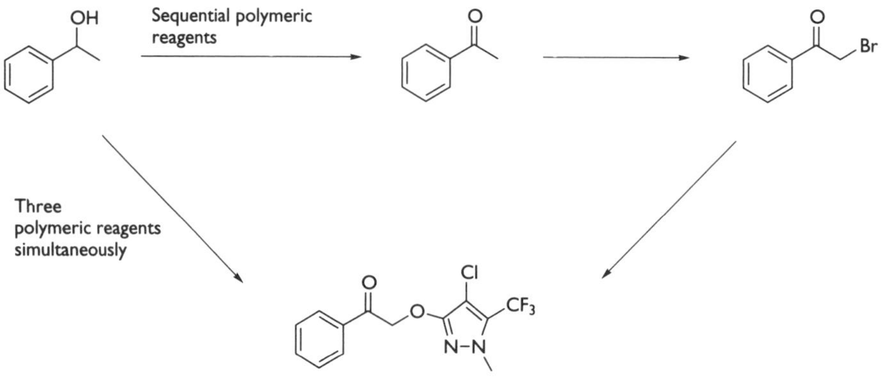

Points of Interest

1. Prior to Parlow's study with three polymeric reagents, only two polymeric reagents had been simultaneously used in one pot.

2. If two separate polymeric reagents (cross-linked) are allowed to be stirred together, the reactive groups do not come into contact, except for the few that are at the surface of the beads. Two or more otherwise mutually incompatible reagents can thus be used in the same reaction flask.

Single Example

Oxidation, bromination, and substitution reactions were performed with support-bound reagents in a single flask.

Literature Summary

Poly(4-vinylpyridinium dichromate) (0.4 g, 0.92 mmol), perbromide on Amberlyst A-26 (1.0 g, 1.0 mmol), and Amberlite IRA-900 (4-chloro-1-methyl-5-(trifluoromethyl)-1*H*-pyrazol-3-ol) (0.5 g, 1.15 mmol) were stirred in cyclohexane. To the mixture of resins was added *sec*-phenethyl alcohol (0.082 g, 0.67 mmol) and the resulting slurry was stirred at 65°C for 16 h. Upon filtration, desired product was isolated in 48% yield.

REFERENCES FOR SECTION 6.3.6

1. Parlow, J. J. *Tetrahedron Lett.* **1995**, *36*, 1395–1396.

6.3.7 Polymer-Supported Phosphonates: Olefins from Aldehydes, Ketones, and Dioxolanes[1]

X = CN, CO$_2$Me

Points of Interest

1. Horner–Emmons Wittig reagents within the pK_a range ca. 6–9 were supported on a basic anion-exchange resin (Amberlyst A-26) by a simple neutralization process. The reagents were prepared immediately before use due to slow decomposition. The loading was estimated from the amount of recovered phosphonate.

2. A 1:1 to 3:1 *E/Z* ratio of alkenes was obtained in high yield for a range of carbonyl compounds (similar to the ratios observed for phase-transfer-catalyzed Horner–Wittig synthesis).

3. Simultaneous ketal deprotection and Wittig reactions were performed with support-bound reagents. Dioxolanes underwent direct olefination by the simultaneous use of the strongly acidic resin Amberlyst 15 H and the polymer-bound α-cyanophosphonates on basic Amberlyst A-26 resin in the same vessel in THF/water (9:1). Such a reaction would be impossible in solution since immediate neutralization of the acidic catalyst by the basic phosphonate reagent would take place.

4. Olefination was performed by either batch or column techniques. Reactions were performed on a macroreticular resin in solvents ranging from hexanes to methanol.

Representative Examples

Aliphatic and aromatic ketones or aldehydes were reacted with support-bound phosphonates.

Literature Procedures

Amberlyst A-26, OH⁻ Form.   Amberlyst A-26, Cl⁻ form (Rohm & Haas Co.; 200 g), was converted into the hydroxide form by washing with 1 N NaOH employing the usual column technique. After rinsing with water and then MeOH, the resin was stored in a refrigerator under MeOH.

Amberlyst A-26, Phosphonate Form.   Amberlyst A-26, OH⁻ form (5 g), and diethylphosphonoacetonitrile (1.78 g, 10 mmol) in methanolic solution were shaken mechanically for 2 h. The resin was then filtered off and thoroughly washed with THF. The solvent was distilled off and from the weight of the residual phosphonate, the amount of the polymer-supported reagent was calculated as 5.6 mmol, corresponding to a loading of 3.5 mmol/g of dried resin (mean value over a number of trials). By a similar procedure, 10 g of Amberlyst A-26, OH⁻ form, supported 5.2 mmol of methyl diethylphosphonoacetate. Both support-bound reagents may also be prepared by a column technique in which an ethereal phosphonate solution is slowly percolated through a packed column and rinsed with ether.

General Procedure for Horner–Wittig Synthesis.   An aldehyde (0.49 g, 3.5 mmol) and Amberlyst A-26-supported phosphonate (5.0 g, 5.6 mmol) in wet form were stirred at room temperature in THF (10 mL). The reaction was complete in 1 h. The resin was filtered off and the solvent was removed by gentle distillation to give the desired alkene in high yield. For deprotection of dioxolanes 0.35 g (1.25 mmol) of Amberlyst 15 H (H⁺ form) was added to the mixture.

REFERENCES FOR SECTION 6.3.7

1. Cainelli, G.; Contento, M.; Manescalchi, F.; Regnoli, R. *J. Chem. Soc., Perkin Trans. 1*, **1980**, 2516–2519.

6.3.8   One-Pot Solution-Phase Cleavage of α-Diols to Primary Alcohols with Polymer-Supported Antagonistic Reagents: Periodate and Borohydride[1]

Points of Interest

1. Support-bound oxidizing and reducing reagents avoid the problems associated with separation of products from the minerals and the instability of the intermediate dialdehydes.

2. Both periodate- and borohydride-supported resins were employed for the *in situ* preparation of the trihydroxy derivatives of adenosine, guanosine, uridine, and cytidine.

Literature Summary

The nucleoside (1 mmol) was dissolved (or suspended) in 10 mL of water and then continuously pumped through a column containing 6 mL of a 1:1 mixture of dry resins. The resins were previously prepared by slowly passing a solution of excess reagent ($NaIO_4$ or $NaBH_4$) in water through a column of

Amberlyst A-27 (chloride form), rinsing with water and ethanol, and then drying *in vacuo* at 60°C overnight. The reactions between the nucleosides and the resins were monitored by TLC. At the end of the reaction the resin was rinsed with 10 mL of water and the combined solutions were evaporated to dryness to afford the corresponding trihydroxy nucleosides.

REFERENCES FOR SECTION 6.3.8

1. Bessodes, M.; Antonakis, K. *Tetrahedron Lett.* **1985**, *26*, 1305–1306.

6.3.9   Investigations on the Stille Reaction with Polymer-Supported Organotin Reagents [1]

Points of Interest

1. The polymer-supported organotin byproduct can be separated by simple filtration and regenerated.

2. The organic target molecule is virtually free of toxic organotin materials. This is particularly advantageous for the formation of products of pharmaceutical or biological importance.

3. Examples where the product of a Stille coupling is produced on support instead of in solution (as in this example with the polymer-supported organotin reagent) have also been reported.[2,3]

Representative Examples

Palladium-catalyzed cross-couplings between polymer-bound allyltin, alkynyltin, alkenyltin, and organic electrophiles (acid chlorides, vinyl iodides, and vinyl triflates) proceeded in 51–95% yield (average 71% yield).

Literature Procedure

To 1.33 g (2.0 mmol of Sn) of derivatized phenylethyltin resin[4] in 8 mL of dry toluene under Ar were added 0.023 g (0.02 mmol, 1 molar % catalyst) of $Pd(PPh_3)_4$ and 2.0 mmol of the electrophile (e.g., benzoyl chloride). The mixture was slowly stirred and heated to 80°C for 1 h. The solution was then analyzed by quantitative gas chromatography relative to authentic samples.

REFERENCES FOR SECTION 6.3.9

1. Kuhn, H.; Neumann, W. P. *Synlett* **1994**, 123–124.
2. Plunkett, M. J.; Ellman, J. A. *J. Am. Chem. Soc.* **1995**, *117*, 3306–3307.
3. Deshpande, M. S. *Tetrahedron Lett.* **1994**, *35*, 5613–5614.
4. References are provided for the preparation of the functional polymer in ref 1.

6.3.10   An Improved Protocol for Solution-Phase Azole Synthesis with PEG-Supported Burgess Reagent [1]

Points of Interest

1. A poly(ethylene glycol)-linked version of the Burgess reagent was developed and applied toward the cyclodehydration of β-hydroxy amides and thioamides. The expected oxazolines and thiazolines were obtained in high yields and excellent purities.

2. The main advantages of this polymer-bound reagent relative to the commercially available Burgess reagent in solution are its improved ease of handling and greatly increased yields in the synthesis of labile oxazolines. Cyclodehydration of serine-derived as well as threonine-derived peptides with polymer-support Burgess reagent cleanly provided oxazolines in ca. 10–20% higher yields than with standard Burgess reagent.

3. Less than 2% epimerization at α- or β-carbons or elimination to dehydroamino acids was detected by NMR spectroscopy.

Representative Examples

Serine- and threonine-derived dipeptides were stereospecifically cyclodehydrated in 76–98% yield (average 86%). A double cyclodehydration provided the corresponding bisoxazoline in excellent yield as well.

Literature Summary

Preparation of PEG-Linked Burgess Reagent.   A solution of 4.88 g (7.8 nmol) of PEG monomethyl ether ($M_w = 750$) in 20 mL of benzene was dried by azeotropic removal of water for 24 h in a Dean–Stark apparatus and then added dropwise to a solution of chlorosulfonyl isocyanate in 20 mL of dry benzene. The reaction mixture was stirred for 1 h at room temperature, concentrated, dried overnight *in vacuo,* and then used without further purification. A solution of the residue in 35 mL of benzene was added dropwise to a solution of 2.5 mL (17 mmol, 2.2 equiv) of Et_3N in 15 mL of benzene. The reaction mixture turned darker with concomitant formation of insoluble Et_3NHCl and was stirred at room temperature for 30 min, filtered, concentrated, and dried *in vacuo* for 2–3 days to yield 6.2 g (82%) of the support-bound Burgess reagent that was used without further purification.

REFERENCES FOR SECTION 6.3.10

1. Wipf, P.; Venkatraman, S. *Tetrahedron Lett.* **1996,** *37,* 4659–4662.

6.3.11   Support-Bound Nucleophiles and Electrophiles for the Purification of Small-Molecule Libraries[1]

limiting reagent	excess reagent	scavenger	solvent/temperature
R^1R^2NH	R^3NCO R^3COCl R^3SO_2Cl	⬤–NH$_2$	$CHCl_3$, rt
(epoxide)–R^3	R^1R^2NH	⬤–NCO	MeOH, rt-65°C
R^3X	R^1R^2NH	⬤–NCO	CH_3CN, rt-65°C
R^2(C=O)R^3	R^1NH_2	⬤–CHO	MeOH/DCM, rt
R^2(C=O)R^3	R^1R^2NH	⬤–COCl	10% AcOH/DCE, rt

Points of Interest

1. Solid-supported nucleophiles and electrophiles were employed to expedite the workup and purification of a variety of different primary and secondary amine alkylations and acylations.

2. Benzylamine was reacted with excess isocyanate. After 1 h, excess (aminomethyl)polystyrene (0.8 mequiv/g) was added as a scavenger for unreacted isocyanate. The reaction mixture was filtered and within the limits of detection by ^{1}H NMR, only the desired urea product was observed.

3. Resin-bound borohydride and resin-bound aldehyde were used simultaneously.

Representative Examples

Thousands of ureas and thioureas were prepared using excess (aminomethyl)polystyrene to quench the excess of electrophilic reagents. The chemistry can also be applied to the construction of amides, sulfonamides, and carbamates if one incorporates the use of a basic resin in the initial step. Support-bound electrophiles were used to quench the excess amine that was employed to drive alkylation reactions to completion. Secondary amines (excess starting materials) could be separated from relatively nonnucleophilic tertiary amines (products) by using a polymer-supported isocyanate[2] as an electrophilic scavenger. Reductive amination was explored as an alternative amine alkylation protocol. Excess primary amine relative to carbonyl component was used. After preformation of the corresponding imine adduct in methanol, reduction was performed using polymer-supported borohydride. Excess primary amine was readily separated from the desired secondary amine product by selective imine formation using a polymer-supported aldehyde.[3] Reductive amination was also utilized to form tertiary amines by the reaction of an aldehyde with an excess of secondary amine and polymer-supported cyanoborohydride in acetic acid/DCM, followed by scavenging of excess secondary amine with polymer-supported benzoyl chloride. Multistep sequences combining protocols with different support-bound reagents were also demonstrated.

Literature Summary

The following is a typical procedure for the reductive amination of primary amines. To a 4-mL screw cap glass vial were added 0.5 mmol of a primary aliphatic amine and 0.33 mmol of an aldehyde in 1 mL of MeOH. The vial was sealed with a Teflon-backed cap, and the solution was then shaken for 2–3 h to allow for imine formation and then treated with ca. 250 mg (2.5 mmol of BH_4^-/g of resin, 0.63 mmol) of Amberlite IRA-400 borohydride resin. The slurry was then shaken an additional 24 h to reduce the secondary amine. Then ca. 300 mg (1 mmol/g of resin, 0.35 mmol) of polystyrene-linked benzaldehyde resin and 1 mL of DCM were added to the vial, and the resulting slurry was shaken overnight, filtered through a cotton plug, and rinsed with MeOH. Evaporation yielded products typically of 90–95% purity in yields ranging from 50 to 99%.

REFERENCES FOR SECTION 6.3.11

1. Kaldor, S. W.; Siegel, M. G.; Fritz, J. E.; Dressman, B. A.; Hahn, P. J. *Tetrahedron Lett.* **1996**, *37*, 7193–7196.
2. Rebek, J.; Brown, D.; Zimmerman, S. *J. Am. Chem. Soc.* **1975**, *97*, 4407–4408.
3. Frechet, J. M.; Schuerch, C. *J. Am. Chem. Soc.* **1971**, *93*, 492–496.

6.3.12 A Polymer-Supported Scandium Catalyst for Solution-Phase Synthesis[1]

R[1], R[2] = aliphatic, aromatic, heteroaromatic, hydrogen, sugars

Points of Interest

1. Scandium trichloride was supported on Nafion (NR-50 commercially available from Du Pont). The authors reference six other examples of different metal Nafions (Hg, Si, Cr, Ce, Al, Ta).[1]

2. The Nafion-Sc catalyst was tested in several synthetic reactions. Many examples of the allylation reactions of carbonyl compounds were provided. In addition, other Lewis acid-mediated reactions were performed with the polymer-supported catalyst including Diels–Alder, Friedel–Crafts acylation, and imino Diels–Alder reactions.

3. Nafion-Sc could be easily recovered and reused. The catalyst was recovered simply by filtration and washing with a suitable solvent, and the activity of the recovered Nafion-Sc was comparable to that of the fresh catalyst.

Representative Examples

Allylation with tetraallyltin proceeded smoothly in both organic and aqueous solvents. Aldehydes, ketones, and nonprotected sugars all react directly to give the corresponding homoallylic alcohols. Diels–Alder, Friedel–Crafts acylpchation, and imino Diels–Alder reaction products were also prepared. The yields for all the polymer-supported reactions reported ranged from 57 to 97% (average 88% yield).

Literature Procedures

Preparation of Nafion-Sc. $ScCl_3 \cdot 6H_2O$ (519 mg, 2.0 mmol) and Nafion (5 g, 1.2 mequiv/g) were combined in acetonitrile (10 mL) under reflux for 40 h. After the mixture was cooled to room temperature, the polymer was filtered, washed with acetonitrile (20 mL $\times$ 3), and then dried under reduced pressure for 24 h. After 40 h, 96% of the $ScCl_3 \cdot 6H_2O$ was consumed, and the polymer thus prepared contained 1.3% Sc according to ICP analysis. Choice of solvents is important at this stage; only 27% of the $ScCl_3 \cdot 6H_2O$ was consumed when 1,2-dichloroethane was used as a solvent. A trial to prepare Nafion-Sc from Sc_2O_3 and Nafion failed (only 0.1% Sc was included in this polymer).

The Nafion-Sc-Catalyzed Allylation Reaction.   To a mixture of Nafion-Sc (250 mg) and aldehyde (0.5 mmol) in H_2O/THF (1:9, 1.5 mL) was added tetraallyltin (0.5 mmol) in H_2O/THF (1:9, 1.5 mL) at room temperature. In contrast, the allylations of sugars and ketones were performed at 60°C. The mixture was stirred for 2 h at the appropriate temperature and then filtered. The Nafion-Sc was washed with ether, and the filtrates were combined. The solvent was removed under reduced pressure, and the crude product was chromatographed on silica gel to afford the corresponding homoallylic alcohol (average yield 88%).

REFERENCES FOR SECTION 6.3.12

1. Kobayashi, S.; Nagayama, S. *J. Org. Chem.* **1996,** *61,* 2256–2257.

6.3.13   Polymer Scandium-Catalyzed 3CC Leading to Diverse Amino Ketone, Amino Ester, and Amino Nitrile Derivatives[1]

Points of Interest

1. The (polyallyl)scandium trifylamide ditriflate (PA-Sc-TAD) catalyst was used to prepare a range of amino compounds. A 1:1:1.1 ratio of aldehyde/aniline/silylated nucleophile cleanly afforded the β-amino compounds after filtration of the catalyst.
2. When silylated esters (ketene silyl acetals) were used as nucleophiles, the yield was dramatically improved in the presence of $MgSO_4$ as a dehydrating agent.
3. Reactions with support-bound catalysts can easily be scaled up.

Representative Examples

β-amino ketones, esters, and Strecker adducts were prepared in 73–99% yield from the scandium triflate-catalyzed reaction between a range of structurally diverse silylated nucleophiles and imines (prepared *in situ*).

Literature Procedure

In the presence of (polyallyl)scandium trifylamide ditriflate (PA-Sc-TAD; see p 277 for the preparation of the polymer-bound catalyst) (56.0 mg), an aldehyde (0.40 mmol), an aromatic amine (0.40 mmol), and a silylated reagent (0.44 mmol) were mixed in DCM/CH_3CN (2:1, 2.4 mL). When ketene silyl acetals were used, $MgSO_4$ (125 mg) was added beforehand. The mixture was stirred at room temperature for 19 h and hexane (20 mL) was added. The catalyst was filtered and the filtrate was concentrated *in vacuo* to afford a crude adduct. After purification by column chromatography (silica gel), the desired adduct was obtained in high yield.

REFERENCES FOR SECTION 6.3.13

1. Kobayashi, S.; Nagayama, S.; Busujima, T. *Tetrahedron Lett.* **1996**, *27*, 9221–9224.

6.3.14   A New Methodology for Solution-Phase Combinatorial Synthesis: Construction of a Quinoline Library Using a Polymer-Supported Catalyst [1]

Points of Interest

1. The (polyallyl)scandium trifylamide ditriflate (PA-Sc-TAD) catalyst is the earliest example of the application of a support-bound catalyst to combinatorial synthesis.

2. For product isolation, the soluble polymer was precipitated from solution by the addition of hexanes to the reaction mixture (2 : 1 DCM/acetonitrile). Combinatorial applications of soluble polymers, such as PA-Sc-TAD, generally require specific conditions under which excess reagents or desired products display distinct solubility characteristics from the polymer for purification.

3. Quinoline derivatives were prepared by a lanthanide-catalyzed three-component coupling reaction[2] in quantities greater than 100 mg.

Representative Examples

Quinolines were prepared (often in high diastereoselectivity) from a range of aldehydes, anilines, and electron-rich alkenes incorporating halogens, ethers, thioethers, and other heterocyclic rings.

Literature Procedures

For the Preparation of (Polyallyl)scandium Trifylamide Ditriflate (PA-Sc-TAD).   Polyacrylonitrile (1.13 g, purchased from Aldrich) was treated with $BH_3 \cdot SMe_2$ (6.8 mL) in diglyme (40 mL) for 36 h at 150°C. After the mixture cooled, 6 N HCl was added and the mixture was stirred for 2 h under reflux conditions. NaOH (7 N) was then added at 0°C, and the mixture was stirred for 1 h at the same temperature. The resulting polypropylamine was filtered, washed with H_2O, dioxane, and ether, and dried to 1.12 g. To a portion of the polymeric amine (121 mg) in 1,2-dichloroethane (5 mL) were added Tf_2O (1.79 g) and Et_3N (0.64 g) at –20°C, and after 5 min of stirring, the mixture was heated for 10 h at 60°C. The mixture was washed with H_2O, dioxane, and ether, and the resulting polymeric sulfonamide was dried to 154 mg. The polymeric sulfonamide (154 mg) and KH (40.0 mg) were combined in THF (3 mL). Sc(OTf)$_3$ (1.0 g) was then added and the mixture was stirred for 48 h at room temperature. After the mixture was washed with water, dioxane, and ether, PA-Sc-TAD (487.6 mg) was obtained.

For the Construction of a Quinoline.   The polymer-bound catalyst PA-Sc-TAD (56.0 mg), an aldehyde (0.40 mmol), an aromatic amine (0.40 mmol), and an alkene (0.44 mmol) were mixed in DCM/acetonitrile (2:1, 2.4 mL) at 40°C for 15 h. After the mixture cooled to room temperature, hexane (20 mL) was added and the catalyst was filtered (the catalyst was recovered quantitatively). The filtrate was concentrated *in vacuo* to afford a crude adduct. After purification by column chromatography, the desired quinoline derivative was obtained. The recovered catalyst could be used in further reactions without loss of any activity.

REFERENCES FOR SECTION 6.3.14

1. Kobayashi, S.; Nagayama, S. *J. Am. Chem. Soc.* **1996**, *118*, 8977–8978.
2. Kobayashi, S.; Ishitani, H.; Nagayama, S. *Synthesis* **1995**, 1195–1202.

6.3.15   Additional Examples of Support-Bound Catalysts

1. Soluble polymer-bound, ligand-accelerated catalysis has been reported for asymmetric dihydroxylation.[1] The soluble polymer-bound catalyst displayed enantioselectivities similar to those of the Sharpless asymmetric dihydroxylation (AD) of olefins in solution. This is in contrast to other studies (generally performed on standard polystyrene–2% divinylbenzene) where the polymer-bound catalyst provided variable yields and lower enantioselectivity.

2. A highly enantioselective polymeric catalyst has been developed for the Diels–Alder reaction.[2] The cross-linking structure greatly affected the performance of the polymeric catalyst.

REFERENCES FOR SECTION 6.3.15

1. Han, H.; Janda, K. D. *J. Am. Chem. Soc.* **1996**, *118*, 7632–7633.
2. Kamahori, K.; Ito, K.; Itsuno, S. *J. Org. Chem.* **1996**, *61*, 8321–8324 and references cited therein.

APPENDIX I

SUMMARY OF FUNCTIONAL GROUP TRANSFORMATIONS FOR COMBINATORIAL SOLID-PHASE SYNTHESIS

DIC HOBt PyBOP HBTU

OAIU Symmetric anhydride Acyloxyphosphonium salt HOBt active ester

3. Amide bond formation with amino acid halides   80

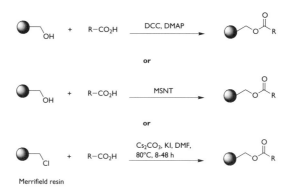

Protected amino acid fluoride   TFFH

4. Azabenzotriazole-based coupling reagents HOAt and
   HATU for solid-phase peptide synthesis   81

HOAt   HATU

5. Alternative procedures for amide bond formation during
   cleavage from resin

II.   Esterification reactions on solid support

1. First amino acid loading by ester bond formation   82

Merrifield resin

III.   Imine formation on solid support

1. Solid-phase imine formation with trimethyl orthoformate   84

Rink or Sasrin
resin

2. Other examples of imine formation on solid support

IV.   Urea formation on solid support

1. A strategy for combinatorial solid-phase synthesis of
   urea-linked diamine libraries   86

V.   Condensations with phosphorus compounds on solid support
1.   Preparation of phosphodiesters in solid-phase oligonucleotide synthesis   87

2.   A combinatorial method for the solid-phase synthesis of
α-amino phosphonates and phosphonic acids   88

3.   A combinatorial method for the solid-phase synthesis of
α-hydroxy phosphonates   89

4.   Other examples of condensations involving phosphorus
compounds

IV. Heck reactions on solid support

1. Heck reactions in solid-phase synthesis 96

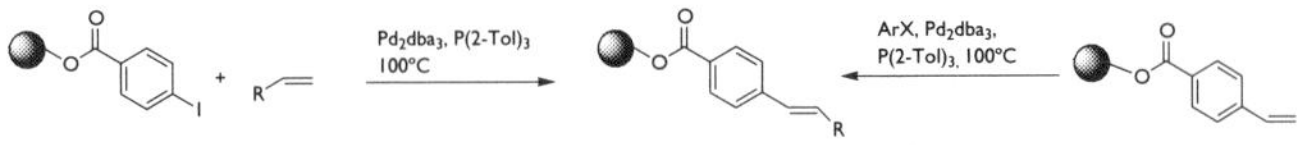
Pd2dba3, P(2-Tol)3
100°C
ArX, Pd2dba3,
P(2-Tol)3, 100°C

2. Solid-phase synthesis using the Heck reaction under phase-transfer conditions at 37°C 97

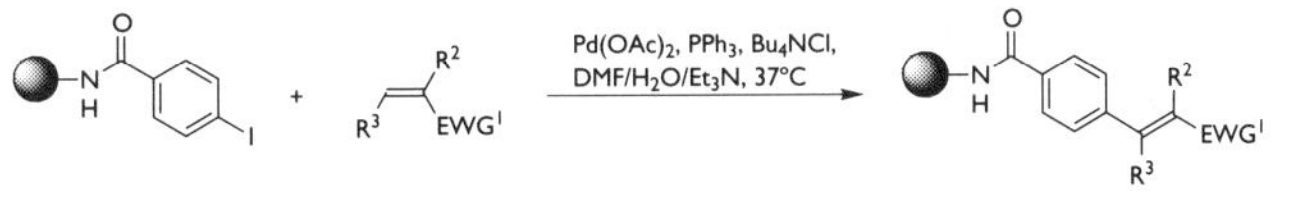
Pd(OAc)2, PPh3, Bu4NCl,
DMF/H2O/Et3N, 37°C

3. Additional examples of solid-phase Heck reactions

V. Enolate alkylation for carbon–carbon bond formation on solid support

1. Enolate alkylation on Kenner's acylsulfonamide linker 98

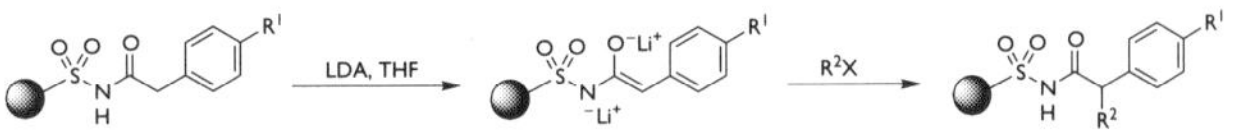
LDA, THF
R2X

2. Aldol and imine reactions with support-bound silyl enol ethers 99

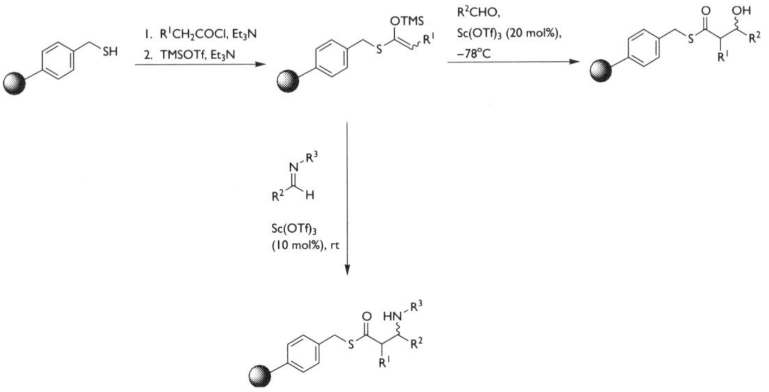
1. R1CH2COCl, Et3N
2. TMSOTf, Et3N
R2CHO,
Sc(OTf)3 (20 mol%),
−78°C
Sc(OTf)3
(10 mol%), rt

3. Other examples of carbon-carbon bond formation with support-bound enolates

VI. Solid-phase Wittig reactions

1. Solid-phase synthesis of alkenes using the Horner–Wadsworth–Emmons reaction 102

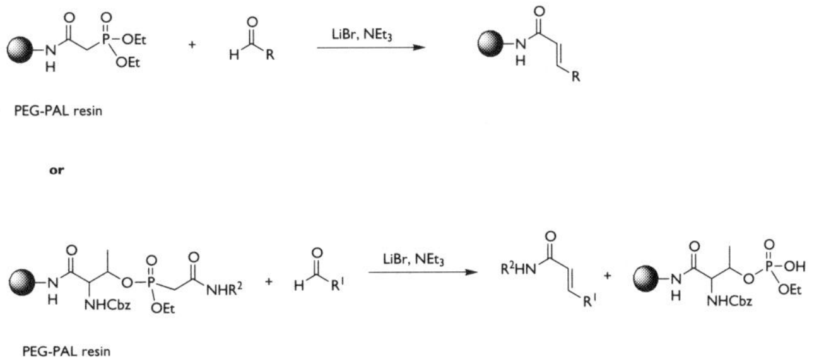
LiBr, NEt3
PEG-PAL resin
or
LiBr, NEt3
PEG-PAL resin

Trityl resin

R = Boc, H

or

Rink amide resin

2. Thiol alkylation in the solid-phase synthesis of β-turn mimetics 113

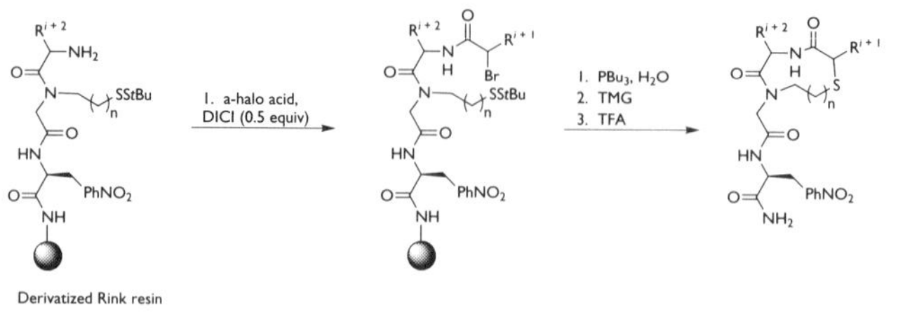

Derivatized Rink resin

3. Anilide alkylation with primary alkyl halides in the solid-phase synthesis of 1,4-benzodiazepines 115

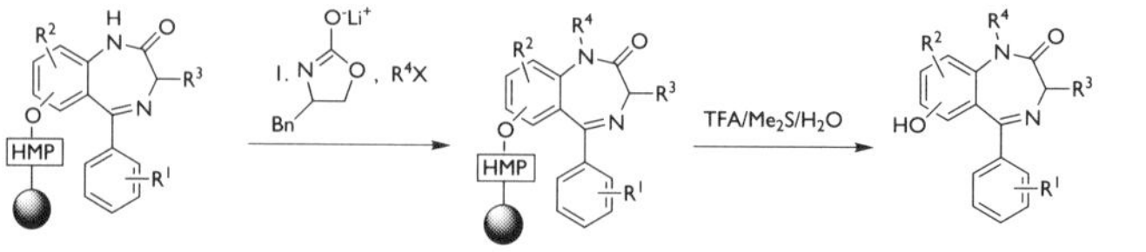

4. One-pot cyclization and anilide alkylation in the solid-phase synthesis of 1,4-benzodiazepine-2,5-diones 116

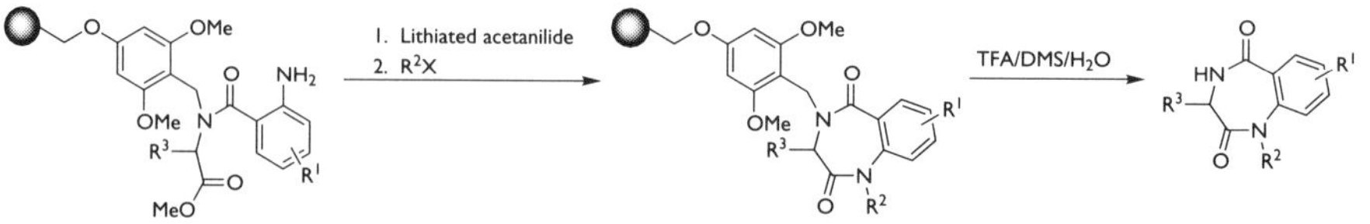

5. Successive amide alkylations: libraries from libraries 117

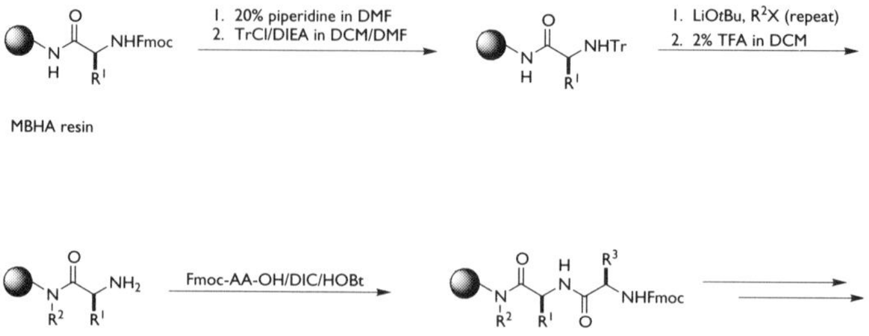

MBHA resin

6. Benzophenone imine α-carbon alkylation in solid-phase unnatural peptide synthesis 118

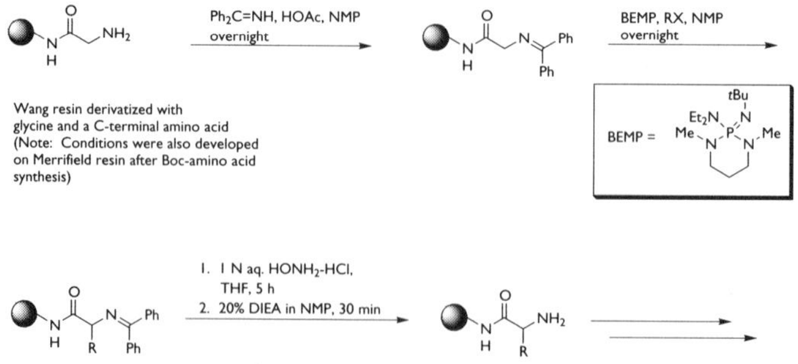

Wang resin derivatized with
glycine and a C-terminal amino acid
(Note: Conditions were also developed
on Merrifield resin after Boc-amino acid
synthesis)

7. Alkylation or sulfonylation of a support-bound phenol 120

8. Enolate monoalkylation for carbon–carbon bond formation on solid support 121

9. Alkylation of support-bound 1,3-diketones in the solid-phase synthesis of pyrazoles and isoxazoles 122

10. Tosyl displacement with primary or secondary amines on solid support 123

11. Grignard additions to support-bound esters   124

THP resin

12. S_NAr reaction on solid support   125

Core structures on Rink resin for S_NAr and S_N2 reactions:

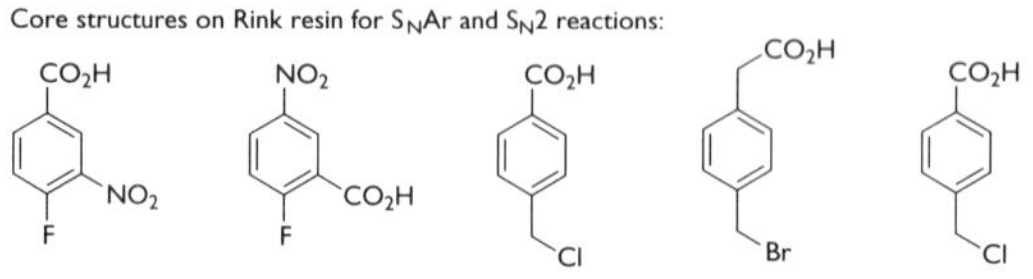

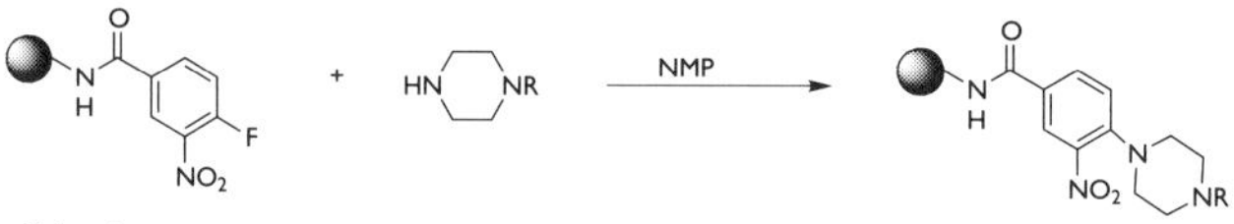

Rink resin

13. Palladium-catalyzed amination of resin-bound aromatic
    bromides   126

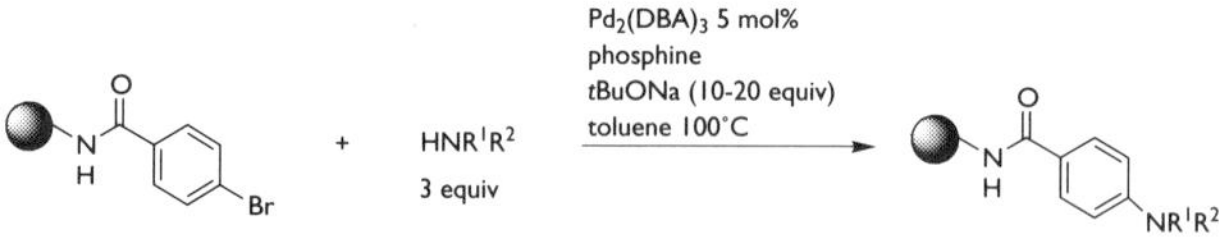

Rink resin

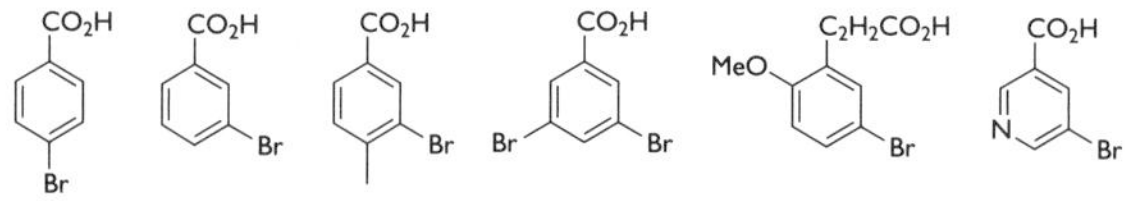

14. 1,4-Michael addition of thiols to support-bound enones   127

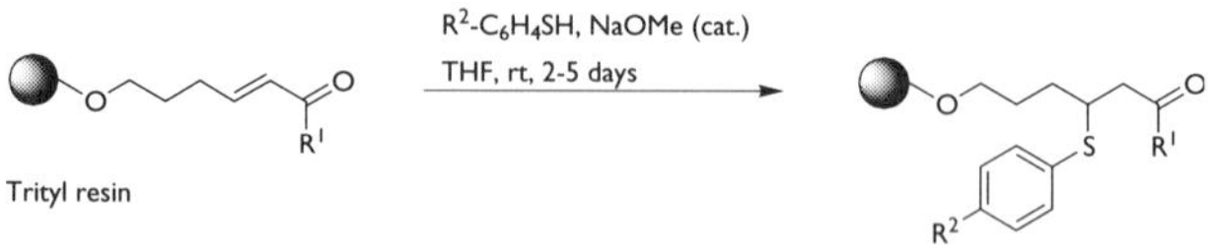

Trityl resin

15. Iodoetherification reaction in the solid-phase synthesis of
    miconazole analogs   128

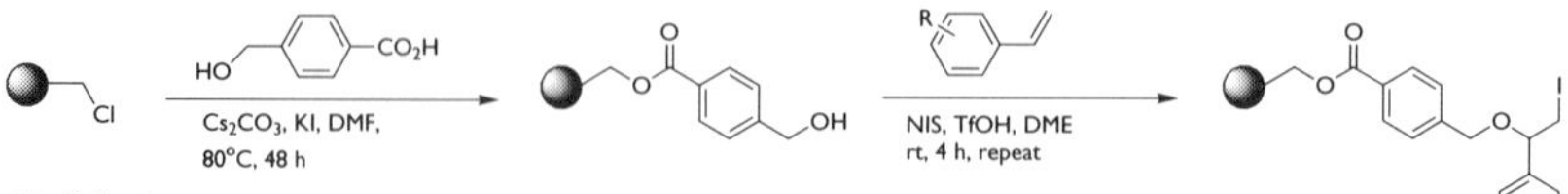

Merrifield resin

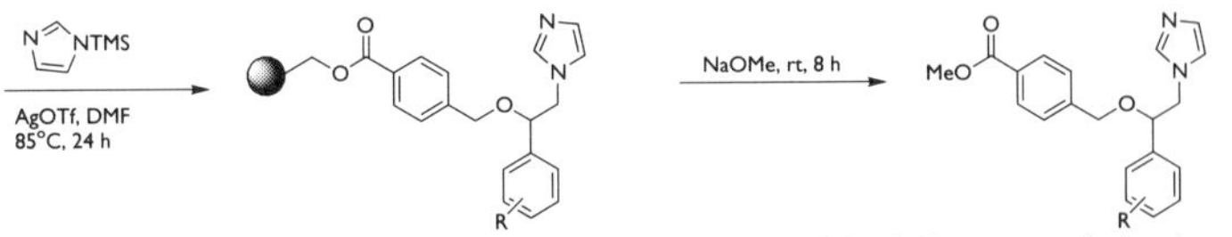

16. Reactions with support-bound alkyl halides in solid-phase peptide and combinatorial synthesis   129

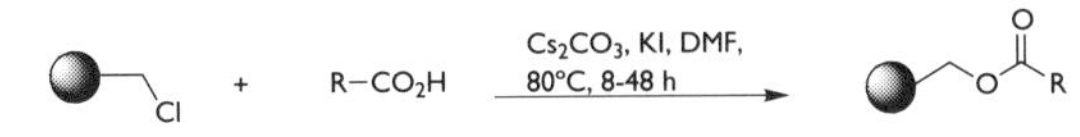

Merrifield resin

E.   Oxidations on solid support
1. Oxidation of alcohols to aldehydes and ketones   131

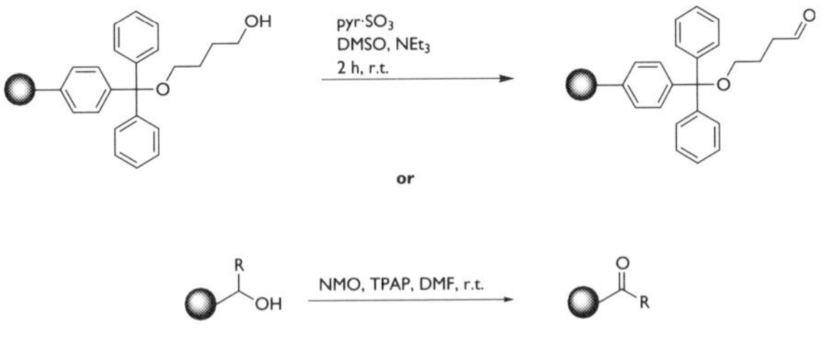

or

R = H, aryl

2. Oxidation of (chloromethyl)polystyrene resin to formyl-polystyrene and carboxypolystyrene resins   132

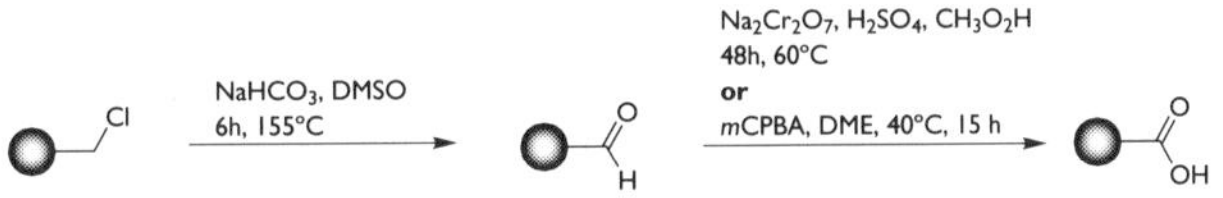

3. Other examples of oxidation on solid support
F.   Reductions on solid support
1. Rapid optimization of oxidation and reduction reactions on a solid phase using the multipin approach: synthe33sis of 4-aminoproline analogs   133

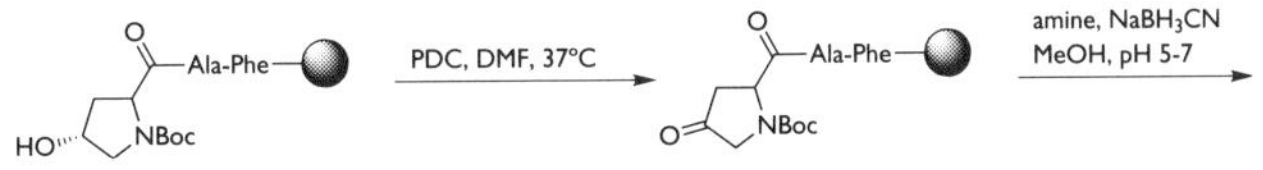

polyethylene pins

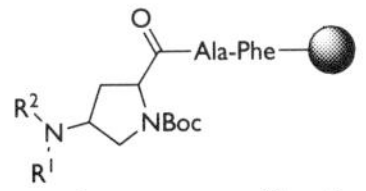

2. Reductive alkylation on a solid-phase: synthesis of a piperazinedione combinatorial library   134

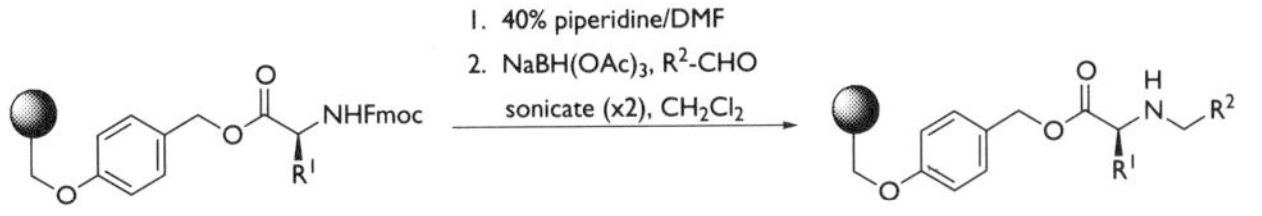

3. Reductive alkylation of Sieber's Xan linker   135

4. Loading amino esters onto resin via reductive amination   136

5. Reductive amination of a support-bound aniline   137

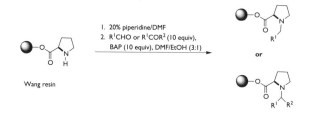

Rink resin:
X = NH, O

6. Solid-phase reductive alkylation of secondary amines using borane–pyridine complex   138

Wang resin

or

7. Reduction of support-bound amides with Red-Al   138

THP resin

8. Reduction of support-bound ketones to alcohols on PEG–PS   139

9. Reduction of a support-bound nitro group to an aniline   139

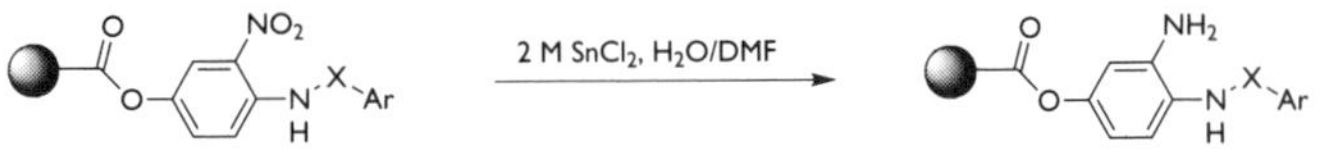

X = CO, CONH

10. Reduction of a support-bound azide to an amine   140

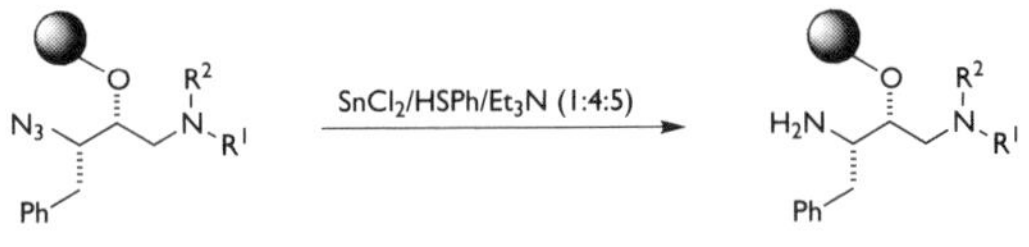

THP resin

11. Solid-phase synthesis of (RS)-1-aminophosphinic acids   141

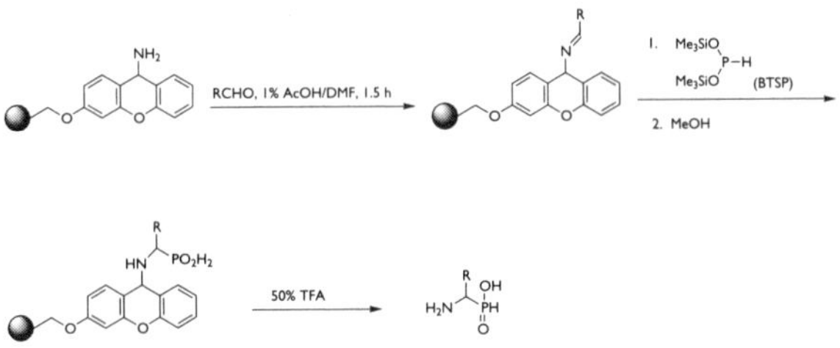

12. Additional studies of reductions on solid support

APPENDIX 2

CLASSIFICATION OF HETEROCYCLIZATION REACTIONS

To assist in locating information in the section describing the solid-phase synthesis of heterocycles, scaffolds are organized by the reaction class, ring size, and page number.

TABLE I. Heterocyclization via lactamizations and related cyclizations

Scaffold	Ring size	Page
1,4-benzodiazepines	7-membered	148
1,4-benzodiazepine-2,5-diones	7-membered	150

1,4-benzodiazepine-2,5-diones	7-membered	152

hydantoins	5-membered	158

5-alkoxyhydantoins	5-membered	159

piperazinediones (diketopiperazines)	6-membered	167

diketopiperazines and diketomorpholines	6-membered	168–169

5,6-dihydropyrimidine-2,4-diones	6-membered	176

1,3-dialkyl quinazoline-2,4-diones	6-membered	181

1,3-disubstituted 2,4(1H,3H)-quinazolinediones	6-membered	182

TABLE 2.   Heterocyclization via imine formation

Scaffold	Ring size	Page
1,4-benzodiazepines 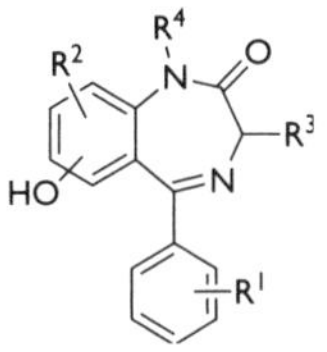	7-membered	145
pyrazoles and isoxazoles 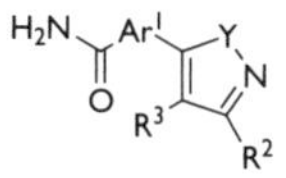	5-membered	160
imidazoles 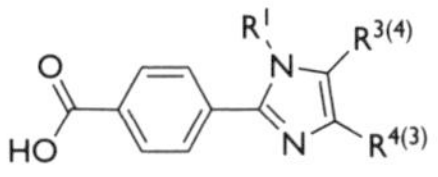 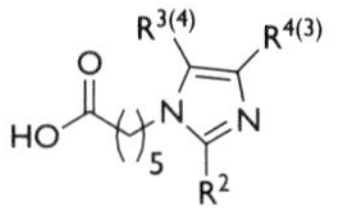 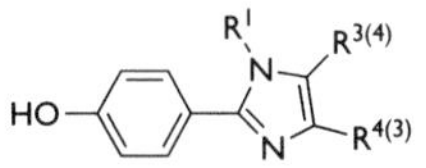	5-membered	164
tetrasubstituted imidazoles 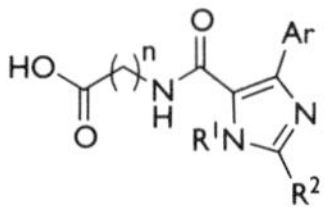	5-membered	165
thiazolidines 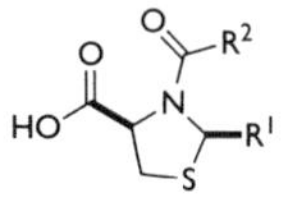	5-membered	188

1,4-benzodiazepines

pyrazoles and isoxazoles

$Y = NR^4, O$

imidazoles

tetrasubstituted imidazoles

thiazolidines

■ **TABLE 3.   Heterocyclization via cyclization on aromatic rings**

Scaffold	Ring size	Page
dihydro- and tetrahydroisoquinolines	6-membered	177
tetrahydroisoquinolines and tetrahydroimidazopyridines	6-membered	184
quinolones	6-membered	185
1,2,3,4-tetrahydro-β-carbolines/Pictet–Spengler	6-membered	200
Fischer indole synthesis	5-membered	204

■ **TABLE 4.   Heterocyclization via cycloadditions**

Scaffold	Ring size	Page
isoxazoles and isoxazolines	5-membered	160
pyrroles	5-membered	166

pyrrolidines	5-membered	189

Z = COR, CN

proline analogs	5-membered	190

β-lactams	4-membered	192

azabicyclo[4.3.0]nonen-8-ones via Pauson–Khand cyclization	5-membered	199

1,2-diazines	6-membered	209

TABLE 5. Heterocyclization via multicomponent coupling reactions

Scaffold	Ring size	Page
1,4-dihydropyridines	6-membered	172

pyridines and pyrido[2,3-d]pyrimidines	6-membered and bicyclic	173

Biginelli dihydropyridines	6-membered	174

4-thiazolidinones and 4-metathiazanones   5-membered   186

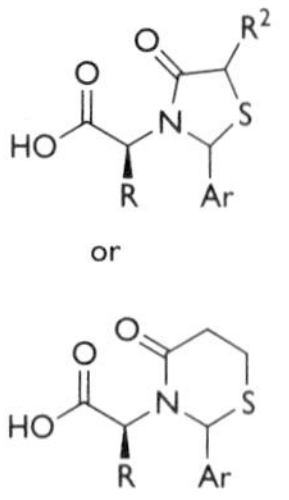

or

TABLE 6. Heterocyclization via other modes of cyclization (i.e. Pd catalyzed, substitution, reductive cyclization, or derivatization of heterocycles preloaded onto resin)

Scaffold	Ring size	Page
β-turn mimetics 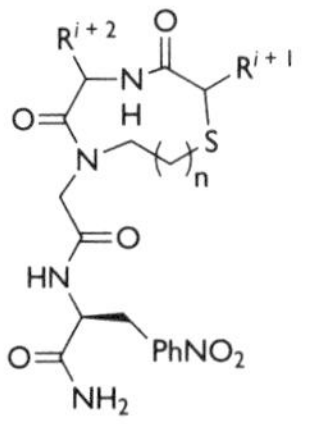 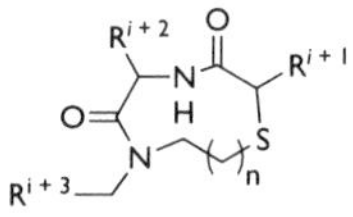	9,10-membered/substitution	154–156
2-oxopiperazines 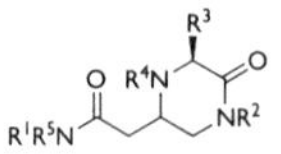	6-membered/1,4-addition	170
peptoid 1(2H)-isoquinolines 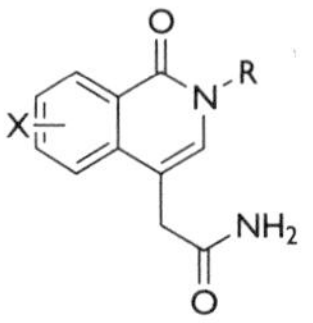	6-membered/Pd catalyzed	178–179
2,6-disubstituted quinolines 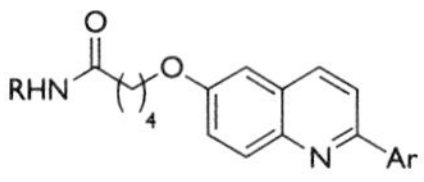	6-membered/reductive cyclization	180
dihydrobenzopyrans 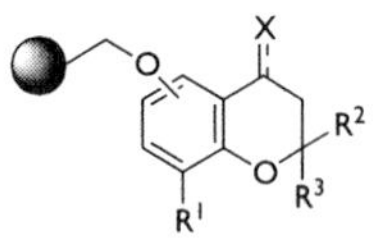	6-membered/condensation	194

$X = O, S_2(CH_2)_2, OH$

bicyclo[2.2.2]octanes | 6/6 bicyclo/1,4-addition | 195

tropanes | 6/5 bicyclo/prepared in solution | 197

(X = CO or CH$_2$)

indoles | 5-membered/Pd catalyzed | 206

olomoucines | purines/prepared in solution | 207

triazines | 6-membered/prepared in solution | 210

UNNATURAL BIOPOLYMERS

The following is a list of methods for preparing unnatural biopolymers.[1] Some of the methods for preparing unnatural biopolymers have been used to generate combinatorial libraries. Procedures and building blocks for preparing unnatural biopolymers can be often applied to the combinatorial synthesis of other small molecules.

1. Libraries from libraries were generated in the postsynthetic modification of peptide libraries by permethylation.[2]

2. Oligo(N-substituted)glycines, or "peptoids," were prepared with the side chains displayed from the nitrogen of an oligoglycine backbone.[3]

3. Oligocarbamates were prepared from N-protected p-nitrophenyl carbonate monomers.[4] N-Alkylcarbamate oligomers were also prepared to increase the density of side chains, remove the main-chain hydrogen bond donors, and decrease the conformational freedom of the backbone.[5]

4. Vinylogous sulfonyl peptides were prepared from N-protected vinylsulfonyl chloride monomers.[6]

5. Oligoureas were prepared from different monomers in separate studies.[7–9]

6. Vinylogous peptides were prepared from both acyclic and cyclic monomers.[10,11]

7. Azatides and peptidosulfonamide peptidomimetics were individually prepared.[12,13]

REFERENCES

1. For a review, see: Liskamp, R. M. J. *Angew. Chem., Int. Ed. Engl.* **1994**, *33*, 633.
2. Ostretch, J. M.; Husar, G. M.; Blondelle, S. E.; Dörner, B.; Weber, P. A.; Houghten, R. A. *Proc. Natl. Acad. Sci. U.S.A.* **1994**, *31*, 11138–11142.
3. (a) Zuckermann, R. N.; Kerr, J. M.; Kent, S. B. K.; Moos, W. H. *J. Am. Chem. Soc.* **1992**, *114*, 10646–10647. (b) Zuckermann, R. N.; Kerr, J. M.; Siani, M. A.; Banville, S. C. *Int. J. Pept. Protein Res.* **1992**, *40*, 498–507.
4. (a) Cho, C.; Moran, E. J.; Cherry, S. R.; Stephans, J. C.; Fodor, S. P. A.; Adams, C. L.; Sundaram, A.; Jacobs, J. W.; Schultz, P. G. *Science* **1993**, *261*, 1303–1305. (b) Moran, E. J.; Wilson, T. E.; Cho, C.; Cherry, S. R.; Schultz, P. G. *Biopolymers (Pept. Sci.)* **1995**, *37*, 213–219.
5. Paikoff, S. J.; Wilson, T. E.; Cho, C. Y.; Schultz, P. G. *Tetrahedron Lett.* **1996**, *37*, 5653–5656.
6. (a) Gennari, C.; Nestler, H. P.; Salom, B.; Still, W. C. *Angew. Chem., Int. Ed. Engl.* **1995**, *34*, 1763–1765. (b) Gennari, C.; Nestler, H. P.; Salom, B.; Still, W. C. *Angew. Chem., Int. Ed. Engl.* **1995**, *34*, 1765–1768.
7. Burgess, K.; Linthicum, D. S.; Shin, H. W. *Angew. Chem., Int. Ed. Engl.* **1995**, *34*, 907–909.
8. Hutchins, S. M.; Chapman, K. T. *Tetrahedron Lett.* **1995**, *36*, 2583–2586.
9. (a) Kim, J.-M.; Bi, Y.; Paikoff, S. J.; Schultz, P. G. *Tetrahedron Lett.* **1996**, *37*, 5305–5308. (b) Kim, J.-M.; Wilson, T. E.; Norman, T. C.; Schultz, P. G. *Tetrahedron Lett.* **1996**, *37*, 5309–5312.
10. Hagihara, M.; Anthony, N. J.; Stout, T. J.; Clardy, J.; Schreiber, S. L. *J. Am. Chem. Soc.* **1992**, *114*, 6568–6570.
11. Smith, A. B.; Guzman, M. C.; Sprengeler, P. A.; Keenan, T. P.; Holcomb, R. C.; Wood, J. L.; Carroll, P. J.; Hirschmann, R. *J. Am. Chem. Soc.* **1994**, *116*, 9947–9962 and references cited therein.
12. Han, H.; Janda, K. *J. Am. Chem. Soc.* **1996**, *118*, 2539–2545.
13. Moree, W. J.; Marel, G. A.; Liskamp, R. J. *J. Org. Chem.* **1995**, *60*, 5157–5169.

OLIGOSACCHARIDES

Oligosaccharide synthesis is a large, complex, and challenging field. The solid-phase synthesis of oligosaccharides has long been recognized as a major challenge. Some recent strides in both the solid-phase and combinatorial synthesis of oligosaccharides are listed below.

1. Random galactosylation of unprotected N-acetylglucosamine was examined as a strategy for the production of oligosaccharide libraries.[1]

2. Glycosylamines were prepared in solution and conjugated to amino-functionalized resins.[2] This work has been extended to the synthesis of N-linked glycopeptides.[3]

3. Solid-phase oligosaccharide synthesis with glycals was reported employing an isopropylsilyl linker.[4]

4. The polymer-supported synthesis of oligosaccharides was reported using dibutylboron triflate to promote glycosylations with glycosyl trichloroacetimidates.[5]

5. Mutliple-component condensation strategies have been employed to generate carbohydrate derivatives.[6-8]

REFERENCES

1. Ding, Y.; Labbe, J.; Kanie, O.; Gindsgaul, O. *Bioorg. Med. Chem.* **1996**, *4*, 683–692.
2. (a) Vetter, D.; Gallop, M. A. *Bioconjugate Chem.* **1995**, *6*, 316–318. (b) Vetter, D.; Tate, E. M.; Gallop, M. A. *Bioconjugate Chem.* **1995**, *6*, 319–322.
3. Vetter, D.; Tumelty, D.; Singh, S. K.; Gallop, M. A. *Angew. Chem., Int. Ed. Engl.* **1995**, *34*, 60–63.
4. Randolph, J. T.; McClure, K. F.; Danishefsky, S. J. *J. Am. Chem. Soc.* **1995**, *117*, 5712–5719 and references cited therein.
5. Wang, Z.-G.; Douglas, S. P.; Krepinsky, J. J. *Tetrahedron Lett.* **1996**, *37*, 6985–6987.
6. Goebel, M.; Ugi, I. *Tetrahedron Lett.* **1995**, *36*, 6043–6046.
7. Park, W. K. C.; Auer, M.; Jaksche, H.; Wong, C.-H. *J. Am. Chem. Soc.* **1996**, *118*, 10150–10155.
8. Sutherlin, D. P.; Stark, T. M.; Hughes, R.; Armstrong, R. W. *J. Org. Chem.* **1996**, *61*, 8350–8354.

LIST OF ABBREVIATIONS

Boc	*tert*-butoxycarbonyl
BOP	benzotriazolyloxytris(dimethylamino)phosphonium hexafluorophosphate
Bzl	benzyl
DCC	dicyclohexylcarbodiimide
DCE	dichloroethane
DCM	dichloromethane
Ddz	[2-(3,5-bis(methyloxy)phenyl)propyl]oxycarbonyl
DEAD	diethyl azodicarboxylate
DIAD	diisopropyl azodicarboxylate
DIC	diisopropylcarbodiimide
DIPEA	diisopropylethylamine
DKP	diketopiperazine
DMF	dimethylformamide
Fmoc	9-fluorenylmethoxycarbonyl
FT-IR	Fourier-transform infrared
HATU	*N*-{(dimethylamino)(1*H*-1,2,3-triazolo[4,5-*b*]pyridin-1-yl)-methylene}-*N*-methylmethanaminium hexafluorophosphate *N*-oxide
HBTU	*N*-{1*H*-benzotriazol-1-yl)(dimethylamino)methylene}-*N*-methylmethanaminium hexafluorophosphate *N*-oxide
HOAt	1-hydroxy-7-azabenzotriazole
HOBt	1-hydroxybenzotriazole

HPLC	high-performance liquid chromatography
Hunig's base	diisopropylethylamine
MALDI	matrix-assisted laser desorption/ionization mass spectrometry
MAS	magic-angle spinning
MS	mass spectrometry
NMM	*N*-methylmorpholine
NMP	*N*-methylpyrrolidinone
NMR	nuclear magnetic resonance
Nvoc	(6-nitroveratryl)oxycarbonyl
PEG	poly(ethylene glycol)
PS	polystyrene
PyBOP	tetrapyrrolidinophosphonium hexafluorophosphate
PyBrOP	bromotripyrrolidinophosphonium hexafluorophosphate
SPOS	solid-phase organic synthesis
SPPS	solid-phase peptide synthesis
TBTU	*O*-benzotriazoyl-*N*,*N*,*N'*,*N'*-tetramethyluronium tetrafluoroborate
t-Bu	*tert*-butyl
TEA	triethylamine
TFA	trifluoroacetic acid
TFFH	tetramethylfluoroformamidinium hexafluorophosphate
TMG	tetramethylguanidine
TMSCl	trimethylsilyl chloride

■ AUTHOR INDEX

Numbers in parentheses are footnote reference numbers and indicate that an author's work is referred to although the name is not cited in the text.

INDEX